W9-BRP-046

Social Science for Counterterrorism
Putting the Pieces Together

Paul K. Davis, Kim Cragin, Editors

Darcy Noricks, Todd C. Helmus, Christopher Paul, Claude Berrebi, Brian A. Jackson, Gaga Gvineria, Michael Egner, and Benjamin Bahney, Contributors

Prepared for the Office of the Secretary of Defense

Approved for public release; distribution unlimited

NATIONAL DEFENSE RESEARCH INSTITUTE

The research described in this report was prepared for the Office of the Secretary of Defense (OSD). The research was conducted in the RAND National Defense Research Institute, a federally funded research and development center sponsored by the OSD, the Joint Staff, the Unified Combatant Commands, the Department of the Navy, the Marine Corps, the defense agencies, and the defense Intelligence Community under Contract W74V8H-06-C-0002.

Library of Congress Cataloging-in-Publication Data is available for this publication.

ISBN 978-0-8330-4706-9

Published 2009 by the RAND Corporation
1776 Main Street, P.O. Box 2138, Santa Monica, CA 90407-2138
1200 South Hayes Street, Arlington, VA 22202-5050
4570 Fifth Avenue, Suite 600, Pittsburgh, PA 15213-2665
RAND URL: http://www.rand.org/
To order RAND documents or to obtain additional information, contact
Distribution Services: Telephone: (310) 451-7002;
Fax: (310) 451-6915; Email: order@rand.org

Preface

This monograph surveys social-science literature relating to counterterrorism. It also takes first steps toward integrating the knowledge reflected in that literature and suggesting theories and methods to inform analysis and modeling. Our project was sponsored by the Modeling and Simulation Coordination Office of the Office of the Secretary of Defense, with oversight provided by James Bexfield, the Director of Planning and Analytical Support in OSD's Program Analysis and Evaluation. Comments and questions are welcome and should be addressed to the editors and project leaders: Paul K. Davis (Santa Monica, California; 310-451-6912; pdavis@rand.org) and Kim Cragin (Arlington, Virginia; 703-413-1100, extension 5666; cragin@rand.org).

This research was conducted within the International Security and Defense Policy Center of the RAND National Defense Research Institute, a federally funded research and development center sponsored by the Office of the Secretary of Defense, the Joint Staff, the Unified Combatant Commands, the Department of the Navy, the Marine Corps, the defense agencies, and the defense Intelligence Community.

For more information on RAND's International Security and Defense Policy Center, contact the Director, James Dobbins. He can be reached by email at James Dobbins@rand.org; by phone at 703-413-1100, extension 5134; or by mail at the RAND Corporation, 1200 S. Hayes Street, Arlington, VA 22202. More information about RAND is available at www.rand.org.

Contents

CHAPTER THREE
Why and How Some People Become Terrorists 71
Todd C. Helmus

CHAPTER SIX

Organizational Decisionmaking by Terrorist Groups. . . . .
Brian A. Jackson

CHAPTER SEVEN

How Does Terrorism End?. . . . .
Gaga Gvineria

CHAPTER EIGHT

Disengagement and Deradicalization: Processes and Programs 299
Darcy M.E. Noricks

CHAPTER NINE

**Social-Science Foundations for Strategic Communications in the
Global War on Terrorism**... 323
Michael Egner

CHAPTER TWELVE

Paul K. Davis and Kim Cragin

APPENDIXES

Benjamin Bahney

Figures

Tables

Summary

Objectives

Social science has much to say that should inform strategies for counter-terrorism and counterinsurgency. Unfortunately, the relevant literature has been quite fragmented and seemingly inconsistent across sources. Our study was an attempt to do better—not only by surveying the relevant literatures, but by "putting together the pieces." This meant taking an aggressively interdisciplinary approach. It also meant representing the knowledge analytically in a new way that enhances communication across boundaries of discipline and organization. Analysts will recognize what we did as constructing conceptual models. We sought also to identify points of agreement and disagreement within the social-science community, to suggest priorities for additional policy-relevant research, and to identify improved ways to frame questions for research and analysis.

Approach

We organized our study around the following questions that transcend particular disciplines:

1. When and why does terrorism arise (that is, what are the "root causes")?
2. Why and how do some individuals become terrorists, and others not?

3. How do terrorists generate and sustain support?
4. What determines terrorists' decisions and behaviors? What are the roles of, for example, ideology, religion, and rational choice?
5. How and why does terrorism decline?
6. Why do individuals disengage or deradicalize?
7. How can "strategic communications" be more or less effective?

For the most part, the monograph's chapter structure follows these questions. However, we added a chapter on the economics of terrorism that reviewed some of the best quantitative empirical research bearing on several of the questions. In addition, we devoted a chapter to thinking about how to represent the relevant social-science knowledge analytically so that it could be readily communicated. Finally, we devoted a chapter that looks across the various papers and highlights particular cross-cutting topics of interest.

Against this background, the following paragraphs summarize our results. The individual papers in the monograph include extensive citations to the original literature and far more nuance than can be captured in a summary.

How Terrorism Arises (Root Causes)

As discussed in the paper by Darcy M.E. Noricks (Chapter Two), "root causes" are not the proximate cause of terrorism. Rather, they are factors that establish an environment in which terrorism may arise. Such factors may be political and economic (that is, "structural"), but may also reflect the pervasive characteristics of culture and relevant subgroups. The subject is very controversial in the literature.

A basic distinction exists between root-cause factors that are *permissive* and those that are *precipitant*. The former set the stage, whereas the latter are the miscellaneous sparks that trigger such developments as insurgency or the use of terrorism. Table S.1 summarizes primary permissive factors.

Table S.1
Permissive Factors

Class of Permissive Factor	Factor
Global systemic factors	Global systemic explanations
State structural factors	Perceived illegitimacy of the regime
	Repression
	Democracy
	Modernization
	Economics
Social and cultural factors	Education
	Human insecurity
	Grievances and anxieties
	Mobilizing structures and social ties
	Ideology, religion, and culture

Figure S.1 arranges the primary root causes in a "factor tree"—a kind of influence diagram discussed in the "analytic" paper by Paul K. Davis. The intention is to include all potentially relevant factors. The relative significance of these factors varies greatly with context, but all of them are thought to be significant sometimes—whether directly or indirectly, and whether as an independent causal factor or part of a combination.

If two nodes on the tree are connected, more of the node at the tail of the arrow leads to more of the node at the point of the arrow. Such trees—diagrammatic versions of top-level conceptual models—allowed us to pull together strands of research from different disciplines and perspectives and at different levels of detail. The factor trees encourage the reader to shift away from single-factor questions toward questions of a more systemic nature—questions that recognize that multiple factors must be addressed simultaneously and that none of the simple explanations are sufficient.

Figure S.1
A Factor Tree for Root Causes of Terrorism

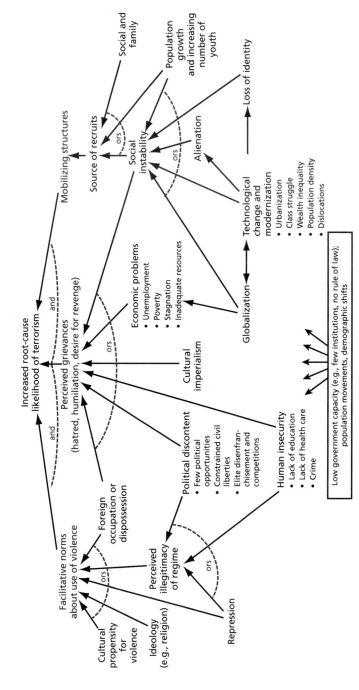

NOTE: "and" conditions are approximate; sometimes, not all factors are necessary.
RAND MG849-S.1

As a whole, Figure S.1 is to be read as saying that whatever role root causes play in the phenomenon of terrorism, the likelihood that terrorism will ensue as a result of root causes will increase if the social group in question believes that violence is legitimate (even if others see it as terrorism), if it has substantial motivations (perhaps stemming from grievances), *and* if social structures exist permitting the terrorist actions. To a first approximation, however, all three factors are *necessary*, as indicated by the "ands."

Reading down the tree, we see multiple arrows contributing to each of these major factors. These are to be read as *alternative* permissive factors. Reading from the left, the acceptability of terrorism may be driven by a cultural propensity for violence, by ideology (including but not necessarily religion), by political repression and regime illegitimacy, *or* by foreign occupation. The operative word is "or." None of these are *necessary*. Any one might be *sufficient*, or it might be that combinations of two or more of them would be necessary. One factor may substitute for another.

As another example, social instability may be due to or exacerbated by alternative factors as diverse as an increase in the youth population, alienation (for example, within an expatriate community), or globalization. Globalization can cause economic problems for those who are displaced and can disrupt traditional societies (for example, by undercutting individuals' sense of identity and by increasing alienation). As indicated at the bottom of the tree, many other systemic or exogenous factors can contribute. For example, an ineffective government and the absence of the rule of law may engender violence, grievances, and the emergence of protest or insurgency movements.

Figure S.1 is our synthesis rather than an extract from the literature. Others would construct the tree somewhat differently (perhaps, for example, treating religion as distinct, rather than as an example of ideology). Some authors would insist that particular items in the tree have been proven unimportant by quantitative studies. We retain the factors in question, however, because there is logic to including them and because the "disconfirming conclusions" sometimes extrapolate unreasonably from particular contexts or levels of analysis. A factor might well not matter "on average," but might matter a good deal to important individuals or groups in particular contexts. Also, a factor might not show up as independently significant from statistical analysis across many cases because it is only one of several contributing factors (that is, its apparent effect is diluted by there being multiple contributors).

Factor trees such as Figure S.1 are schematic, qualitative, analytical models. Because they juxtapose different pathways upward, their use in discussions can help avoid fruitless arguments about which factors matter and which do not. When experts argue on such matters, they are often talking past each other because they have studied terrorism in different contexts and with different disciplinary paradigms.

Despite the considerable literature on root causes of political violence and terrorism more narrowly, we found serious shortcomings. Table S.2 sketches what might be done to improve the situation. First, because context matters so greatly, data analysis needs to distinguish better among (1) classes of political violence (for example, terrorism that is or is not part of an insurgency), (2) the types of terrorist

Table S.2
Shortcomings in the Current Knowledge Base on Root Causes

Step Needed	Example
Distinguish better among classes of political violence	Terrorism versus rebellion, ethnic conflict, social movements, and civil war
Distinguish types of terrorism	Separatist versus religious and left-wing movements
Distinguish different levels and components of terrorist system	Leaders versus lieutenants, foot soldiers, facilitators, financiers
Improve methodology and measurement	Datasets skewed toward Irish Republican Army (IRA) and Israel-Palestine cases; excessively aggregated measures (for example, national gross domestic product)
Address understudied causal factors	Rule of law, strength of related institutions
Address discrete knotty problems	Better characterization and measurement of the roles of ideology, religion, and culture; assessment of whether, for example, some religious tenets are better vehicles for terrorism than others

organizations (for example, separatists versus extremist religious or left-wing movements), and (3) the levels and components of the terrorist system (for example, the leaders rather than the lieutenants or foot soldiers). These distinctions need to be recognized by those posing questions and commissioning research or analysis.

Second, existing quantitative analysis depends heavily on datasets skewed toward the data-rich cases of the IRA and the Israeli-Palestinian conflict. Much of the existing analysis is also highly aggregated, which introduces measurement error. For example, economic factors prove not to be a causal factor of terrorism in the large, but we know that individuals sometimes move toward or away from terrorist organizations in part according to whether personal-level opportunities exist. Third, a number of important causal factors have not been adequately studied. These include whether an area enjoys the rule of law and whether it has strong related institutions. Finally, a few knotty problems need to be addressed more carefully and rigorously. Some of these involve the roles of ideology, religion, and culture.

In considering how to address the shortfalls, we note that

- A good deal of existing data should be reanalyzed and recoded with the distinctions suggested by Table S.2.
- However, much more data is needed, especially the kind obtained only by scientific fieldwork, rather than merely mining readily accessible materials or collecting anecdotal material.

In some important cases, relevant data exist but are treated as classified or are otherwise restricted. Declassification or sanitation should often be possible.

Why People Become Terrorists

Root-cause factors affect terrorism indirectly by contributing to an environment, but how do we conceive causes at the level of individuals? Why, given the dangers and moral issues, do some people become terrorists? Here, the relevant literatures include psychology, social psychology, sociology, and religious studies.

Some of the important research conclusions are "negative": It has simply not proven feasible to identify terrorists by general characteristics, as discussed in the papers by Todd C. Helmus and by Claude Berrebi. Terrorists tend to be males, aged 17–30 (although sometimes women do become terrorists). Notably, however:

- Terrorists are *not* particularly impoverished, uneducated, or afflicted by mental disease. Demographically, their most important characteristic is normalcy (within their environment). Terrorist leaders actually tend to come from relatively privileged backgrounds.

These conclusions are firmly supported by empirical analysis, although there are many nuances, as discussed by Berrebi.

What, then, are the factors at play? As in the research on root causes, a myriad of factors have been identified and discussed. To make sense of them, we can use the factor tree shown in Figure S.2, which comes from the Helmus paper.

The first-order factors in this figure (listed in red) are group socialization processes, expected rewards, a felt need to respond to grievances, and a passion for change.

The first factor is well established: Abundant evidence indicates that socialization processes are a necessary precondition for radicalization (by which we mean the process of becoming willing to conduct a terrorist act). Group processes assure individuals that their chosen path is correct, build up socially motivated courage, and help to dehumanize selected targets.

Another factor that is usually necessary is the perception of rewards for participation in terrorism. Three examples are the friendships and camaraderie solidified in the terror cell or organization, the social status derived from membership (for example, the respect shown to members of Hamas and Fatah), and the heavenly gains of martyrdom. Group processes and rewards ultimately combine with one of two key motivational factors (the right side of the factor tree) that are

Figure S.2
Factor Tree for Individual Willingness to Engage in Terrorism

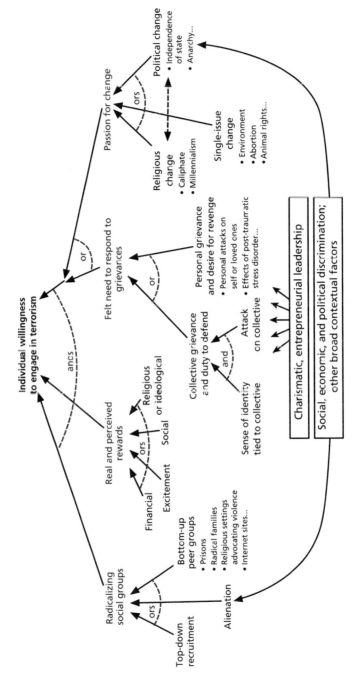

different psychologically. The first is a strong sense of necessity, as with a perceived duty to defend a people or achieve revenge for either personal or collective wrong. The second is a passion for change, which might be religious (as in establishing an Islamic caliphate) or political (as in revolution against repression). These two factors form the ideological basis for terrorism and constitute overt reasons for terror action. Neither is in itself a necessary factor but at least one is likely required.

Dipping more deeply into the factor tree, a number of observations are possible. First, group processes take place in essentially one or both of two ways: There may be top-down recruitment strategies initiated by a terror organization or cell, there may be bottom-up processes dominated by peer bonds and other social influences, or both (note the "ors"). Within the bottom-up trajectory, groups of individuals meet and interact in any of various settings that include prisons, radical families, religious houses advocating violence, and the Internet (bulleted items in the figure are examples but do not include all the possibilities).

Groups within either top-down or bottom-up processes may be influential as a result of perceptions of social and religious alienation. Feelings of alienation in Muslim communities throughout Europe and the Middle East draw individuals to places where they can meet and identify with like-minded people. This alienation is likely fed by perceptions of social, economic, and political discrimination.

One motivational set (the third of four branches) involves perceived grievances. These may be collective, as in defending one's people or rejecting an occupier, or personal, as in a desire for revenge against those who killed or imprisoned friends or family. Personal traumatization, often manifested in post-traumatic stress disorder, may exacerbate motivations for revenge.

An alternative motivational set involves the passionate desire for change (the right branch), which may be related to political change (for example, independence), religious change (for example, establishing an Islamic caliphate with Sharia law), or even single-issue change (regarding the environment or abortion, for example).

Finally, as shown at the bottom, below the tree, some factors affect most or all of the items above. These include many contextual factors, but also the existence of charismatic and entrepreneurial leaders.

How Terrorists Obtain and Maintain Support

Given that individuals are willing to become terrorists, how does a terrorist organization gain support, what support is needed, and how is the support sustained? Christopher Paul's paper addresses these questions, drawing heavily on the sociology literature among others. Figure S.3 indicates the types of support needed and from where it may come. Of the support types, some is provided by "active" support, whereas some is provided merely by a population or state looking the other way or perhaps sharing information. Not all organizations are equal. Some terrorists are able to obtain much of what they need through straightforward purchases, from wealthy members, or through criminal activities. Also, small self-organizing groups need less support than does, say, an insurgent army. However, some things must always be obtained from external sources. The issues of sanctuary and toleration are especially important. When the population turns against a terrorist organization, intelligence tips increase markedly. Further, the terrorist

Figure S.3
Support of Terrorist Organization

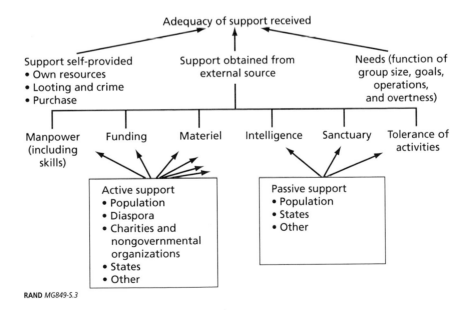

organization itself must be more cautious and must worry increasingly about internal penetration. *That public support is very important to terrorist organizations is perhaps the strongest conclusion from social science affecting the counterterrorism policy issue.*

The factor tree for this discussion is shown as Figure S.4. The top-level factors include the perceived *need for resistance and action, identification* with the terrorist organization, and *pressures* to support that organization. Usually, all the top-level factors are needed (note the approximate "and" relationship), although there are exceptions, such as when intimidation by the terrorist organization may be sufficient to force support. Identification with the terrorist organization is especially important, as indicated by the larger arrow.

In examining support-related counterterrorism possibilities, it becomes clear that one size does not fit all. The most important implication here is that

- *Policymakers should first ascertain the specifics of the particular case they are dealing with.*

That is, rather than applying a generic concept (perhaps one in vogue in Washington), they should identify the type of group (size, goals, nature of operations, and covertness), the extent of support needs (manpower, funding, materiel, intelligence, sanctuary, and tolerance of activities), and how the group's needs are being met. It then becomes much easier to specify interventions to reduce support motives. To put it differently, although ideal cases have a long and valued role in academic studies, applying the lessons of social science is another matter. Not all details matter, certainly, but which details *do* matter differ with the case. The conclusion might seem banal but for the fact that this principle of starting with context is often violated.

A second implication of the review is that, given knowledge of the case-specific matters, *it is wise to focus on factors that matter and that can be changed.* Cultural characteristics change over decades or centuries, not weeks, but whether a state provides essential social services, whether a state can protect a population from intimidation, and

Figure S.4
Factors Affecting Support of Terrorism

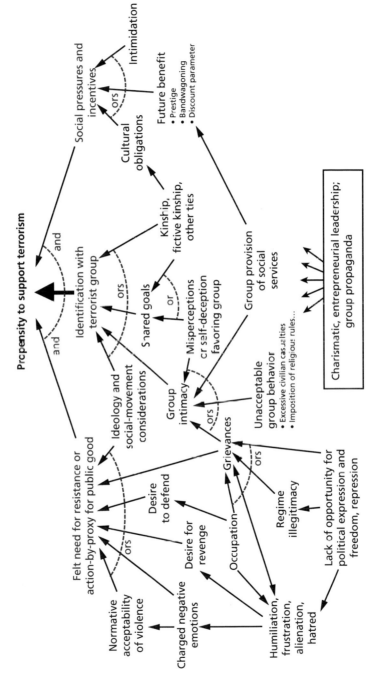

whether a terrorist organization and its actions are considered legitimate and effective may all be better targets for change efforts.

How Terrorists Make Decisions

A Rational-Choice Framework

Given that terrorist organizations exist, how do they behave, to include making decisions? This is the subject of papers by Claude Berrebi and Brian A. Jackson, which drew on literatures from organization theory, political science, game theory, and economics, among others.

To make sense of the many factors and processes at work, we adopted a structure described in detail in Jackson's paper and summarized in Figure S.4. As with the earlier factor trees, we show and/or conditions (always to be regarded as approximate). The overall framework for organizing is one of rational choice, although a better terminology is perhaps "limited rationality," for reasons discussed below.

Despite these caveats, much of what terrorist organizations do *can* be understood well in a rational-analytic framework, so long as allowance is made for misperceptions and cognitive biases. The structure in Figure S.5 describes such a framework. We believe that this is a useful way to organize and collect intelligence and to understand behaviors at different levels of detail. The major factors shown are perceived benefits, acceptability of risks, acceptability of expending the resources required for success, and the sufficiency of information in making a judgment. This modest set of four factors is influenced, however, by many subtle lower-level factors. For example (left side of the figure), a decision may reflect judgments (perceptions) about the degree to which a contemplated action will cause positive reactions among the relevant population. That judgment can be quite wrong: If the group overreaches and kills too many of the wrong people (such as al-Qaeda in Iraq's attacks in Amman, Jordan, against other Muslims) or innocent civilians generally, the reaction may be quite negative even if highly successful. But public reaction is only one of many concerns. Even if morally debatable within the organization, would the action advance the organization's interests? Or, to change the language somewhat to

Figure S.5
Factors Influencing Terrorist Decisions and Behavior

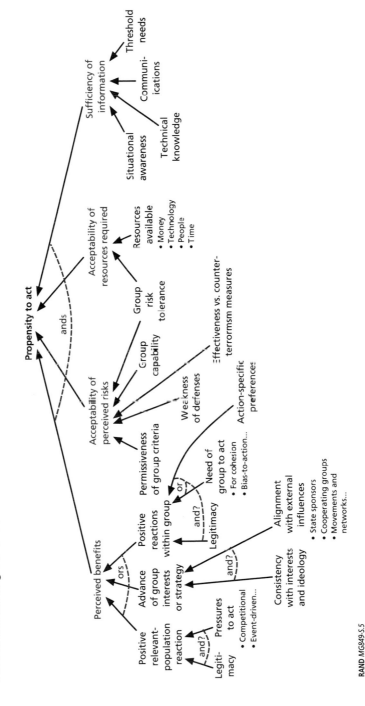

RAND *MG849-S.5*

correspond with other perspectives, is the action *demanded* by God or Allah or *demanded* by one's sense of honor in redressing past grievances, as discussed above? Would the action benefit the group itself positively, or would it cause dissension and splintering? Moving rightward in the figure, we see a mixture of objective and subjective subfactors at work. For example, a group's willingness to accept perceived risks is to some extent subjective, whereas assessment of a target's success, of operational security, or of the group's capabilities may be rather objective.

Explaining Empirical Results with a Rational-Choice Model
Some of the best quantitative research on terrorism and counterterrorism has been accomplished by economists applying rational-choice models to empirical data. Some such work has gone well beyond the usual statistical analysis of heterogeneous data with uncertain significance to analysis of special datasets that can be seen as reflecting "natural experiments"—that is, phenomena under a range of conditions akin to the range that an idealized social experiment would call for. The result is an ability to get closer than usual in social-science research to being able to infer causality or disprove claims of causality.

Claude Berrebi's paper describes several recent studies in which such techniques have been used to assess apparent rationality. The results support the rational-choice model for interesting cases that draw on experiences in the long-running conflict between Israel and Palestine. Some selected results are the following:

- At the group level, tactical- and operational-level rationality explains where and when Palestinian terrorists chose to attack. That is, attacks were not random but rather quite "sensible" when considering such issues as target value, attack cost, and risk.
- Attack timing was also explained, but only with inferences about the value terrorists place on targets of different types. For example, terrorists are not content to leave certain high-profile areas untouched, even though it would be easier and, in a narrow sense, more fruitful to attack others.
- Terrorist use of suicide bombers is well explained by understanding suicide bombers to be special assets with particular value

against "hard targets" (that is, targets difficult to attack in more conventional ways). Perhaps even more interesting is evidence of how "human capital" considerations matter. Not all volunteers for suicide attacks are equal and mounting a suicide-bombing attack involves a large operation. As would be expected from rational-choice theory, "better" suicide bombers (older and more-educated terrorists who are, according to the data, more likely to succeed) are used preferentially against larger, important, and lucrative civilian targets.

• Strategic-level rationality has been manifested in explicit statements by Osama bin Laden, among others (for example, he crowed about the positive exchange ratio between the cost of the September 11 attacks and the cost of its consequences to the United States). More generally, the rationality of terrorist objectives relating to imposing economic hardships on the targeted countries is supported by several studies. That is, there is an empirical basis for terrorists' imagining that they will be able to achieve many of their objectives through violent action.

Such solid evidence of rationality is both encouraging and discouraging. On the one hand, we can expect terrorists to be clever and to make good operational choices that exploit target weaknesses. More positively, however, it means that with good intelligence and analysis, we can expect to understand their calculations and how to affect them. Further, at least some terrorists should be expected to respond to incentives. It is not just wishful thinking to imagine this.

Limitations of the Rational-Choice Model

There are limits to the rational-choice model. These involve bounded rationality (for example, the inability to gather the information needed for idealized rational-analytic calculations, and misperceptions), the many cognitive biases that afflict human decisionmaking (for example, the consequence of selecting data that reinforce preferences, of demonizing adversaries), the character of individual leaders (such as their risk-taking propensity), emotions (for example, the fervor that commanders seek to build before battles or the fears that can paralyze), physiological circumstances (such as exhaustion and, variously, paranoia or paralysis), and leaders' idiosyncrasies (for example, those of Shoko Ashahara of Aum Shinrikyo). The study of such considerations has led to Nobel prizes and is reflected in the relatively new field of behavioral economics. Despite these limitations, the rational-choice models fare better than erroneously assuming that terrorists and terrorist organizations behave chaotically. Their rationality may be "limited" or "bounded," but it is understandable and needs to be understood. The primary admonition here is simple: In applying the "rational-choice model," analysts should take pains to use realistic assessments of *terrorist* perceptions and values rather than our own.

How Does Terrorism End?

Interestingly, the historical evidence describing how terrorism ends uses somewhat different terms than descriptions of how it arises. Naively, one might think to bring an end to terrorism simply by working to reduce all the factors causing it in the first place. However, as noted above, many factors matter to different relative degrees in different contexts (including aspects of context dependent on "random" events). Which pathway through the factor trees will prove to be most relevant?

As discussed in the paper by Gaga Gvineria, it is possible to summarize the modes by which terrorism declines as in Table S.3, which includes examples:

- Terrorist movements often decline as the result of at least partial success and partial accommodation reflected in state policy. Also, new alternatives may arise as a result of political compromise, civil war, or economic prosperity. This is arguably playing itself out today: It is difficult to imagine negotiations with al-Qaeda central, but states can address local grievances with a diminution of terrorism by al-Qaeda affiliates.
- Sometimes, terrorist movements are defeated by direct counterterrorism activities, which may be repressive or which may at least walk a tight line and sometimes transgress.

Table S.3
Classes of Cases and Historical Examples

Dominant Mode of Terrorism Decline	Notable Historical Examples
Substantial success (primary objectives met, by whatever means)	Original Irish Republican Army (IRA) (circa 1921) EOKA (Cyprus) Croatian Ustasha African National Congress (ANC) Nepalese Maoists Irgun/Stern Gang (Israel)
Partial success	Palestinian Liberation Organization (PLO) Provisional Irish Republican Army (PIRA)
Direct state counterterrorism activities (sometimes repression)	Revolutionary Armed Task Force (RATF) Symbionese Liberation Army (SLA) George Jackson Brigade Narodnaya Volya Uruguayan Tuparamos Muslim Brotherhood
Disintegration through burnout	Weather Underground Front de libération du Québec (FLQ) Red Brigades
Loss of leaders	Shining Path Real Irish Republican Army (Real IRA) Aum Shinrikyo
Unsuccessful generational transition	Red Brigades The Second June Movement The Japanese Red Army Weather Underground Symbionese Liberation Army Baader-Meinhof group (Red Army Faction)
Loss of popular or external support	Weather Underground Front de libération du Québec Real IRA Red Brigades Shining Path Armenian Secret Army for the Liberation of Armenia (ASALA)
New alternative to terrorism	Front de libération du Québec Provisional IRA Palestinian Liberation Organization Revolutionary Armed Forces of Colombia Khmer Rouge Armed Islamic Group Maoists in Nepal Guatemalan Labor Party/Guatemalan National Revolutionary Unit

- Often, decline occurs because the organization itself weakens—through loss of leaders, "burnout," unsuccessful generational transitions, and so on.
- Loss of popular support has often been very important. As support wanes, intelligence tends to increase on terrorist activities, penetrations occur, and operations become more difficult. The effects are across the board and may not be easy to measure.

The examples listed in Table S.3 sometimes exhibited more than one of the failure modes listed.

Having reviewed a considerable literature on the ascent and decline of terrorism, we selected two aspects of decline for additional study: deradicalization and strategic communications. We also added an appendix (not summarized here), by Ben Bahney, which takes a first cut at the literature on metrics to suggest ways in which factors arising in the various chapters can be measured.

Disengagement and Deradicalization

As discussed in a paper by Darcy M.E. Noricks (Chapter Eight), deradicalization has not yet been adequately studied by scholars. A number of valuable observations are possible, however. These include the following:

- Disengagement is often a more realistic goal than deradicalization. People often disengage from the activities of terrorism without rejecting their cause or beliefs (although their passion for those may also wane over time).
- The pathways for radicalization and deradicalization are different, which has important implications for policy interventions (Table S.4).

Interestingly, although aggregate-level quantitative research has not found economic factors to be predictive of radicalization, a number

Table S.4
Pathways for Radicalization Versus Those for Deradicalization or Disengagement

Radicalization	Deradicalization/Disengagement
Individual economic factors not predictive	Majority of programs provide economic support for targeted individuals and their families
Ideology/religion sometimes predictive; other times not	Many programs based in ideological re-education 　Delegitimize use of violence 　Reinterpret theological arguments
Supportive peer group	Isolation from peer group Role model important Saudi program: self-esteem counseling Saudi and Singapore: target broader family network
Traumatic event catalyzes	Traumatic event catalyzes
Failure of nonviolent strategies	Failure of violent values and beliefs

of countries have deemed it important to provide economic support for both individuals and their families in their deradicalization programs. Also, although ideology and religion are only sometimes root motivators for joining a terrorist cause, most of the deradicalization programs include ideological "re-education." This may be necessary even if the dangerous ideas were picked up as part of indoctrination rather than having had deeper roots. Peer-group issues loom large in both radicalization and deradicalization. Extracting individuals from the terrorist group is important, as is providing new role models. In some programs, self-esteem counseling and counseling in a family-network context is included. The last two items of the table are parallel: Traumatic events can catalyze radicalizing *or* disengagement, and people learn: Just as a failure of nonviolent protests can lead to violence, so also can failure of violent activities lead to disengagement.

A final conclusion from the social science so far is that "pull factors" are more effective than "push factors." That is, people are more likely to disengage from terrorist activities because they are positively attracted to a "normal life," new employment, or a new social group than they are to disengage because of the threat of punishment, counterviolence, or a negative reputation.

Strategic Communications

It is now widely recognized that it is essential (necessary, if not sufficient) to reduce public support for terrorism. "Strategic communications" is one of the primary mechanisms discussed for that purpose. It is also controversial. We use the term here, even though some associate it with careless and heavy-handed propaganda. Any term that we might choose would likely also be tainted. In any case, we have in mind "good" strategic communications. Michael Egner's paper reviews relevant literatures and reaches conclusions that would seem innocuous and obvious except that they are so often not heeded.

First, we should distinguish sharply between short-, medium-, and long-term aspects of a strategic communications campaign. Second, once again, context matters. Here, however, it is *audience* that matters. A frequent error in strategic communications has been to develop messages that are suitable for one audience but counterproductive for attempting to influence another. The implication is that messages should be targeted and built by people with a close understanding of those particular audiences. Further, close monitoring and rapid adaptation are important because perceptions and concerns change rapidly. All of this argues against highly centralized message construction, especially when driven by American headquarters intuition rather than local knowledge.

A third observation is that a core issue in strategic communications is the simple reality that actions speak louder than words (although words matter as well). What the United States actually does in the international arena weighs heavily on results. Sometimes those actions are helpful to strategic communications (Tsunami relief) and sometimes they are not (U.S. failure, for some years, to provide basic security for the population after occupying Iraq; or the appearance, for some years, of having tilted excessively toward Israel).

Analytic Representation of the Social Science

As discussed in the paper by Paul K. Davis, we developed new analytical methods to bring a certain amount of order out of chaos. The relevant social science is fragmented and discipline-bound, with researchers tending to study one or a very few factors in great detail but not addressing the whole. Further, their work involves different levels of analysis, which makes for difficult communication across projects. To make things worse, some of the social science is primarily observational, and other parts are quantitative and rigorous but narrow. Finally, there is the problem that most of the quantitative social science depends heavily on statistical analysis, which has shortcomings for understanding and explaining phenomena and reasoning about intervention alternatives.

We also concluded that the current *quantitative* social science was too heavily imbalanced toward statistics-heavy atheoretical empirical work. Both theory-driven and atheoretical approaches are crucial, but the current situation is out of balance.

System Theory

The approach we took in our study, reflected ultimately in all of the papers, was to organize thinking with causal system models, i.e., theory. In mature sciences, "theory" is good; it is the means by which the whole can be seen and the strands pulled together. "Theory" in this sense is anything but ad hoc speculation or simplistic "It's all about X" assertions. In a good theory-informed approach, one uses data to test the validity of a theory, to identify its shortcomings, and—when the theory appears to be valid—to calibrate its parameters. A good theory provides the integrated framework within which to recognize principles and mechanisms. Alternative theories may be necessary, but that also is good because they sharpen the issues for debate and inquiry.

The factor trees used throughout the monograph illustrate how a great deal of confusion can be eliminated by viewing matters in this way. The approximate use of "and" and "or" relations clarifies many unnecessary disagreements and, at the same time, distinguishes between individually critical factors and factors that can substitute for one another to achieve the same effect. Counterterrorism that attacks only one of several "or" branches will likely prove ineffective because of the substitutions. On the other hand, successful attacks on any of the "and" branches might prove to quite effective.

Humility About "Prediction"

Another theme of our work is that, even where the social science is "strong," the paradigm of reliable prediction is usually inappropriate. Too many factors are at work, many with unknown values and some not even knowable in advance. Except in rare cases in which matters are over determined, there will be a substantial "random" component in social behavior. Strategy, then, should be developed with an eye toward achieving flexibility, adaptiveness, and robustness.

A Systems View

Figure S.6 illustrates another aspect of taking a systems view. In this particular depiction, the terrorist organization (red ovals) has capabilities that depend on its organization and its resources. There is a "demand" at any given time for terrorist actions, whether generated internally or by relevant publics. The resulting attacks will be more or less effective depending on such factors as target vulnerability and counterterrorism efforts. The consequences of attacks will feed back, affecting everything else in the system. Public support for the terrorists may then rise or diminish; targets may be hardened further or become more vulnerable than previously, and so on. Support for terrorism (orange oval to left) includes that from states and relevant publics. To decrease support requires separate attacks on each such source of support. Some of our factor trees can be seen as "zooms" into the macro factor of support in Figure S.6 (excluding state support).

Looking for Leverage

An important concept of analysis amidst uncertainty is the need, as mentioned above, to forgo pursuit of certainty in favor of an approach to improve odds. Where is the "leverage"? Figure S.7 illustrates the kind of analytic display we have in mind. In this, the issue is whether to expect high, medium, or low levels of public support for the terrorist organization and its case. Level of support is indicated by color: Red is

Figure S.6
An Illustrative High-Level Systems View

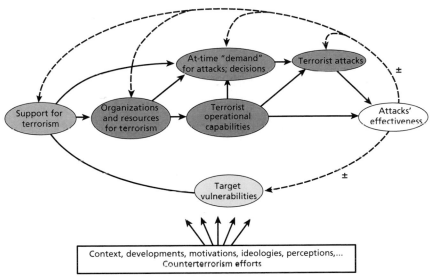

RAND *MG849-S.6*

high support (bad from our perspective); green is low support (good). Consistent with the earlier factor-tree discussion, the expectation is based on assessments of the motivation for such public support (such as a felt need to redress grievances or a passion for violent jihad; vertical axis), the expected price that will be paid personally or by one's society as the result of supporting the terrorism (horizontal axis), and the degree to which the violence is regarded as legitimate (left versus right panel). If the baseline situation is Point A, then if legitimacy is deemed high (left panel), motivation must drop substantially or the price of support must increase to very high levels if support is to be low. In contrast, if legitimacy is deemed low (right side), the baseline situation is already less dire, and relatively small changes in motivation or perceived price will shift support level to low. The drop in perceived legitimacy can come about naturally, as when terrorists become too bloodthirsty and indiscriminate.

Figure S.7
Support for Terrorism as Function of Motivation for Supporting, Price of Supporting, and Perceived Legitimacy of Terrorist Tactics

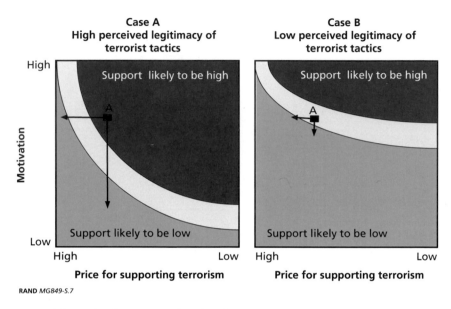

RAND *MG849-S.7*

Although the particular display in Figure S.7 is notional, it may suffice to illustrate how social-science knowledge could be displayed to clarify issues of relative leverage.

Future Research and Analysis

With respect to future analysis and increasingly ambitious modeling, we recommend a process as described in Table S.5. A core element of the process is exploiting factor-tree representations to define modules for the research community to address separately—but always with the larger perspective in mind. With multiple research thrusts for each of the crucial modules, it should be possible to sharpen the trees, to define how to "measure" the factors (even if subjectively), to sharpen the understanding of how the various factors combine and when they are more or less important, and to work toward *approximate* causal descriptions that improve the odds of correct diagnosis and prescription.

Table S.5
Procedural Elements of an Approach

Tier One
Collect factors, focusing on concepts, not proxies
Define factor levels meaningfully
Organize in multiresolution factor trees
Consider alternative trees for different perspectives
Translate trees into influence diagrams with feedback loops, dotted lines
Annotate diagrams to indicate first-order combining logic
Review, debate, iterate, refine

Tier Two
By module, characterize combining functions with diagrams, logic tables, "operator math," and pseudo code
Implement simple module-level models, exercise, refine
Compare representations and conduct live model-observation exercises to elicit comments, insights
Do, as above, for system-level model

Doing so would include conferences, peer-reviewed work, and convergence-focused activities, with the goal of reaching integrated conclusions (even though expressed more in terms of odds than confident prediction).

We emphasize the importance of in-depth work on each module, but the higher-level system perspective is crucial as well. On the one hand, little credence should be placed on detailed system models dependent on a myriad of poorly understood details and uncertain inputs in a myriad of subcomponents. Such system models can certainly be constructed and will "run," but the social science (and experience with big-model analysis under uncertainty) does not justify confidence in their results. Thus, the system models should be seen not as an answer machine, but as an integrative framework, with most of the debate and analysis being conducted at the module level.

The framework can also be valuable for higher-level analysis. As should be evident from the factor trees, it will often be possible to make more progress working at higher levels of particular modules than by immersing oneself in the myriad of details of the subbranches. Exploratory analysis with a low-resolution overall system model may prove quite useful. Such analysis varies all the key inputs systematically rather than fixing on alleged best estimates. It looks for conclusions that are relatively insensitive to uncertain assumptions.

Selected Cross-Cutting Observations

As we conducted our review of the social sciences and their relevance to studies of terrorism and counterterrorism, we identified some key points of agreement and disagreement bearing on counterterrorism policy. These are discussed in the paper by Kim Cragin.

Points of Agreement

Context. As mentioned above, context matters so much that those contemplating a counterterrorism campaign should *start* by understanding and characterizing their context, rather than looking to generic principles or conventional wisdom. This may seem banal, but in fact it is a principle often violated. Context varies drastically, even within what at first appears to be a single nation, area, or people. As merely one example, attempting to refute extremist religious arguments might be entirely appropriate with one subgroup and a waste of time and credibility with another that is concerned about matters of security, politics, or even economics.

Root Causes Versus Basis of Support. The root causes of terrorism are many, complicated, and subtle. However, they are not the key to either the sustainment of terrorism or to short- and mid-term counterterrorism. Table S.6 contrasts some of the factors identified above for root causes with those for maintaining support. There are some overlaps, but many of the most important—and actionable—factors are arguably less root causes than they are proximate causes of grievance or intimidation that can actually be addressed.

Table S.6
Root-Cause Factors Versus Sustaining-Support Factors

Root Causes	Maintain Support
Perceived illegitimacy of state	Perceived illegitimacy of state
State repression	State repression
Lack of opportunity	Lack of opportunity
Constrained civil liberties	Humiliation and alienation
Elite disenfranchisement	Resistance as public good
Ethnic fractionalization	Defense of self or community
	Identification with group
	Kinship and fictive kinship
	Intimidation by group
	Group provision of services
	Perceived group legitimacy

Ascent and Decline Are Different. Interestingly, the decline of terrorism does not mirror its ascent. For counterterrorism purposes, it is important to keep in mind the various modes of decline so that they can be recognized and accelerated where possible. To illustrate this point, Table S.7 compares root causes with the modes of decline. It demonstrates the disconnect between the various factors in our analysis. This disconnect makes sense, since root causes do not account for terrorist decisionmaking or the relationship that emerges between the terrorist group and support populations. In fact, it can be argued that the most likely situations of decline relate to terrorist group decisionmaking. That is, terrorism seems to decline in situations where terrorist leaders assess the risk of counterterrorism activities or the loss of popular support as greater than the benefits of the fight, or when they take such extreme actions as to lose popular support. The root-cause factors need not have been resolved.

Popular Support Matters Greatly, But Is Only One Consideration. A strong point of consensus in the social sciences is that terrorist

Table S.7
Descent Does Not Mirror Ascent

Root Causes	Modes of Decline
Constrained civil liberties	Success or preliminary success
Elite disenfranchisement	Burnout, poor succession, loss of leaders
Ethnic fractionalization	N/A
Illegitimacy of state	Success or preliminary success
State repression	Success or preliminary success
Lack of opportunity	Success or preliminary success
N/A	Loss of popular support
N/A	Counterterrorism activities
N/A	Loss of state support

NOTE: N/A = not applicable.

groups rely on popular support to sustain their activities and membership. Nevertheless, popular support is not the only factor that terrorist leaders weigh in their decisionmaking. For example, some evidence suggests that al-Qaeda leaders have sometimes viewed popular support as a preeminent consideration. Yet, at other times, al-Qaeda leaders have forsaken popular support to accomplish immediate operational objectives. The most well-known examples relate to the attacks against fellow Muslims by al-Qaeda in Iraq under Abu Mus'ab al-Zarqawi's leadership, including against hotels in Amman, Jordan. Our findings also suggest that terrorist leaders at the operational and tactical levels must meet—virtually or physically—to weigh factors more regularly than those terrorist leaders providing broad guidance to their followers. Thus, the relative value of these factors for terrorist groups are more likely observable at an operational level. Moreover, changes in those values are more likely to take place first at the operational level and then filter upward to those leaders separated from the day-to-day survival of the group.

Points of Controversy

We end by discussing briefly three nettlesome issues.

Which Dominates, Supply of or Demand for Terrorists? Our research revealed an interesting apparent conflict. On the one hand, economists (and some others) have noted that terrorist organizations typically operate in contexts that include very large numbers of potential recruits, when only much smaller numbers are needed. Al-Qaeda, for example, may need hundreds or thousands, but not hundreds of thousands. Since it is normal, not unusual, for a society to include many individuals that are angry, disaffected, or otherwise potential recruits, and since even volunteers for suicide attacks appear to be plentiful, it might seem that efforts to reduce supply are doomed to fail. At the same time, Helmus, Paul, and Jackson find that ideology matters to radicalization and support, that *high-quality* recruits are in shorter supply than others, and that specialized skills matter. These findings seem to suggest that it is worth focusing heavily on the supply problem.

Synthesis is needed. We accept the conclusion that the supply of raw volunteers or recruits far exceeds demand. However, our analysis suggests that the payoff is likely to be in attenuating the absorption rate. If recruiters find it difficult to operate, and if opportunities for systematic indoctrination and training are minimal and tenuous, then the *flow* of effective new recruits into al-Qaeda operations will be reduced. That is, the flow is determined not by the raw supply but by bottlenecks in the process of recruiting, radicalizing, indoctrinating, training, and employing. If so, then the goal should not be to "drain the swamp" (however desirable that might be) but rather to disrupt operations enough to minimize flow. This also suggests that affecting motivations is likely to be less important in determining the flow of recruits, but it is very important for other reasons, including influencing popular support for terrorism and encouraging deradicalization.

Are We Dealing with a Centrally Controlled Terrorist Organization or a Distributed, Bottom-Up Network? Over the last decade, we have seen al-Qaeda move to more decentralized networked operations. Some of the discussions about that phenomenon disparage a more classic organizational view and—rather frequently—convey a sense of hopelessness. Other discussions argue that, despite the decentralizaton of operations, al-Qaeda Central continues to have great significance: motivational, strategic, and facilitative. We pondered how to deal with the controversy but concluded that it was very much a controversy at a snapshot in time and that we should not allow such current specifics to determine the structure of our intendedly more general social-science depiction. After all, with a major al-Qaeda success, the central leadership could gain more power again; and with a major failure (or the loss of its leaders) it could shrink further in its importance. Thus, Helmus's chapter allows for both bottom-up and top-down processes, and Jackson's chapter describes how terrorist organizations make decisions with a structure that does not depend explicitly on the degree of decentralization. Whether decisions are made in a single room or much more indirectly, the factors he identifies still apply. Details will differ a good deal, but no generalizations are likely to hold up and—in this case—it is better to start with the generic structure and then interpret it for the context.

Arguably, an even more important point is that the appropriate perspective is not really one of a classic hierarchical organization or of a network but rather of a system. A system has many components and many functions, each of which is subject to attack. If one takes a system view, then the natural approach is to identify the major functions and related components and to then mount attacks on them thematically. For example, a campaign to disrupt recruiting or financing would be global from the outset rather than focused geographically (as in merely attacking al-Qaeda in Iraq). This would correspond to a network view in that one would not imagine a single node to be a "center of gravity," but the focus would not be something ethereal such as "the network generally." The focus would be on the specific operations (in this case, recruiting or financing). Such a system perspective leads naturally to thinking in whole-of-government terms.

What Is the Role of Religion in Current Struggles? Most of the literature that we reviewed avoids or skirts the issue of religion (except in studies that purport to show that it is not an important factor in terrorism). Most of our own monograph is restrained in discussing the role of religion. Why is this? A basic reason is that the subject is uncomfortable. Intellectually, scholars are uncomfortable highlighting religion because they see it as a mere subset of ideology (or, at least, as heavily overlapping with ideology). They know well the many instances throughout history in which terrorism has been driven by motivations having nothing to do with religion. A second reason is that the short-hand of referring to "religion" is troublesome because religions can be powerful agents of either the positive or the negative. Other reasons come into play as well. Social-science terrorism literature tends not to draw on the religious-studies literature, especially the relevant Islamic litera-ture. This is a straightforward shortcoming but a rather dramatic one.

This said, the issue of religion arises in numerous places through-out our monograph, albeit in a muted way. Noricks notes that religion can contribute to a "facilitative norm for the use of violence," especially when people see external threats with sacred meaning. Helmus notes that religion contributes to individual-level radicalization, perception of rewards, and a passion for change. Paul notes that religion can be used as a *tool* of validation for terrorist organizations garnering public support (and as an important part of developing a common identity). However, Berrebi observes that religion correlates poorly with terror-ist violence in the Israeli-Palestinian conflict. That, arguably, demon-strates how context matters and can be more subtle than is sometimes appreciated.

Some of the conclusions that we draw are as follows:

- Militant religion sometimes matters a great deal and sometimes not at all.
- Level of analysis matters (for example, leaders may be more affected by religious extremism than the foot soldier).
- The effects of religion may be "original" or subsequent, as when not-particularly religious young males join a terrorist organiza-tion and then—as part of bonding and indoctrination—adopt the religious trappings of the overall story.
- Because the role of religion differs so much, both policy and on-the-ground activities, such as counterradicalization and deradi-calization activities, should be locally tailored rather than dictated by generalizations.

Conclusions

Some overarching principles proved valuable in establishing our approach and making sense of results:

- Many factors contribute to terrorism phenomena and it is counterproductive to argue about "the" key factor: An interdisciplinary *system approach* should instead inform thinking from the outset.
- Existing social science identifies many relevant factors, but a multiresolution analytic approach is needed to make that information coherent.
- The answer to "Which factor matters most?" is, in most cases "It depends." Centrality of context is a first principle and establishing context should be the first order of business in organizing thought. The issues of the Taliban in Pakistan are simply not very comparable to those of the Irish Republican Army or to those of Hizballah and Hamas in Palestine, or even to the current activities in Baghdad.
- A combination of logical thinking and empiricism that draws on the social-science base allows us to go well beyond the dismissive "It depends," characterizing the circumstances in which one or a combination of particular factors is likely to dominate. This amounts to systematizing what experts already do inside their minds. Distinguishing such cases can go far toward explaining or resolving apparent contradictions or heated disputes in social science.
- In social science, it is seldom possible to make strong predictions: Many key facts are not known and "random factors" intrude. A better aspiration is to "improve the odds" of correct diagnosis and prescription and to lay the groundwork for rapid adaptations in response to more information, including that from experience at the time. He who "bets the farm" on the predictions of a model purporting to be based on social science is likely to lose that farm—if not the first time, then the second or third.

Overall, a good deal of structure and coherence can be found in the existing base of social science for terrorism and counterterrorism. Gaps exist in our understanding of the "It depends" contexts that might provide better guidance to policymakers. And many of these "It depends" contexts, such as al-Qaeda's decisionmaking and the prioritization of Afghanistan or Iraq, have significant implications in the near- to mid-term. Several nettlesome issues, such as the relative importance of supply versus demand, need further research and analysis. Even beyond the call for more and better data (and reanalysis exploiting improved distinctions among cases), much remains to be done in going beyond the "factor tree" descriptions and developing tighter and more analytic subject-area by subject-area (module-by-module) characterizations. Doing so will require a combination of theory-informed and data-driven research, as well as the systematic collection and dissemination of empirical data to researchers.

Acknowledgments

This monograph benefited from comments and reviews by Bruce Hoffman (Georgetown University and RAND consultant), Eli Berman (University of California, San Diego), Martha Crenshaw (Stanford University), Douglas McAdam (Stanford University), John Horgan (University of Pennsylvania), Brian Jenkins (RAND consultant), Steve Simon (Council on Foreign Relations), Mark Stout (Institute for Defense Analyses), and Ben Wise (RAND). We also benefited from seminar discussions with many of our RAND colleagues at the outset of work as we developed our plan of approach. Their comments were greatly appreciated, but—as always—the authors are solely responsible for the final manuscript.

CHAPTER ONE

Introduction

Paul K. Davis and Kim Cragin

Background

The Challenge Posed

U.S. defense planning has been changing dramatically in an effort to adapt to new threats and realities. The Department of Defense (DoD) has changed its strategic emphasis (Rumsfeld, 2006; Gates, 2009) and, as part of that, has put a priority on improving the usefulness and quality of its analysis of the overlapping subjects of irregular warfare, counterterrorism, and counterinsurgency. Doing so is challenging because the phenomena at issue are so different from those relevant to analyzing weapon systems or military forces in major combat. Analysis requires addressing issues in multiple dimensions.[1] Most of the dimensions deal with *social-science* phenomena, rather than, say, the physics of precision weapons or global navigation. They involve *people*, whether individuals, groups, organizations, interactions, or processes.

This distinct feature of the new analytic challenge led DoD to ask RAND for a critical survey of what is known from social science that should be reflected in analysis and supporting models of social-science phenomena in counterterrorism and counterinsurgency. Further, DoD specifically wanted the study to focus on the academic and otherwise scholarly literature and to address issues relevant to national counterterrorism strategy.

Approach

Challenge and Objectives

The challenge we faced was considerable, for reasons worth recounting here. First, the relevant social-science literature is highly fragmented in at least four ways: by academic discipline, by the divide between theory and empiricism, by methodological approach, and by level of analysis.[2] Second, much of the research is reported in what might be called model-hostile terms. That is, the research may provide interesting and important facts but not help someone who seeks to *reason* about terrorism and counterterrorism in cause-effect terms, or to extrapolate the insights from research in one area in establishing strategy for another.

To illustrate the challenge, consider that we were initially asked to address questions such as (1) What are the relationships between political reform and terrorism? (2) What are the relationships between economic opportunity and terrorism? (3) What social and cultural factors are important in terrorist recruiting? (4) What psychological factors and influences affect terrorism? and (5) What are the relationships between Muslim public opinion and al-Qaeda activities? These are all excellent questions and might have formed the basis for structuring our research. However, it was foreseeable that the answer to each such question would be "It depends." Such a conclusion would not be very useful but would be inevitable because the questions themselves are discipline-bound, which is a problem because in almost all cases multiple factors are at play.[3] We needed to be able to "put the pieces together" and go beyond "It depends."

The approach we settled on, then, would have a number of objectives: (1) providing a *system* perspective allowing discussion of either parts or the whole; (2) capturing the insights of scholars working at different levels of analysis, from different disciplinary and methodological perspectives, and with different mixes of theory and empiricism; and (3) being able to *communicate* our results to people from diverse backgrounds.

Organizing Questions

As a practical matter, it was important to choose the questions on which to focus. We found that we could accomplish our goals reasonably by organizing around the questions in Table 1.1. Some of the questions (1, 2, 3, 5, and 6) relate to the life-cycle stages of a terrorist organization: its genesis, growth, and decline. Some relate to the individual level of analysis (2 and 6) and some to the group level (3 and 5). Question 4 asks about the decisionmaking and behavior of terrorist organizations. All of the questions have implications for counterterrorism, but Question 7 addresses a particularly cross-cutting counterterrorism issue on which social science has something to say. More questions could always be added (and we will note some along the way), but—taken together—addressing these questions would cover a great deal of ground.

Scope and Character of Inquiry

Definitions

Each of the organizing questions in Table 1.1 presupposes an understanding of what constitutes terrorism, but a myriad of definitions exist and the scope of our inquiry would depend on the definition that we

Table 1.1
Organizing Questions

When and why does terrorism arise (that is, what are the "root causes")?
Why and how do some individuals become terrorists and others not?
How do terrorist organizations generate and sustain support?
What determines terrorist organizations' decisions and behaviors? What are the roles of, for example, ideology, religion, and rational choice?
How and why does terrorism decline?
Why do individuals disengage or deradicalize?
How can "strategic communications" be more or less effective?

adopted. To maximize the range of literatures we would tap, we used the most fundamental definition of terrorism:

> *Terrorism is the use or threatened use of violence for the purpose of inducing terror.*

Under this definition, terrorism is a tactic or strategy—one that can be employed by a state as well as by substate, nonstate, or individual actors. Research on such terrorism occurs in numerous literatures on political violence—literatures dealing with, for example, insurgency, rebellion, civil wars, and urban gangs. We wanted to be able to draw on all of them.

Although casting our research net broadly, our ultimate interest was the kind of terrorism most troubling to the United States today—that of subnational or nonstate actors. The usual characteristics of that type of terrorism include (Hoffman, 1998) (1) the existence of a terrorist organization with a chain of command or cell structure, (2) threats or acts of violence against noncombatants, (3) intended repercussions beyond immediate targets, and (4) pursuit of political goals. This subset of terrorism activity is what governments and readers ordinarily have in mind when referring to terrorism. It excludes state-supported terrorism, which is addressed by the laws of war, international condemnation of repression, and other activities outside what is usually considered to be counterterrorism.

Disciplinary Scope

The organizing questions in Table 1.1 are such that we would take an aggressively *interdisciplinary* approach, rather than, for example, viewing the issues through alternative disciplinary lenses and then juxtaposing the results.[4] In doing so, we would draw on the traditional social-science disciplines and their subfields, but also on such cross-cutting fields as terrorism studies, criminology, organization theory, and policy analysis (Table 1.2).

Table 1.2
Fields Drawn Upon

Traditional Social-Science Fields and Subfields	
Anthropology	Social and cultural, ethnology . . .
Economics	Microeconomics and macroeconomics . . .
Geography	Physical, human
History	Social, military . . .
Political science	Political economy, political theory, philosophy, . . .
Psychology	Cognitive, behavioral, social . . .
Sociology	Political, gender, demography, criminology, organization theory...
Cross-Cutting Fields	
Terrorism studies, criminology, organization theory, policy analysis . . .	

NOTE: The fields shown are rather standard (National Research Council, 1997), but some argue, for example, that history is part of humanities rather than social science.

Structuring the Research and Monograph

With the above considerations in mind, we decided that each of the questions in Table 1.1 would be the basis of a chapter-length paper. Each paper would be written by someone with a strong disciplinary background suitable to the particular question but someone who would also be able to survey literatures from the different perspectives. Initially, these papers would be written independently and with distinct characters reflecting author orientations and assignments, so as to capture differing perspectives. However, there would be overall guidance, considerable cross-talk along the way through meetings and sharing of drafts, and some integration to improve coherence of the whole. We then assembled the authors of this monograph—authors with disciplinary backgrounds in cultural anthropology and history, psychology, sociology, political science, economics, and the sciences, as

Table 1.3
Macro-Structure of the Monograph

Summary

Introduction

1. Introduction (Davis and Cragin)

Part One

2. Root causes (Noricks)

3. Radicalization: why individuals become terrorists (Helmus)

4. Popular support: how terrorists gain and sustain support of populations (Paul)

5. Economic analysis: what matters empirically and how rational-choice theory helps (Berrebi)

6. Decisionmaking of terrorist organizations (Jackson)

7. How terrorism ends (Gvineria)

8. Disengagement and deradicalization: how individuals cease to be terrorists (Noricks)

9. Implications of social science for strategic communications (Egner)

Part Two

10. Summary insights from social science (Cragin)

11. Representing and integrating social-science knowledge analytically (Davis)

12. Conclusions (Davis and Cragin)

Appendix A. Author biographies

Appendix B. Potential metrics (Bahney)

well as experience in system thinking and modeling (see Appendix A for brief biographies). Table 1.3 describes the chapter structure of the monograph.[5]

Darcy M.E. Noricks discusses the controversial subject of terrorism's root causes; Todd C. Helmus describes what is known about individual-level radicalization; and Christopher Paul covers the issue of how terrorist organizations gain and sustain public support and what support they in fact need. Claude Berrebi's chapter is cross-cutting. He reviews the hard empirical evidence on the significance or nonsignificance of some factors treated in other chapters and then comments on the usefulness of a version of the rational-actor model extended to recognize noneconomic utilities (for example, support of a cause). Brian A. Jackson draws on organizational theory to discuss the decisionmaking of terrorist organizations. Gaga Gvineria surveys what is known about how terrorism ends, identifying processes and factors different from those of earlier chapters. In the first of two chapters written with an eye on counterterrorism, Darcy Noricks asks what is known about what influences individuals to disengage from terrorism or even to deradicalize—an understudied subject. Michael Egner then pulls together lessons from the literature about strategic communications intended to influence potential or current terrorists, or the populations that may or may not support terrorist organizations.

Kim Cragin's chapter begins the integrative portion of the monograph by looking across the earlier chapters for cross-cutting themes or issues to be highlighted. Paul K. Davis's chapter discusses ideas and methods for integrating and representing social-science knowledge. Some of these guided work on earlier chapters (see below); others discuss what might be done to extend and tighten knowledge and to move toward more concrete models. The conclusions chapter (by Davis and Cragin) is brief because the monograph includes a lengthy executive summary. One appendix provides information on the authors; the other, by Ben Bahney, describes possible metrics to be used in measuring the factors identified in earlier chapters.

Analytical Guidance

As discussed earlier, we sought to take a causal-system-model approach so as to be able to see both the whole and parts.[6] That is, we sought to be able to address such analytical question as

- What *combinations* of factors stimulate terrorism (or its demise)?

- *When* are various factors or combinations most important (that is, in what circumstances defined by the many other factors at work)?
- Of these, which are potentially subject to influence through strategy, policy, and tactics?
- And, assuming efforts to influence *everything* that can be influenced, what degree of success might be anticipated and with what uncertainties and side effects?

Such questions are not the norm in social science, although researchers often address them along the way to a greater or lesser degree.

Paul K. Davis elaborates on this theme in Chapter Eleven, but a few particular concepts discussed there influenced the way other chapters were written. To impose an interdisciplinary view that would assist in framing questions addressing combinations of factors, and to reduce the cognitive complexity created by having scores of factors, we drew on the theory and methods of multiresolution modeling and gave the following guidance to chapter authors:

- Since you are likely to identify a great many factors affecting the question you are addressing, develop hierarchical abstraction trees ("factor trees") to organize the factors of your chapter to be useful in understanding the whole of your topic and pointing toward what might constitute a model and an approximate theory. Attempt to construct the trees as "influence diagrams," but be willing to suppress some real-world complexities in doing so (for example, feedbacks and weak interactions).
- Distinguish between factors that are contributory to a phenomenon and factors that are more or less *necessary.* Distinguish between contributory factors that are and are not *substitutable* for one another.
- Where feasible, describe the approximate combining logic relating multiple factors to the given phenomenon or effect.
- Alternatively (a function of the chapter), identify the "processes" most relevant to your topic, thinking about them where possible in causal-model terms.

In summary, the guidance asked chapter authors to consider multiple variables simultaneously, to organize the variables at different levels of resolution, to modularize, to approximate, and to describe first-order combining relationships. As discussed in the integrative chapters, a good deal of sense-making is possible by looking across the chapters and exploiting the structures resulting from the instructions. The value of this approach was amply demonstrated in discussions and debates within the project itself and in subsequent briefings of results to a variety of audiences.

The guidance reflected recognition that the existing literature cannot fully answer the analytical questions listed above. Thus, the chapter authors were asked to take first steps toward answering them, first steps that could be communicated readily to and discussed among people with highly varied backgrounds. Chapter Eleven discusses potential next steps in extending and tightening social-science knowledge.

With this background, then, let us proceed to discussion of root causes and then the other questions outlined in Table 1.1.

Bibliography

Davis, Paul K., and Brian Michael Jenkins, *Deterrence and Influence in Counterterrorism: A Component in the War on al Qaeda*, Santa Monica, Calif.: RAND Corporation, 2002. As of March 5, 2009:
http://www.rand.org/pubs/monograph_reports/MR1619/

Gates, Robert, "A Balanced Strategy," *Foreign Affairs*, Vol. 88, No. 1, January/February 2009.

Geddes, Barbara, *Paradigms and Sand Castles: Theory Building and Research Design in Comparative Politics*, Ann Arbor, Mich.: University of Michigan Press, 2003.

Gupta, Dipak, *Understanding Terrorism and Political Violence*, New York: Routledge, 2008.

Hoffman, Bruce, *Inside Terrorism*, New York: Columbia University Press, 1998 (2nd ed. issued in 2006).

National Research Council, *Research-Doctorate Programs in the United States: Data Set*, Washington, D.C.: National Academy Press, 1997.

Rumsfeld, Donald, *Quadrennial Defense Review Report*, Washington, D.C.: Department of Defense, 2006.

Stout, Mark E., Jessica M. Huckabey, and John R. Schindler with Jim Lacey, *The Terrorist Perspectives Project: Strategic and Operational Views of al-Qaeda and Associated Movements*, Annapolis, Md.: Naval Institute Press, 2008.

Endnotes

[1] In DoD parlance, the dimensions are sometimes abbreviated by the acronym PMESII, for political, military, economic, social, infrastructure, and information. Another common abbreviation is DIMEFIL, standing for diplomatic, information, military, economic, financial, intelligence and law enforcement—major elements of national power.

[2] This fragmentation also includes viewing issues through very different paradigms, making comparison and integration difficult, but nevertheless important (Geddes, 2003; Gupta, 2008).

[3] Because this is so, it is particularly frustrating when debate trends toward simple-formula answers, such as "It's all about poverty," "it's all about repression," or "it's all about Islamic extremism."

[4] The distinction we draw is sometimes referred to as interdisciplinary versus multidisciplinary.

[5] We should also mention omissions. Despite the breadth of our effort, we did not include (1) case histories or other in-depth information on particular terrorists or terrorist groups, (2) a critical survey of the counterterrorism literature per se, although we drew on that literature, (3) in-depth information on terrorist groups or terrorist individuals (such as bin Laden), or (4) discussion of existing models for counterterrorism and counterinsurgency. We also relied exclusively on the unclassified literature and did not use, for example, sensitive information arising from the ongoing conflicts in Afghanistan and Iraq, except for material that has been released in the open literature (e.g., Stout et al, 2008).

[6] The desirability of doing so had been suggested in earlier RAND work (Davis and Jenkins, 2002).

CHAPTER TWO

The Root Causes of Terrorism

Darcy M.E. Noricks

Introduction

Objectives

The "root causes" of terrorism are not the proximate causes of terrorism, but rather factors that help establish an environment in which terrorism is more likely to occur. In this review paper I focus primarily on societal- or state-level causes, both structural and sociocultural. In doing so, I draw on the scholarly literature of terrorism and political violence more generally. I consider political, social, cultural, and ideological factors; I also touch on organizational factors and social ties.[1]

The structure of this paper is as follows. The first section identifies the varieties of relevant literatures; the next sections review those literatures, rather much on their own terms; the next section then summarizes points of agreement and disagreement. The following section describes my effort to integrate the material. Finally, I describe some implications for how policymakers should think about root causes, and about research that still needs to be conducted.

The Political Violence Literature

One difficulty with the root-causes literature is the conflation of the terrorism literature with the rebellion/revolution, ethnic conflict/ethnic riots, and civil war literatures—not to mention the social movement literature. Each of these has something to offer in understanding root causes, but there has been no structured effort to determine which lessons from the other literatures are or are not relevant, and why. The conflation is particularly significant, because many terrorism studies assume relationships among factors based on, for example, empirical work about revolutions. They do so despite the argument of many scholars of social revolutions that their set of theories does not apply to terrorism.[2] Goodwin (2005) notes, in contrast, that there is a growing literature on Islamist movements as a "revolutionary phenomenon" (p. 404).

There is a good argument for considering the literatures more holistically; both earlier and newer scholars have suggested relationships among the various types of political violence (Gurr, 1970; Blomberg, Hess, and Weerapana, 2004; Bjorgo, 2005). Bjorgo suggests that terrorism is often "an extension and radicalization of various types of conflict"; he cites ethnic conflict and conflict between rival ideological groups in particular and goes on to conclude, "Obviously the root causes of such conflicts are also root causes of terrorism" (p. 4). Blomberg, Hess, and Weerapana developed a model that predicts when political violence will take the form of terrorism rather than the form of a civil war, coup, or revolution. These authors follow closely in the footsteps of Gurr, who hypothesized that the type of political violence was related to the balance of coercive control between the regime and the dissidents but was also determined by the level of organizational membership of the government and the dissident group; the scope and opportunities for protest against the regime; and the geographic concentration of dissidents in isolated areas. Different combinations of these factors resulted in one of three types of violence: turmoil, conspiracy (including small-scale terrorism), and internal war (including large-scale terrorism and guerrilla wars).

It seems apparent that similar root causes likely play a role in different levels and types of political violence. The precise nature of that relationship is still unclear and would require a much more extensive effort than this paper allows. At the risk of undermining my own critique of conflating the political violence literatures, however, I have elected to include the literature on related types of political violence (for example, ethnic conflict and civil wars) where relevant. Because these literatures are much more extensive than the terrorism literature in hypothesizing links between root causes and political violence, I refer to them at times to round out the discussion of relevant factors or to emphasize areas where conclusions conflict.

Factors

In the sections that follow, I attempt to categorize and review the various root-cause explanations for terrorism. These theories about terrorism are multilayered and the categories are not neatly bounded, so there are overlaps and duplications across topical areas. Some of the studies about root causes test multiple variables as well as their interaction.

Precipitant Versus Permissive Factors

The fuzzy boundaries of the root-causes concept partly result from the important distinction made in the literature between permissive and precipitant factors (Crenshaw, 1981; Newman, 2006). Permissive factors—also called preconditions—are the more traditional causes that set the stage for terrorism over the longer term. They help to establish a context in which opportunities for terrorism are created. A precipitant factor, in contrast, is an event or incident that helps catalyze or trigger a change in behavior, particularly a move toward violent action. Precipitant events may also feed into grievances. The concept of root causes includes both permissive and precipitant factors, although the latter are conceived more as symptoms of the problem than as sources of the illness. Newman (2006) explains, "certain conditions provide a social environment and widespread grievances that, when combined with certain precipitant factors, result in the emergence of terrorist organizations and terrorist acts" (p. 750). He also suggests that it is the structural factors—as well as underlying grievances—that determine the operational base and that provide recruits and ideology. Precipitant factors, in contrast, provide a window of opportunity, determine leadership and organization, and help to shape the political agenda.

Among the factors considered permissive are lack of political opportunity, perceived illegitimacy of the regime, economic inequality, social instability resulting from the processes of modernization, and cultural and ideological factors, such as cultural acceptance of violence.

Precipitant factors are also seen as important in the broader literature about political violence, particularly in the literature on ethnic violence (Varshney, 2001; Horowitz, 2003). Precipitant factors can include a wide range of events and phenomena (for example, repression of the targeted group, rumored or threatened disruption of a targeted group, and failed elections); the important thing is that the participants in violence perceive the precipitating event as significant. This section focuses more closely on permissive factors and then discusses precipitant factors.

Categories of Permissive Factors

Permissive factors can be divided into global systemic explanations that emphasize such factors as the pressures of the international system, state structural explanations that emphasize political and economic conditions internal to the state, and social and cultural factors (see Table 2.1).

 Global Systemic Explanations. A number of scholars have suggested global systemic explanations for the development of terrorist groups (Cronin, 2003; Rapoport, 2001; and Sedgwick, 2004, 2007). This stream of literature gives less emphasis to various root "causes" of terrorism than to seeing terrorism as a consequence of event dynamics in a period of time. The contrast in perspectives is quite interesting. Rapoport identifies four, overlapping, historical waves of terrorism: anti-empire; anti-colonization; anti-Western; and the current, religion-based fourth wave, prompted by the Iranian revolution and the defeat of the Soviet Union in Afghanistan. Each wave is defined by a particular set of tactics, goals, and ideologies. Each begins with a precipitating event or events, lasts about 40 years before breaking, and dissolves as another wave rises to take center stage. The waves also build on one another in the vein of the social movements literature on contagion, demonstration effects, and the diffusion of tactics. For example, the success of North Vietnamese guerrilla tactics against the United States in Vietnam provided a model for other revolutionary movements in the third wave and modern proof-positive that asymmetric tactics could be effective (pp. 420–421).

Table 2.1
Permissive Factors

Class of Permissive Factor	Factor
Global systemic factors	Global systemic explanations
State structural factors	Perceived illegitimacy of the regime
	Repression
	Democracy
	Modernization
	Economics
Social and cultural factors	Education
	Human insecurity
	Grievances and anxieties
	Mobilizing structures and social ties
	Ideology, religion, and culture

Cronin builds on Rapoport's model and expands on the conditions that undergird what she calls the "jihad era" of terrorism. She is particularly interested in the effects of globalization and the facilitative effect of increased linkages between terrorist groups during the 1970s and 1980s. She concludes that "Terrorism is a by-product of broader historical shifts in the international distribution of power in all its forms—political, economic, military, ideological, and cultural" (p. 53); terrorism is a means through which individuals can exert control over their globalized environments. Sedgwick agrees with both Rapoport and Cronin that local causes cannot explain the appearance of global waves of terrorism. Nor is the idea of a prevailing ideology or the idea that major global events trigger waves of terrorism a satisfactory explanation. Instead, Sedgwick posits a "diffusion effect." The recent success—or apparent success—of terrorism as a strategy, anywhere in the world, is what explains others' embracing it, and the resultant global waves of terrorism.[3]

The literature on "protest cycles"—also called "cycles of contention" (Tarrow, 1995; Koopmans, 2004), "protest waves" (Karstedt-Henke, 1980), and "moments of madness" (Zolberg, 1972)—is a prominent strain in the social movements literature, although it largely considers clusters of reform movements in western democracies that saw only modest violence at the very end of these cycles. This literature argues that we can empirically observe the rise and radicalization of social movements as well as the occurrence of protests and demonstrative actions as clustering in time and space (McAdam, 1994). Tarrow (1994) specifies that a cycle of contention is "a phase of heightened conflict across the social system," with "intensified interactions between challengers and authorities which can end in reform, repression and sometimes revolution" (p. 153). He explains that these cycles begin when the government is revealed to be vulnerable to social change. During this time, collective action diffuses from more mobilized groups (for example, students) to those who are traditionally less so (for example, peasants, workers in small industry), new "master frames" of collective action are created and diffuse when the early groups are seen to be successful, and repertoires of contention expand (such as innovation and diffusion of new tactics) (pp. 92–94).

Tarrow's protest wave is similar in form to Rapoport's but with more detail at the organizational level of analysis. The zenith of Tarrow's cycle is when new organizations flourish and existing organizations struggle to maintain membership and relevance. This leads to competition between groups, some of which attempt to differentiate themselves by adopting more radical tactics. Some of the groups move toward institutionalization to maintain mass support, but this often leads their competitors to radicalize "to gain the support of the militants and prevent backsliding" (Tarrow, 1994, p. 148). Another factor influencing the twin response of institutionalization and radicalization is government action. When government pursues a combination of co-option (with the moderates) and repression (of the radicals)—which is most often the case—this tends to result in even more violence on the part of the radicals. Tarrow's last phase is demobilization.

For Karstedt-Henke (1980), terrorism is the inevitable result of a protest wave. The four phases of the wave include

1. **Mobilization:** After protesting begins, the authorities repress the outburst but do so in an "inconsistent and undifferentiated way" that leads to public anger and provokes additional protests (pp. 200–209).[4]
2. **Differentiation:** The authorities continue to use repression but also attempt to reach out to groups they think will be amenable to conciliation. Because the authorities are still unfamiliar with the players however, they sometimes confuse the groups and misapply their intended measures, leading to continued growth in both the radical and moderate wings of the movement (pp. 209–213).
3. **Integration or radicalization:** Increasing differentiation between moderate and radical groups allows the authorities to co-opt the moderate wing, which becomes integrated into the political system. This allows the radical wing to have more control over decisions about tactics and allows the authorities to accurately identify members of the radical wing whom they target for full-scale repression. This results in a tit-for-tat of violence and counterviolence, which produces terrorist groups (pp. 213–217).
4. **Latency:** The last phase of the wave is a decline in protest activity as moderates become institutionalized and radicals lose members because of the high costs of participation and radical groups' need to go underground (pp. 217–220).

Koopmans' (1993, 2004) theory builds on both Karstedt-Henke's and Tarrow's but again focuses on tactics. According to Koopmans, the initial phase of a protest wave focuses on novel strategies, such as confrontational protests, which will be covered by the media and will help to mobilize people in large numbers (1993, p. 654). Once the novelty wears off, as the protest wave continues, movements tend to lose adherents. Movements then decide between a strategy of pursuing additional participants or increased violence. The former is more likely if preexisting, co-optable groups are willing to ally themselves with the movement's goals. Similar to both Tarrow's and Karstedt-Henke's theories, this is where radicalization occurs. These preexisting groups are usually moderate. Incorporation of these groups leads to intergroup conflict between moderates and radicals, leading the radicals to become increasingly militant and sometimes violent to ensure that their views are heard. Increased radicalism may also usher in a period of decline in protests, either because internal conflict among activists redirects their energies away from external activities or because of increasing repression and marginalization by the authorities. Koopmans (1993) warns that extreme violence or terrorism can also provoke extreme levels of repression, which then "undercuts the general legitimacy of protests" (p. 655).

The broader social movement literature has also increasingly been used as a lens through which to view and better understand the activity of Islamist groups, because they are seen as part of a larger movement that has swept both Africa and Asia in recent times. Charrad (2001) emphasizes the "wave" of Islamic fundamentalism that gathered throughout the 1980s (p. 170). Maddy-Weitzman (1997) elaborates, "Signs of an Islamic revival outside of authorized state structures were widespread during the 1970s and 1980s" (p. 11).

Discussion and Cautions. Tarrow, Karstedt-Henke, and Koopmans all insist on the role of government repression in catalyzing later stages of the wave, but it is difficult, in these accounts, to unlink the choice of violence from competition with (or an attempt to differentiate themselves from) moderate groups who are securing mass support through the institutionalization process. These issues come up again in separate sections on repression and democracy, below.

Global systemic explanations for terrorism have been criticized for their lack of historical accuracy (Gelvin, 2008), and "wave theories" tend to be inconsistent and more interesting than compelling (Gvinernia, 2009). The difficulty of defining the boundaries of any particular wave a priori are obvious. However, a powerful theme across the political violence literature is the importance of historical context (rather than just global influences). Skocpol (1979), for example, specifies that her theory of revolution is most relevant in the context of a specific "world historical time": pre-colonial agrarian autocracies forced to meet the challenges of modernization. Piven and Cloward (1979), too, specified the importance of historical context, noting that "Popular insurgency does not proceed by someone else's rules or hopes; it has its own logic and direction. It flows from *historically specific circumstances*: it is a reaction against those circumstances, and it is also limited by those circumstances" [emphasis added] (p. xi).

Perhaps the most interesting fact about wave theories is the one that is least often discussed: that wave theories' own parameters imply that at some point the current wave will crash and peter out and will be followed by the rise to prominence of a new, historically specific wave. A few authors have hypothesized the next wave: Kaplan (2007) suggests that the fifth wave includes ethnic and tribal groups, which seek to realize a utopian vision of society at a local level and within a compressed time frame. The defining tactic of this wave is rape. Some futurists have suggested that environmental terrorism is next on the docket, with the book *The World Without Us* as the guiding ideological framework. Still others hold that Islamist-inspired terrorism began as a ripple in the broader religious fourth wave, which gained enough momentum to take on a life of its own. One fear is that just as the returnees from the Afghanistan conflict contributed to the problems of radicalization in the various home countries of the Mujaheddin, the returnees from the Iraq conflict will pose a similar problem.

State-Level Structural Explanations. Responding to the contention that political violence is shaped by historical circumstances, a number of scholars have attempted to isolate common variables that might be shaping situations most vulnerable to political violence. The largest body of literature associated with root causes deals with explanations that emphasize the social, political, and economic characteristics of a society. This group of explanations is extremely broad. It includes variables as high-level as the economic system and the language or languages spoken in a given state. But it also includes such factors as variation in social mobility across societies. Specific factors emphasized in the terrorism literature include political factors, such as perceived legitimacy and state repression; economic factors, such as wealth disparity and poverty; social factors, such as demographics and human security; and cultural and ideological factors. I begin the next section with a discussion of political factors, about which there is substantive agreement, including the importance of the perceived legitimacy and strength of a regime. Then I discuss other structural factors, about which there is much less agreement, including the role of economic factors. Finally, I discuss social and cultural factors, which sometimes act as permissive causes (for example, education, religion, human insecurity) and at other times function as precipitant causes (for example, grievances, mobilizing structures).

Perceived Illegitimacy of the Regime. Many scholars emphasize the delegitimation of the state to help explain the appearance of violent behavior directed at the state (della Porta, 1995; Moore, 1966; Sprinzak, 1990; Weinberg, 1991; Weinberg and Pedahzur, 2003; Weinberg, Pedahzur, and Perlinger, 2008). Variations on this explanation focus on a contending group's perceptions of the legitimacy and weakness of a regime. As a regime is perceived as increasingly illegitimate, the likelihood that an oppositional group will use violence increases. Sprinzak suggests that this is because the group extends delegitimation of the regime to every individual associated with the regime through dehumanization and depersonalization (Sprinzak, 1990, pp. 80–82). Weinberg (1991), Weinberg and Pedahzur (2003), and Weinberg, Pedahzur, and Perlinger (2008) found that political parties are most likely to use terrorism when the group has grandiose goals—such as the establishment of a new social order—and when party doctrine emphasizes the illegitimacy of the existing regime (Weinberg, 1991, p. 437). The combination of a regime that is perceived as both illegitimate and weak is a ripe permissive condition for terrorism.

Some experts see an accumulation of permissive conditions that degenerate the relationship between citizens and the state until a point of crisis occurs. Delegitimation most often occurs during a period of political or social change within the state, although it might also occur in reaction to international events—such as foreign occupation (Pape, 2003, 2005). The point of crisis might occur when a state's efforts to reform or modernize are blocked by one or more competing groups of elites. The link between elite disenfranchisement and terrorism is another political factor commonly pointed out by experts, although mainly in literature concerning revolutions. The state's failure to mobilize sufficient resources for reform leads to administrative and military collapse, thereby creating a political opportunity for rebellion (Skocpol, 1979).[5]

Repression. A similarly agreed-on substantive factor linked to terrorism is the role of government repression in moving a group from nonviolent to violent tactics (Callaway and Harrelson-Stephens, 2006; Crenshaw, 1995, 2001; della Porta, 1995; Gurr, 1970; Weinberg, 1991). Delegitimation of the state may occur over time through repeated interactions between the state and a nonviolent group, but the most significant action a state can take is to use violence against its own citizens. Della Porta suggests that the use of state violence against mobilized groups is a key mechanism for delegitimation. Researching German and Italian terrorist groups from the 1970s, della Porta found that the state's use of excessive violence not only delegitimized the state but at the same time legitimized the use of violence by activists.[6] This is in line with Gurr's earlier observation that a group's initial reaction to the use of government force is to perceive the government's use of force as a legitimization of the use of force overall (Gurr, 1970, p. 17). Weinberg (2001) notes that repression not only increases the likelihood that a mobilized group will turn to terrorism but that failed repression also emphasizes the state's vulnerability to being overthrown (p. 437). In della Porta's cases, activists perceived the state's excessive use of force (for example, police brutality, the death of activists in prison) as a rejection of the democratic compact. The activists concluded that the only way to oppose an authoritarian state was through violence (della Porta, 1995, p. 158). Callaway and Harrelson-Stephens (2006) go so far as to assert that state repression ("denial of security rights" in their terminology) is a necessary condition for the creation and growth of terrorism, although that is not rigorously true.[7] Moreover, state repression creates "martyrs and myths" and encourages "secondary deviation—the individual's even stronger commitment to his or her deviant behavior" (della Porta, 1995, p. 191). In addition, a government's excessive or perceived excessive use of force can sometimes send moderates into the ranks of the extremists (Crenshaw, 1995, 2001; Gupta, 1990). In the context of a delegitimized regime, state repression catalyzes a violent counter-response from the opposition group.

Other processes that have been suggested to lead to regime dele-gitimation include (1) regime support for unpopular economic, social, or cultural institutions; (2) evidence of corruption; (3) weak infrastruc-tural power (for example, policing, provision of services); (4) exclusion of mobilized groups from political participation or access to resources (sometimes reflected in class struggle or elite disenfranchisement); and the specification of the repression hypothesis to include (5) regime use of *indiscriminate* violence against oppositional groups or political repre-sentatives. Although revolutions scholar Jeff Goodwin (2001) does not refer explicitly to regime legitimacy, he suggests that all of these prac-tices, when employed by the state, have a cumulative effect that can lead to the development of strong opposition movements. Moreover, delegitimation of the regime occurs in the context of rising inequality and increasing resistance, during which frustrated members of politi-cal parties or social movements become alienated and militarized in response to fraud and repression, concluding that violence is "the only way out" (pp. 25–26).

Democracy. If regimes are viewed on a spectrum of legitimacy, surely democratic regimes are at the high end of that spectrum (assum-ing a degree of competence and security). However, there is disagree-ment about the effects of democracy and political inequality on ter-rorism. Empirically, the results are mixed, although there is more agreement as the variable is increasingly disaggregated. Some schol-ars, taking a page from democratic peace theory (Kant, 1795; Doyle, 1983), argue that, since democracies provide increased opportunities for both participation and nonviolent resolution of conflict and griev-ances, democracies should be less likely to produce terrorism (Schmid, 1992; Gurr, 2003; Li, 2005; Kaye, Wehrey, Grant, and Stahl, 2008). Although Eisinger studied cities rather than states, his early (1973) work distinguishing those American cities that did or did not have rioting in the 1960s concluded that cities with more available avenues of political participation tended to preempt riots by offering an oppor-tunity to redress grievances. Along these lines, Engene (1998) found that successful unionization was negatively correlated with domestic terrorism (Lia and Skjolberg, 2004).

Eubank and Weinberg's (1994, 1998, 2001) studies surprised many by finding the opposite: that democratic regimes are actually *more* likely to both host terrorist groups and experience terrorist violence than are authoritarian regimes, and that political and civil liberties are positively correlated with terrorism. Scholars who align themselves with this side of the debate argue that democracy is positively correlated with terrorism because democracies offer increased opportunities for terrorism thanks to freedom of movement and association and access to such potential targets as government buildings, thereby lowering the costs of conducting terrorism. They also argue that democracies have a more difficult time convicting terrorists because of efforts to protect civil liberties, which result in constraints on government action (Crenshaw, 1981, p. 383; Gurr, 1998; Wilkinson, 2001).

Eubank and Weinberg's initial (1994) findings were challenged on the grounds that they should have used incident data—rather than the number of terrorist groups within a country—for more methodologically rigorous identification of the countries affected by terrorism (Sandler, 1995). Their subsequent efforts (1998, 2001) did just that but achieved similar results. Li (2005) pointed out in a subsequent study that, since incident data are collected from open sources, there is a natural upward reporting bias of terrorism in democratic countries because of greater press freedom. Hoffman (1998) suggests that it is precisely this press freedom, as well as "the unparalleled opportunities for publicity and exposure that terrorists the world over know they will get from the extensive U.S. news media" (p. 137) that makes American targets so attractive.

Further efforts attempted to clarify the issue by disaggregating the variables. Eyerman (1998) found that new democracies were the most likely to experience terrorist violence, whereas established democracies were less likely than nondemocracies to experience terrorism. Li (2005) differentiated democracies in terms of the type of electoral system: proportional representation (such as Spain, Germany, and Norway), majoritarian (the United States and United Kingdom, for example), or mixed (for example, Ecuador, Russia, and Taiwan). He found that proportional representation systems experience fewer incidents than the other two. This aligns with similar findings in the civil-war and ethnic-conflict literatures. Reynal-Querol (2002) finds that civil war is less likely in states with proportional systems. Moreover, proportional representation systems are a more stable solution for ethnic conflict–prone regions because majoritarian systems are less amenable to representing minority interests (Carnegie, 1997). A post-9/11 National Research Council report, edited by Smelser and Mitchell (2002), found that terrorism "was discouraged by policies of incorporating both dissident and moderate groups responsibly into civil society and the political process" (p. 2). Proportional systems are believed to do this more effectively than majoritarian (sometimes called "winner-take-all") systems.

In some respects consistent with Eubank and Weinberg's findings, Pape (2003, 2005) argued that suicide bombers are particularly likely to target democracies that are perceived to be foreign occupiers and that democracies are particularly vulnerable. Pape argues that terrorists, rather than selecting suicide terrorism as a tactic because of extreme religious views, are driven by nationalism, along with a belief that suicide bombing works as a *strategy*. (Also see Bloom, 2005).

Pape's work demonstrated that the suicide-bombing phenomenon need not be the result of Islamist extremism; rather, it could be more deeply rooted in secular considerations and strategic logic (the feasibility of coercing an occupying power to leave). However, some aspects of Pape's 2003 article were challenged, including his undervaluing the role of religion (particularly Islam) and his claim that democracies were particularly likely targets (see, especially, Moghadam, 2005; Piazza, 2008; Wade and Reiter, 2007). Wade and Reiter (2007) reassessed Pape's data and expanded the dataset; they found that democratic states experience more suicide terrorism but that this is correlated with the number of religiously distinct minority groups within the country. More religious minority groups correlates with more suicide terrorism in democracies. In addition, they determined that "the size of [the] country, whether it [was] a majority Muslim state, and its past experience with terrorism" had a larger effect than democracy (p. 330). And in contrast, although Muslim states are more likely to experience suicide terrorism, this effect is mitigated by the degree to which the Muslim state is democratic. This finding led Wade and Reiter to suggest that "democracy may be a partial palliative for suicide terrorism among Muslim states" (p. 342).

Recent RAND research on the Middle East also found some support for the idea that potentially violent actors can be successfully co-opted into the political system (Kaye, Wehrey, Grant, and Stahl, 2008).

Discussion and Cautions. All of the studies on the relationship between democracy and terrorism have difficulties, in part because of their level of aggregation and reluctance to deal with country-specific contexts. Further, one may ask whether the research is helpful to strategy and policy. After all, no one is seriously proposing to recommend against democracy because of some difficult-to-interpret correlational data relating democracy to terrorism. Arguably, the value of the debate for decisionmakers has to do more with increasing wisdom than with identifying policy levers. One consequence of wisdom may be more differentiation and subtlety, rather than broad-brush assumptions that more democracy is good.

Despite the points of contention, there are some areas of substantive agreement. Neither the most free nor the most authoritarian states experience more terrorism. Rather, states with an intermediate level of political freedom are more prone to terrorism (Abadie, 2004, 2006). Violence also often increases immediately after democratic government is instated, particularly when newly democratizing states are also in the midst of market liberalization (Chua, 2002). This may be because the institutions commonly thought to facilitate peaceful protest in a democracy are too immature to either function efficiently or to garner sufficient public trust. Or it may be because "Turbulence is an inevitable by-product of democratic principles and processes" (Rapoport and Weinberg, 2001, p. 3). In either case, social instability is commonly linked to terrorism. It is also clear that terrorism requires some degree of political space to operate. Repression may undermine a state's legitimacy, but total repression can effectively stifle dissent. Tilly (1978) explained that authorities can decrease the likelihood of protest by offering activists less-costly routes for achieving their desired ends, but he also conceded that authorities can decrease the likelihood of protest if they use a sufficient level of repression (see also Callaway and Harrelson-Stephens, 2006). Perhaps most important, disaggregating the variable of democracy by type of electoral system brings the findings in the terrorism literature (Li, 2005) into alignment with those of the ethnic conflict and civil war literature (Smelser and Mitchell, 2002)—emphasizing the value of a proportional system that best incorporates both moderate and dissident groups.

Modernization. In line with some of the hypothesized links between new or transitioning democracies and terrorism, modernization is thought to result in a similar kind of turbulence and social instability. The general link between modernization-related turbulence and political violence is another area of substantive agreement. The classic argument is that the process of modernization destabilizes society and may ultimately weaken the perceived legitimacy of the state as it undergoes a rapid expansion and centralization (Crenshaw, 1981; Kegley, 2002; Lia and Skjolberg, 2004). Modernization can also be conceived of as processes of social and political change that accompany economic evolution and that generally include the breakup of the traditional family—in part because of increased mobility and the related migration of wage earners (Wilkinson, 1974). On an individual level, modernization is often associated with job loss and weakened family and community ties.

The concept of modernization is a bit too broad to be empirically useful, but it is based on the idea that societies transition from more traditional patron-client relations toward market relations, which leads to an interrelated set of factors, each constituting a potentially significant permissive cause of terrorism (for example, urbanization [Massey, 1996], population density, and advances in transportation and communications infrastructures [Crenshaw, 1981; Jenkins, 1980; Hoffman, 1998], and other technological changes [Jackson, 2001]. The latter have tended to be more important in terms of tactics, targets, and organization than in terms of root causes of terrorism. However, as suggested elsewhere (Helmus, 2009; Paul, 2009), the relationship between terrorism and modernization might best be understood viscerally in terms of the effects of feelings of desperation, loss of valued traditions and relationships, and general anxiety that often accompany modernization.

Population. Other factors produced by the modernization process and potentially related to an increased likelihood of terrorism are population growth—the "youth bulge" in particular (Urdal, 2006; Ehrlich and Liu, 2002) and uneven population growth across ethnic and religious groups within a given state (Stern, 2003). Goldstone (1991) compares revolutions to earthquakes and asks, What conditions create a permissive structure in the first place? His answer is demographics. He argues that population growth alters the relationship between population size and the state's ability to provide services, causing new social stresses and pressures for change. Population pressures lead to increased demand for resources, an expansion of the army and related rising costs, an expansion in the size of the elite population jockeying for positions, and urban migration and falling real wages—particularly as a result of increasing numbers of young unemployed (pp. 24–25). After some resistance to change, a precipitating event (such as bankruptcy, rioting) occurs, weakening that resistance and leading to subsequent state breakdown if the state's institutions are not flexible enough to accommodate these changes (pp. 35–37). Revolutions occur when a state experiences three simultaneous challenges: (1) state fiscal crisis because of an inability to respond to demographic changes, (2) intra-elite competition and division, and (3) high mobilization potential within the population as a result of an increased sense of deprivation combined with conditions such as a large numbers of unemployed youths and growing urbanization with its attendant pressures (high rents, low wages) (p. xxiii). In most of these arguments, however, it is state weakness and societal instability that really creates the political "space" for political violence; population growth is merely one factor that triggers a series of changes that may lead to state weakness and may complicate the state's ability to provide services to the population. The period of state crisis "increase[s] the salience of heterodox cultural and religious ideas; heterodox groups then provide both leadership and an organizational focus for opposition to the state" (p. xxiv).

Economics. Berrebi (2009) covers economic factors in greater depth, especially the issue of the rational-choice model and its usefulness, but I would be remiss to ignore them completely in a paper that purports to discuss root causes. Much as with the issue of political freedom and democracy, there is decided disagreement on the importance of economic factors in the development of terrorism. Some of this disagreement is due to conflating micro and macro levels of analysis. The micro evidence is consistent that leaders or organizers often come from fairly privileged backgrounds (for example, Mao Tse-tung, Fidel Castro, and Osama bin Laden). They act, or claim to act, on behalf of relatively impoverished segments of society. In contrast, there is no reason to expect (and no evidence to suggest) that typical participants in a terrorist organization will have similar backgrounds. Another significant problem in parsing this issue is determining what aspect of "economic" we should be, and are in fact, measuring. The majority opinion seems to be that poverty, at least, is not at all predictive of terrorism. The article most cited as evidence of the lack of relevance of economic factors is Krueger and Maleckova (2003). Using several different methods and types of data (Hizballah militants, Palestinian suicide bombers, Israeli Jewish Underground members), they find no evidence for a poverty-terrorism connection. In contrast, they find some evidence that individuals with higher incomes and higher education levels are slightly more likely to join a terrorist group. One of the datasets on which they rely (Berrebi, 2003) found a positive correlation between a higher standard of living and participation in Palestinian terrorism in Israel. Recent work by Berrebi (2007) corroborates and expands on his earlier work. Krueger and Maleckova find stronger evidence that a lack of civil liberties is more directly correlated to participation in terrorism. Abadie (2004, 2006) replicates these results using a different dataset and also controlling for political rights and geography.

Somewhat contradictory research (Li and Schaub, 2004) found that economic development in a given country and its trading partners decreases the likelihood of terrorism, and that there is some evidence of a positive correlation between very high unemployment and political violence (Gupta, 1990). Robison, Crenshaw, and Jenkins (2006) also find that levels of foreign direct investment correlate with reduced transnational terrorism over time. Paxson (2002) uses Richard Rose's survey research in Northern Ireland and finds that Protestants with higher incomes and higher levels of education profess more moderate views and less support for terrorism. Among Catholics, however, income does not seem to matter—although more education is also associated with rejection of "hard-line" views. Lia and Skjolberg (2004) test the contention that terrorism happens least in the world's poorest countries (for example, Sub-Saharan Africa). Using the RAND-

MIPT Terrorism Incident Database[8] rather than the more commonly used ITERATE dataset[9] (Krueger and Maleckova assembled their own event dataset from State Department reports), Lia and Skjolberg find Africa to be the continent with the highest number of terrorism-related injuries in the seven years before 2004, even though there is widespread underreporting of terrorism in less-developed countries.

It is interesting to note that successful programs to deradical-ize terrorists often involve economic incentives (Cragin and Chalk, 2003; Ibrahim, 1980; also see "Preachers to the Converted," 2007, and Noricks, 2009). Moreover, a recent report on detainees in Iraq (Bowman, 2008) noted that the U.S. military is currently releasing more detainees than it is bringing to detention centers. According to National Public Radio's Tom Bowman, this is because the United States discovered that the majority of those detained were "young, poorly educated men without jobs who accepted money from al-Qaida in Iraq (AQI) to serve as lookouts, or to build or plant roadside bombs." In support of this contention, Major General Doug Stone, the head of American detention facilities in southern Iraq, conceived a plan to keep newly released detainees from returning to AQI's control. Detainees are monitored for a period of six months after release. For each month that they return to the detention center to check in, they receive a sti-pend of about $200 a month, roughly equivalent to what they were previously receiving from AQI (Bowman, 2008).

Although the contradictory nature of these findings about the links between economic variables and terrorism can sometimes be attributed to differently measured concepts, the type and quality of datasets used, and the failure to distinguish different types of terror-ism, the "more murder in the middle" thesis (Fein, 1987) also has long-standing roots in terrorism studies. Several scholars have remarked on the number of educated, middle-, and upper-class participants in terror-ist organizations (Sageman, 2004; Friedman, 2002)—particularly in the Palestinian case (Hassan, 2001; Berrebi, 2003) and in 1970s groups such as Baader-Meinhof (Aust, 1988; Combs, 2003). Callaway and Harrelson-Stephens (2006) posit that the relationship between subsis-tence and terrorism is an inverted "U": Those at the low end are too busy trying to survive to rebel and those at the high end are fairly satisfied with their lot. It is those in the middle who have the greatest number of unmet expectations. Explanations for this commonly observed phenomenon are based in Gurr's seminal (1970) work on rebellion and his conception of relative deprivation. Citizens rebel when their expectations—political, social, or economic goods to which people believe they are entitled—exceed their opportunities to meet these expectations. Education as a factor is also posited to follow this inverted "U" pattern. Angrist (1995) noted that the period preceding the first *intifada* was marked by a doubling of Palestinian men with 12 years or more of education but also by a sharp increase in unemployment for college graduates. Friedman (2002) also suggests that underemployment is a factor in al-Qaeda's ability to recruit from the Arabian Peninsula.

For comparison, the greed-grievance debate is also alive and well in the civil war literature, but the fact that civil war is concentrated in the poorest countries is empirically well supported (Collier and Hoeffler, 2000, 2004) and has generated somewhat broader support than is the case with the terrorism literature. Although the primacy given to economic explanations varies in this subfield as well, the importance of such factors is less contested. Collier and Hoeffler's "greed theory" of civil war identifies low income, low rates of growth, and a large degree of dependence on primary commodity exports as the key explanatory variables for civil war. Other economists, such as Elbadawi and Sambanis (2000), identify low levels of per capita income and dependence on natural resources as two key factors, but they also note the causal significance of demographic factors (too many young, poor, uneducated men), as well as a failure to develop strong democratic institutions, which they believe compounds all the other problems. Other scholars have expanded Collier and Hoeffler's dataset and found that it is income inequality, rather than merely low income, that increases conflict risk (Nafziger and Auvinen, 2002). This may equally be the case for dependence on natural resources; what matters may not be the degree of dependence on, but the distribution of, the revenues earned from natural resources.

Addison (2001) refines this argument further, examining key variables with economic effects not traditionally measured (or captured) as such: individual access to productive assets (land, water, other "natural capital"), infrastructure, and stock of human capital. These variables are important in that they help determine how states are connected to the growth process and whether some groups (with ethnic, regional, and religious characteristics) are marginalized from its benefits. Addison hypothesizes that if aid directly improves the lives of disadvantaged groups—by financing community projects and encouraging pro-poor expenditure reform—then it will raise their participation in growth and will reduce conflict by more than Collier and Hoeffler have allowed.

Discussion and Cautions. The disputes exist not only because of different datasets and methodology but also because there are important subtleties involved (for example, aggregate versus distributional effects). One such subtlety is the difference between a broad factor acting on people (such as poverty or rate of development) and more specific factors, such as whether opportunities and choices exist (as when incentives are provided for deradicalization). Another difficult factor to capture is the effect of the "Robin Hood" impulse. Although Lia and Skjolberg's (2004) findings challenge Krueger and Maleckova's (2003) on this account, both sets of authors rely on macro-level data. Neither is able to really measure the significant motivator that redressing economic (as well as social) inequality seems to have been for all manner of leftist terrorist groups from the Narodnaya Volya in seventeenth century Russia to the Red Brigades. The anti-globalization movement is another example of the complex relationship between economic variables and, at the very least, sympathy for terrorism (see Karmon, 2005). Ehrlich and Liu (2002) point out that certain structural conditions lead easily to "moral indignation" (p. 187).

Social and Cultural Factors. *Education.* A number of studies also examine the relationship between education—sometimes measured in terms of illiteracy (Krueger and Maleckova, 2003)—and terrorism, mostly finding that the only relationship is a positive one: Terrorists turn out to be more rather than less educated than the general population. An assessment of Jemaah Islamiyyah terrorists determined that more jihadis than not had either some college or advanced technical training (Magouirk, Atran, and Sageman, 2008). Even with this training, however, a majority still worked in unskilled jobs. Krueger and

Maleckova (2003, p. 142) propose that if we think of terrorism as a violent form of political engagement, then it is natural for the more educated to be the more engaged.

The literature on the role of education still needs some additional work to reach the standard of "substantive agreement." Methods that evaluate education in terms of national literacy levels and compare that information with the number of attacks or terrorist groups within a given country are too high-level to be useful to the policymaker. More recent studies (including my own research on behalf of the John Jay and ARTIS Transnational Terrorism Database [JJATT], 2009) have been able to code demographic data for degrees earned and type of education or school. These types of distinctions are important. There is a fairly large chasm between the type of education that Americans imagine when they think of a bachelor's degree in liberal arts and the education received by a graduate in Islamic studies from a Saudi university.

Education can encourage terrorism in several ways. One is that schools may be used simply as convenient recruiting hubs or, in some cases, even as "mobilizing structures" with the right mix of youth, insulation from social control, and opportunities. Another is that schools may propagate violent ideology and expand the context in which the use of violence is considered appropriate and desirable. Pursuing the latter of these first, education as a root cause of terrorism is clearly related to the variables of culture, ideology, and religion, discussed in greater detail below. I introduce the issue in this section because it relates to the importance of distinguishing *level* of education from *type* of education. There is an ongoing debate in the literature about the role of madrassahs in breeding terrorists, which became particularly fierce after 9/11. Bergen and Pandey (2006) argue against their significance by noting that madrassahs (Muslim schools) do not teach the technical skills necessary to connect the dots between reciting the Koran by heart, on the one hand, and planning a coordinated terrorist attack, on the other. Madrassahs have not been important in the case of al-Qaeda in the Middle East. But madrassahs turn out to be particularly important in the case of Jemaah Islamiyyah as a base for recruitment and the formation of ties that lead to a desire to join the jihad (Magouirk, Atran, and Sageman, 2008)—something to be discussed in greater detail in the section on social ties, below. Moreover, Stern's (2000) early research on Pakistan's madrassahs suggested that a small percentage of madrassahs were not only teaching a radical version of Islam but were also exhorting their graduates to fulfill their spiritual obligations by participating in ongoing jihad conflict such as that in Kashmir. Many of these same schools prepared their graduates for such an undertaking by sending them to jihad training. Pakistani officials estimated that 10 to 15 percent of the country's 40,000 to 50,000 madrassahs were extremist in nature.

Discussion and Cautions. As with the discussion of democracy, above, the usefulness of this research probably relates more to wisdom and subtlety than to broad implications for strategy. No one is going to suggest a strategy of deemphasizing education merely because of some correlational data suggesting that increased education could lead to more terrorism. However, there are implications for distinguishing among types of education.

Human Insecurity. Although this is something of a catch-all category, factors broadly related to human insecurity include low levels of civil liberties, high levels of crime, low levels of education and health care, and a lack of subsistence rights. Each of these variables has been suggested as an additional factor that contributes to terrorism. Callaway and Harrelson-Stephens (2006) contend that the prime breeding ground for terrorism is the nexus between poor political rights and poor human rights conditions. A study of terrorism in Latin America (Feldmann and Perala, 2004) also found that terrorism was more common in states with widespread human rights violations. Krueger and Maleckova (2003, p. 139) also identified a relationship between civil liberties—as defined by Freedom House—and terrorism.

Ehrlich and Liu (2002) test a series of potentially relevant human factors, including gender equity, health and population growth, education, and peace and order. They conclude, "these interacting and largely structural factors can be important to the motivations and recruitment of terrorists, even when those terrorists are relatively prosperous individuals . . . [since] . . . the socioeconomic and political conditions in their nations provided a good basis for both moral indignation and grassroots support" (pp. 186–187). Ehrlich and Liu subsequently attempt to answer the critical question: If these factors are key root causes, why do we not see anti-U.S. terrorism in Latin America, where there are such similar conditions to those in the Middle East? They suggest that U.S. support for Israel and its policies related to its oil dependence have been triggers in the case of the Middle East that are absent in other regions (p. 189).

Grievances and Anxieties. This brings us to the issue of grievances. Although obviously difficult to measure, isolate, and test, the notion is widespread that there is some sort of grievance driving the use of violence. Grievances are sometimes related to permissive factors and sometimes to precipitating events. Long-standing or historical grievances are part of a larger permissive context but recent events can quickly become grievances as well. Experts stress that political grievances, such as inequality, can be real *or* perceived. Permissive factors alone do not explain the prevalence or absence of terrorism. There is wide agreement that permissive factors must be combined with precipitant factors—usually events—that occur immediately before the act of terrorism. Horowitz (2003) explores the role of precipitating events, also called "exogenous shocks" (Varshney, 2001). These are seemingly minor events, or a series of events, that act as a trigger for large-scale violence; the important thing is that the participants in large-scale violence perceive the precipitating events as significant. Bandura (1998), Weinberg (1991), della Porta (1995), Crenshaw (1998a), and others have cited the failure of nonviolent strategies as a necessary condition for the turn from nonviolence to violence. This kind of failure—whether in some single dramatic stroke or slowly over time—may also function as a precipitating event. Because the use of violence is not, in most cases, a socialized norm, the decision to use it is not immediate, even in the case of those socialized to view it as a legitimate tool. The use of violence is understood to be an escalatory step, and one with repercussions.

Humiliation is one of the most commonly cited grievances (Kristof, 2002; Stern, 2003; Hoffman, 1998; Newman, 2006), followed by the related emotions—caused by a wide range of possible grievances—of revenge, despair, and impotence. A 2008 Pentagon study of foreign fighters in Iraq identified alienation and a desire to "make their mark" as two of the traits most common to suicide bombers there.[10] An attempt to find or establish an identity is also a common theme among British Muslim radicals studied recently by the British Intelligence Service MI-5. The study found that "'Membership in a terrorist group can provide a sense of meaning and purpose. It can lead to enhanced self-esteem, and the individual can feel a sense of control and influence over [his] life. . . .'" (See also Helmus, 2009, for additional discussion on alienation, identity, and so forth.)[11]

The relationship between grievances and terrorism may be either direct or indirect. A number of surveys have documented U.S. foreign policy in the Middle East as a source of grievances. Atran (2004) notes that the second Iraq war is just the most recent affront, and he quotes a Defense Science Board report produced after the 1996 terrorist attack on U.S. military housing at Khobar Towers in Saudi Arabia: "Historical data show a strong correlation between U.S. involvement in international situations and an increase in terrorist attacks against the United States" (p. 74). An alternative interpretation suggests that it is the perception of U.S. weakness in Lebanon and elsewhere that led to the increase in attacks against the United States (Schachter, 2002). The literature on "political opportunity" similarly emphasizes that challenges to the state are most likely to develop when state authorities are seen as newly vulnerable or particularly receptive to calls for change.[12] These are not actually inconsistent: The United States may be seen both as intervening (attempting to use its strength) but as lacking the willpower to be effective. Indirectly, U.S. support to regimes that engage in human rights abuses might be part of the human security–terror link discussed above. Of course, determining which is the chicken and which the egg, in relating the issue of grievances to the decision to use terrorism, is the tricky part. Popkin (1987, p. 9) argues that political entrepreneurs are needed to organize the masses to protest at the local level and that, to do so, these entrepreneurs must be able to identify which material and ideological incentives will attract the greatest number of individuals to the cause. Political entrepreneurs can and do use grievances as ideological incentives for action.

Speckhard and Akhmedova's (2006) study of Chechen fighters found a link between the decision to join the rebels and the death of a family member. But grievances are just as likely to be communally rather than individually experienced. An individual who experiences relative deprivation with respect to his individual preferences is actually less likely to engage in violent activities against the state than an individual who experiences frustration on behalf of the group to which he belongs, according to Gupta (1990, p. 250). He explains the rational actor quandary (on which he elaborates further in his recent book; see Gupta, 2008b). Economic theory predicts that rational actors pursue courses of action that maximize their utility but does not explain why these same individuals might pursue courses of action that maximize the utility of a group to which the individual belongs when those courses of action risk a less than maximal outcome for the individual (and indeed may cause considerable loss for that individual).[13] Gupta hypothesizes that an individual's self-perception consists of himself both as an individual and as a member of some social group. The social group can be one to which the individual is born or one of his choosing. An individual's utility, therefore, consists of both his individual and his collective utility. Individual utility is achieved through economic activities, whereas collective utility is achieved by taking part in group activities or in activities that achieve collective goods on behalf of one's identified group. Given that terrorism is fundamentally a group activity (Crenshaw, 1998b)—occasional "lone wolves" excepted—the possibility that group grievances might be a more powerful organizing force than individual grievances is not too surprising.

Mobilizing Structures and Social Ties. Another important factor sometimes included in discussions of precipitants is mobilizing structures (see also Jackson, 2009). Tilly (1978), for example, has argued that the degree to which different preconditions exist is irrelevant without an organization structure. Much of what we believe to be true about the mobilization of terrorist groups originates in the literature on social movements (della Porta, 1995; Fernandez and McAdam, 1988; Karstedt-Henke, 1980; Koopmans, 1993; McAdam, 1986, 1999; Tarrow, 1995, 1998) and takes its current form in Sageman's (2004, 2008) work. The greatest substantive agreement is on the idea that committed individuals bring their friends and family members into terrorist groups using the strength of their relationships first—as opposed to the strength of their grievances or their faith. In the case of some groups, such as Jemaah Islamiya, these relationships were formed predominantly through common attendance at madrassahs, mosques, and religious study groups, in addition to military training in Afghanistan or the Philippines and, later, through kinship ties (Ismail, 2006). But in other cases—Australia, for example—these relationships were also formed at places of work and at social organizations, such as soccer clubs. The history of other groups, such as the Irish Republican Army, emphasizes the important role of kinship ties in establishing relationships that bridge the gap between family group and terrorist group.

Many scholars (della Porta, 1995; McAdam, 1986, 1988; Sageman, 2004; Snow, Zurcher, and Ekland-Olson, 1980) argue that there is more bottom-up "enlistment" than top-down recruitment; individuals join activities because their friends are already members rather than because they are initially committed to either the activities or the organization's greater goals. McAdam (1986) argues that an initial receptivity to the idea of participating in "low-cost/low-risk activism" is facilitated by socialization through family or other agency ties. When these individuals subsequently come into contact with political activists, they are receptive to participation in low-cost/low-risk activism. Participation in low-risk/low-cost activism then makes it more likely that those same individuals will be "drawn into more costly forms of participation through the cyclical process of integration and resocialization" (pp. 68–71).

This finding was confirmed by both della Porta's research on the Red Brigades (1988) and Sageman's (2004) work on the global salafist movement. Della Porta found that 45 percent of the 1,214 Italian militants she studied had personal ties to eight or more group members before joining a terrorist organization. Similarly, Sageman found that 75 percent of the 172 Mujaheddin he identified had prior relational ties to jihadis already involved in training for, planning, or conducting terrorist activities. Snow, Zurcher, and Ekland-Olson (1980) further found that individuals with the fewest or weakest social ties to alternative networks were more likely to join a movement than were individuals with strong ties to countervailing networks. These individuals were more "structurally available" for participation.

Ideology, Religion, and Culture. Perhaps the area of greatest disagreement in the root-causes literature is the relationship between culture, ideology, religion, and terrorism. Although there is disagreement over the degree of importance of these various factors, there is more general agreement that "Violent behavior is a consequence of violent socialization" (Rhodes, 2000, p. 1093). The early literature on terrorism tended either to use the concept of violence in a cavalier fashion, as though violence were comparable to any other tactic in the tool kit of an organization,[14] or to suggest that organizations were violent because they attracted individuals who were predisposed to, or intrinsically interested in, violence (Chai, 1993). But the debate has since evolved and the current state of knowledge tends to begin with the premise that individuals are not born predisposed to violence (as distinct from aggression).[15]

The idea of socialization to violence is not entirely new of course. Gurr (1970, p. 155) posits that whether or not an individual or group will turn to political violence depends, first, on the degree and scope of normative justification for political violence within the collective. That is, at what level is political violence socially acceptable and to what degree does it occur regularly within the society? The second factor is the degree and scope of utilitarian justification for political violence within a society. That is, to what degree has political violence succeeded in achieving specific ends in the past? Horowitz (2003) posited that ethnic violence, for example, is less common in the West because of a "considerable rethinking about the legitimacy of interethnic killing," on a societal level (p. 491).

The most widely cited reference exploring the links between religion and terrorism is Juergensmeyer's (2001) study on terrorism in five religious traditions. He concludes that although religion is not completely innocent (in many cases, religion provided the ideology, the motivation, and the organizational structure), it generally does not—by itself—lead to violence. Only when religion is combined with movements for social or political change, in which norms about the use of violence have been reinterpreted, does it lead to violence. Juergensmeyer introduces the idea of a "culture of violence" in which perpetrators are enmeshed. Believing that they are acting on behalf of a larger supportive community of activists, and that they are protecting the community from an existing outside threat, they use violence against outsiders merely as a response to this threat and their actions also take on sacred meaning in the context of their religious affiliation. Krueger and Maleckova (2003, p. 140) found that having a higher proportion of the population affiliated with any of the religions for which they coded—Islam, Christianity, Buddhism, and Hinduism—was positively associated with terrorism.

Stern (2003) emphasizes the role of a bifurcated worldview in which a tight knit "ingroup" focuses its hatred on an "outgroup." She emphasizes the similarity of this bifurcation across different types of religions and ethnic groups and numerous other self-defined interest groups: blacks versus whites in America; Jews and Christians versus Muslims; antiabortion crusaders versus gynecologists; traditional societies versus contemporary America and women's rights; and the list goes on. Stern also links issues of humiliation to ideology, and Hafez (2003) observes a relationship between foreign occupation and ideology.

Some commentators have hypothesized that Islam is particularly vulnerable to being hijacked for terrorism because of its essentially political character; the lack of a reformist period similar to that undergone by Christianity (Friedman, 2002, 2005; Manji, 2004, Rushdie, 2005); or because Islam has an unusually bellicose historical context, which contributes to its misuse (Lewis, 2002; Auster, 2005). Pope Benedict XVI famously roused the ire of the Muslim community in 2006 when he contrasted the rationality of Christianity with the violence of Islam—focusing on the Islamic concept of jihad.[16] In contrast, a recent British study conducted by the behavioral science unit within MI-5 concluded, "There is evidence that a well-established religious identity actually protects against violent radicalization" (Travis, 2008).[17] The study involved several hundred case studies of violent Islamist extremists in the UK, a large number of whom were discovered to be little better than religious novices. The study reported that few of those involved in terrorism were brought up in religious households; there were a large number of converts involved; and even the nonconverts were surprisingly illiterate about Islam. Numerous comparisons have also been made between the current wave of Islamist terrorism and the wave of anarchist terrorism in the 1970s (Crenshaw, 2007; Gelvin, 2008), suggesting that the phenomenon may not be as unique as some scholars believe.

Albert Bandura's famous (1998) essay on moral disengagement explains that self-sanction regulates an individual's moral conduct. Individuals "refrain from behaving in ways that violate their moral standards, because such behavior would bring self-condemnation" (p. 161). Moral standards are not controlled by autopilot, however, and can be disengaged through a process of "cognitive reconstrual." Bandura explains that cognitive reconstrual may occur through unconscious cognitive processes, as well as through intentional training (such as military training, religious indoctrination) or through social learning in which aggression may be observed and imitated. In addition, moral reconstrual is facilitated "when nonviolent options are judged to have been ineffective . . ." (1998, p. 164).

As Bandura suggests, the norms emerging from an organization need not be an intentional outcome; it can also be a secondary result of association and activities. This is illustrated in the case of the civil rights movement as McAdam (1999) describes it. McAdam explains that recruitment into the civil rights movement was not just direct recruitment from the ranks of churchgoers; rather, "it was a case of church membership itself being redefined to include movement participation as a primary requisite of their role" (p. 129). McAdam quotes John Lewis, a former SNCC[18] president, who said, "People saw the mass meetings as an extension of the Sunday services."[19] Another observer agrees, "To the [black church member] of Montgomery, Christianity and boycott went hand and hand" (Walton, 1956, p. 19). McAdam (1999) also says that the same thing happened with respect to the student protests: "Participation in protest activity simply came to be defined as part and parcel of one's role as a student" (p. 130).

Norm development can be an iterative process that progresses from both a group's experiences interacting with other groups as well as from members' influence on their organization. In the case of small, independent Islamist prayer groups, for example, creating a "culture of violence" often includes regular viewing of carnage tapes from Chechnya, Ambon, and Iraq as well as discussions about the approved parameters of violent jihad in the context of broader religious discussions; and specifying appropriate rationale, targets, and means. When these discussion groups are supplemented with pseudo-military or actual military training, locally or abroad, it reinforces the intellectual culture of violence and complements the theory with development of practical skills. This supplementing, of course, may have inculcated violence as an explicit objective.

Anthropologists and historians seem particularly comfortable with the link between culture and violence, as are regional scholars in such countries as Algeria. Mousseau (2002–2003) noted that certain norms and historical traditions render terrorism more socially acceptable in some societies than others. Martinez (1998) suggests that it is Algeria's "war-oriented *imaginaire*" (emphasis in original) that explains the ready acceptance of violence and the subsequent spiral into mass violence by the many armed groups in Algeria. A war-oriented *imaginaire* is a worldview in which "the use of violence is respected as a means of social advancement," which includes the acquisition of power, status, and the economic benefits that accrue from both (p. 11). Martinez refers to historical experiences that underscored the "virtues of violence as a means of accumulating resources and prestige," but he also emphasizes the fact that many members' most recent quasi-political experience was participation in the Afghan jihad. Martinez does not argue that Algerian culture is violent but rather that the use of violence has become respected by some groups as a means of advancement because of historical models that included successful advancement of the corsair under the Ottoman Empire, the Caïd under French rule, and the military officer of the Armée de Libération Nationale (ALN), the armed wing of the socialist party's Front de Libération Nationale (FLN), during and immediately following the successful war for independence from France (p. 10). Ehrlich and Liu (2002) also highlight such cultural factors as religious fundamentalism and attitudes toward globalization as root causes of terrorism, citing Barber's (1996) work *Jihad vs. McWorld*.

Callaway and Harrelson-Stephens (2006) hypothesize that states that suffer more brutal colonial experiences will have more terrorism than those with no, or less violent, colonial experiences.[20] Again for comparison, a robust finding in the civil war literature is the "conflict trap," essentially the observation that violence begets violence. States that have experienced civil war are more likely to fall into it again.

Conclusions: Making Sense of the Factors

Despite a series of sometimes contradictory outcomes, particularly in the areas of political freedom/democracy and poverty/economic factors, there are a few clear areas of significant consensus in the root-causes literature. No one permissive condition or even combination of permissive conditions is thought to have sufficient power to predict the emergence of terrorism on its own. But areas of consensus include the criticality of regime illegitimacy, the (almost) necessary condition of repression, and the inverted "U" effect of political freedom on terrorism. There is broad consensus that perception of regime illegitimacy and strength is a key factor in creating a political opportunity for terrorism. There is also broad support for the contention that repression is, in the majority of cases, a necessary condition for terrorism. Problematic is the fact that this relationship breaks down when we distinguish between domestic and international terrorism. There is also broad support for the purported relationship between weak or transitioning democracies and the increased likelihood of terrorism—this is related to the effects of modernization and social instability produced by such processes. This is also true for the relationship between curtailed civil liberties and the increased likelihood of terrorism. In parallel, there is strong case study evidence that both moderate and formerly violent groups can be co-opted within a more democratic political system. It is interesting to note that there does seem to be a common theme to the factors about which there is the most consensus. They can all be linked to the broader concept (and importance) of "rule of law."

The concept of rule of law is not often directly addressed in the terrorism literature, although it does come up in the ethnic conflict literature. Although the concept is somewhat fluid,[21] the rule of law in its simplest form is the idea of equal justice and protection under the law. Laws are public, independently applied, and enforced without prejudice (implying an independent judiciary). In its more progressive form, the rule of law includes the idea that national laws are fair, meet international standards, and are consistent with human rights principles. Establishing and enforcing the rule of law requires strong and capable institutions and very little corruption. Rule of law is relevant to the political freedom variable, since new democracies are less likely to have the institutions capable of providing what we traditionally see as democratic forums for problem-solving—hence, they can be said to have weaker rule of law.

Freedom from corruption is endemic to the rule of law. In the ethnic conflict and riots literature, Horowitz (2003) emphasizes the critical role of law enforcement. He finds that ethnic violence essentially *never* occurs without the support (explicit or implicit) of the authorities or law enforcement. He concludes, "If the instruments of public order are capable and determined, they may not be able to prevent all forms of violence, but they can have a profound effect on the course it takes" (p. 489). Horowitz groups the conditions that facilitate violence into three categories: (1) uncertainty: an already tense situation between two groups; (2) impunity: a belief that the perpetrator either will not be caught or will not be punished; and (3) justification: a belief that the perpetrator has a valid rationale for acting against the opposite group and, furthermore, that violence is a legitimate means (Horowitz, 2003, p. 326).

Simon (1994) notes that, for most of the nineteenth and early twentieth centuries, America's primary experience with terrorist violence was internal and was predominantly focused on the anarchists and the U.S. labor movement. The 1910 dynamiting of the *Los Angeles Times* building by John J. McNamara, the secretary-treasurer of the International Association of Bridge and Structural Iron Workers, followed a protracted period of union-management disputes and strikes at the newspaper. The explosion killed 21 nonunion workers, and the subsequent arrest and trial of McNamara and his brother James led to the establishment of a Presidential Commission of inquiry on violence in labor-management relations (pp. 38–42). By all accounts, the labor movement today is comparatively violence-free. Although this does not mean that labor-management relations are free of conflict, the end of overt violence is still a remarkable change and begs the question, What accounts for this rather dramatic difference in the acceptability of violence over time? I hypothesize that violence was successfully institutionalized and underpinned by the rule of law. Clearly, norms about the appropriate use of violence in the context of labor disputes changed. If it turns out to be true that violence was successfully institutionalized, then a prime policy recommendation would be to work toward stronger, more-capable institutions and the rule of law as a way to address certain types of terrorism. Unfortunately, the existing literature does not have enough to say about these factors for us to be certain about these recommendations.

Areas of continued dissension include the role of education, poverty, and other human security factors, as well as the role of ideological and cultural factors. Some of the disagreement centers on the degree to which these factors play a role, but other studies have concluded that the factors do not play a role at all. There does appear to be an inverted "U" relationship between terrorism and the factors of education and wealth, although that relationship might be contested in terms of measurement validity. Some of this complexity probably stems from the conflation of the revolutions/rebellion literature and the terrorism literature, because much of the former focused on peasant rebellions or the role of the "masses" in fomenting revolution. More-recent demographic research has revealed that individual participants in terrorist groups and in terrorist violence are both more educated and more financially well off than was previously believed—although this was no surprise to scholars who studied anarchist and other social revolutionary terrorist groups in the 1970s. However, the emerging picture of foreign fighters and suicide bombers in Iraq suggests that they fit the old model of the undereducated, unemployed, alienated terrorist far better than the new model. This contrast, too, might be better understood by distinguishing types of terrorism. I discuss this in greater detail in the next section.

In an effort to better understand the way that various factors interact and are related to one another and consistent with the general approach adopted in the larger volume of which this paper is part (Davis and Cragin, 2009), I developed a causal path diagram (shown in Figure 2.1). This diagram seeks to include "all" of the factors discussed, whether or not there is agreement on them and whether or not the empirical evidence appears to confirm or disconfirm them. The reason for this is that any of the factors can be important in at least some circumstances. Further, a given factor can be important as part of a phenomenon even if it plays an intermediate role rather than what an empiricist would regard as causal role. Such subtleties are discussed further in the companion paper (Davis, 2009).

Figure 2.1 is a visual representation of the way that root causes might be connected to one another. It does not take into account the degree of agreement or disagreement in the literature about the importance of any one factor. Instead, it presents multiple possible contributing pathways—any or all of which may operate simultaneously. Rather than the variables at lower levels "leading" to the variables at higher levels, variables at higher levels represent larger (more abstract, higher-level) categories, whereas variables at lower levels represent factors that would likely be included in these larger categories. Factors at any level can exist independent of the factors below it. A final point is that the lower-level variables are not comprehensive, although they include the most commonly discussed factors from the literature.

Things to note in Figure 2.1 are the three highest-level variables with arrows that point directly to "Increased root-cause likelihood of terrorism." These three variables (Facilitative norms about use of violence; Perceived grievances; and Mobilizing structures) are linked by the word "and," indicating a threshold. Although many of the permissive conditions represented in the diagram could be combined in numerous ways to create a volatile situation, terrorism is most likely the result when all three "gateway" conditions are present: facilitative norms about violence in general and terrorism specifically; grievances to serve as motivation; and mobilizing structures to provide the organization. These three variables also represent, to some degree, the relationship between permissive and precipitant conditions and the way traditional root causes (those below the "gateway" variables in red) are remote causes rather than direct causes of terrorism. When root-cause scholars talk about precipitant factors, they tend to mean events that feed into the perceived grievances variable, but sometimes they also

Figure 2.1
Relationships Among Root Causes

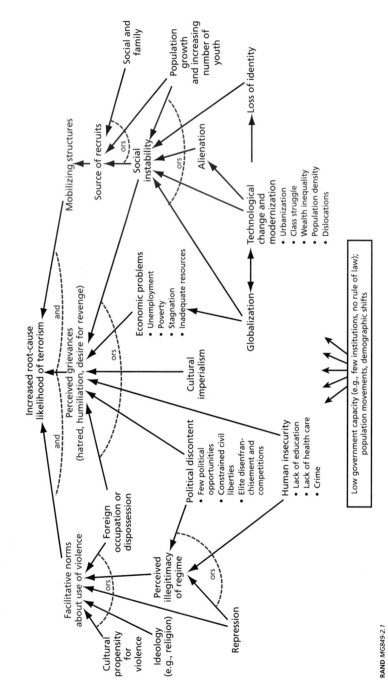

mean an event that creates a larger source of recruits. The fact that root causes are remote, rather than direct, has important policy implications, of course, since not only is it difficult to affect root causes, it is also difficult to measure how, and the degree to which, root causes *are* affected through policy changes.

Another interesting element is the factor "Low government capacity" at the bottom of the figure. Although low government capacity (that is, weak institutional capacity) is rarely called out as a potential root cause in and of itself, many variables listed as potential root causes seem to be derived from, or related to, a situation in which the government has a reduced capacity to govern, provide services, or make and enforce effective policy. This is true in the case of the rule of law, discussed above, as well as in the case of economic inequality or social instability. Many factors identified as root causes could be ameliorated, or even removed, if the government in question had sufficient capacity to effect change.

Implications for Strategy, Policy, and Research

Table 2.2 evaluates the combination of presence, importance and mutability of various root causes. Unfortunately, root causes are by their nature some of the factors *least* amenable to policy influence—particularly if we focus on the need to influence these factors within the sovereign realm of another nation. The country that is (however unwillingly) host to terrorist groups would have a greater ability to influence root causes, although the time frame might be long. The first column assesses whether or not a root cause is likely to be a relevant situational factor. The second column notes whether or not there is substantive agreement in the literature about the importance of this factor. The third column assesses the degree to which the root cause is amenable to policy influence by the United States if the terrorist group is hosted by a third party. The fourth column assesses the degree to which the root cause is amenable to policy influence by the host country. An "X" indicates whether the factor is likely to be present, is agreed to be important, and is amenable to policy influence. A slash indicates some degree

Table 2.2
Root-Cause Presence, Importance, and Mutability

Factor	Likely Present	Substantive Agreement	Amenable to Policy Influence (Third Party)	Amenable to Policy Influence (Host)
Facilitative norms about use of violence	X		/	/
Cultural propensity for violence	/			/
Ideology or religion	/			/
Perceived illegitimacy of regime	X	X	/	X
Foreign occupation		X	X	
Repression	X	X	/	X
Political inequality	X		/	X
Constrained civil liberties	X	X	/	X
Lack of political opportunity	/		/	X
Elite disenfranchisement		X	/	X
Reduced government capacity	X		/	X
Human insecurity			/	X
Crime				X
Lack of education			/	X
Lack of health care			/	X
Migration			/	/
Grievances (real or perceived)	X	X	X	X
Economic inequality			/	X
Modernization/massive social change		X		
Modernization (technologies)		X		
Class struggle		X		
Wealth inequality			/	X
Urbanization				/

Table 2.2—continued

Factor	Likely Present	Substantive Agreement	Amenable to Policy Influence (Third Party)	Amenable to Policy Influence (Host)
Population density				/
Economic stagnation			/	X
Poverty and unemployment			/	X
Mobilizing structures	X	X	X	X
Relationships and social ties	X			
Population growth (youth)	X	X		
Social instability	X	X	/	/
Humiliation	/		/	/
Alienation				/
Dispossession				/
Loss of identity				/

of presence, importance, and amenability that is less than total. Other points to note: In cases where there is substantive agreement but no "X" in the "likely present" column, this indicates that when the factor is present, there is substantive agreement that it matters. However, the factor is not relevant in every, or even in most, situations (for example, foreign occupation, elite disenfranchisement).

As Table 2.2 indicates, the only factors likely to be present, substantively agreed on as important when present, and also amenable to policy influence are grievances and mobilizing structures. Perceived illegitimacy of regime, repression, and curtailed civil liberties are similarly evaluated but are somewhat less amenable to international policy influence. Again, the fact that root causes are remote rather than direct causes of terrorism means that is it difficult to affect them. More important, it is difficult to measure how, and the degree to which, root causes are actually affected through any specific policy change.

Critical Tasks for Future Research: What Should We Tackle First?

Throughout this paper, I identified a number of lacunae in the existing literature on terrorism. Some of these weaknesses are simply the result of the relative youth of terrorism studies as an academic field of inquiry (as compared with, for example, the study of war). This includes the failure to appropriately categorize different types of terrorism and to maintain methodological consistency across studies. Overreliance on too few datasets and too few cases is another problem that is partly due to the youth of the field and partly due to the difficulty of data collection given the subject matter. Horgan (2007) suggests that the persistent gaps in our ability to answer some of the most basic questions about terrorism might be due to this "paucity of reliable data on all but the most well-researched terrorist groups" (p. 106). He also contends that, despite the piquing of scholarly interest that accompanies a major attack such as that on September 11, the number of researchers who pursue terrorism studies full-time has not risen overall. Both the paucity of data and the weak growth in the number of scholars committed to advancing the field reinforce my sense that terrorism studies is immature as an academic field of inquiry.

In some cases, the conflation of different types of terrorist groups within a single dataset or methodological inconsistency across studies helps to explain contradictory findings in the terrorism literature. Other instances, however, are as much a result of weak measurement validity. In the next section, I discuss a set of solutions that could help us to reinterpret the existing literature in a more useful manner, as well as move the field forward conceptually by emphasizing certain critical paths for future research.

Methodological and Measurement Problems

As mentioned above, some contradictory findings are probably a result of the relatively data-poor nature of the terrorism field. Researchers are forced to rely on only one or two large datasets; and even the smaller data-collection efforts and case studies are biased toward long-standing conflicts, such as those involving the Palestinians and the Irish Republican Army. However, other methodological problems can be attributed to invalid or weak measures of the independent variable and to failure to treat multiple variables simultaneously. One example is the issue of education discussed above. One way to evaluate the education variable has been to compare national literacy levels to the number of attacks or terrorist groups within a given country. These measures have obvious aggregation weaknesses. More recent studies have been able to code demographic data for academic degrees earned and type of education or school. These types of distinctions are important. As mentioned above, there is a large distinction between, for example, an American curriculum leading to a bachelor's degree in history and a Saudi curriculum in Islamic studies. Whether and how much that distinction matters remains to be studied.

Poverty is another area in which valid measurement is critical for determining whether and how important this variable is. There was initially too much emphasis in the literature on poverty measured in terms of the gross domestic product of a nation. Economic concepts are more usefully tested at a disaggregated level. Purchasing power parity, for example, allows a more finely tuned measure of economic factors. In addition, demographic data show that although a large percentage of al Qaeda members have some college education, these same members are also largely unemployed or underemployed and not working in the field for which they were educated. It is important to pursue the means to better measure and test each of the higher-level root cause variables at a disaggregated level (JJATT, 2009).

Accurately measuring the role of ideology and culture poses particular challenges. This may explain the unusual degree of contention extant in attributing causality to the relationship between culture, ideology, religion, and terrorism. Because culture is a process more than it is a static variable, figuring out how best to measure its effect takes imagination. The qualitative literature on culture is rich and cooperation between qualitative and quantitative scholars in this area has real potential.

Leaders Versus Followers

Terrorism research would also benefit from a greater effort to distinguish between leaders and followers or, indeed, among the many different types of actors in a terrorist system (Davis and Jenkins, 2004); National Research Council, 2002, Chapter 10; Stern, 2003), which include leaders, lieutenants, foot soldiers, sources of ideology and inspiration, facilitators of finance and logistics, and the portion of the population that either condones or supports the terrorist organization. Although the organizational theory portion of the literature does tend to focus on this distinction to a greater degree (Crenshaw, 1981; Chai, 1993), it is mentioned much less often in the broader terrorism literature—although it is not totally ignored (McCauley, 1991; Reader, 2000). But it is a distinction that matters. From a practical perspective, group leadership tends to be more stable than group membership, for example. Moreover, Victoroff (2005) observes, "Leaders and followers tend to be psychologically distinct. Because leadership tends to require at least moderate cognitive capacity, assumptions of rationality possibly apply better to leaders than to followers" (p. 33). Crenshaw and Chai distinguish between leaders and followers in terms of differing levels of commitment, different interests, and even different goals. Lipsky (1968), Popkin (1987), and others emphasize the important role of movement leadership to groups of the "relatively powerless [and] low income." Lipsky also suggests that "groups which seek psychological gratification from politics, but cannot or do not anticipate material political rewards, may be attracted to [more] militant protest leaders" (p. 1148).

The distinction between leaders and followers is particularly important when it comes to policy prescriptions. The solutions for eliminating funding for terrorist groups or activities will be different from the solutions for preventing grassroots radicalization. Foreign fighters and suicide bombers in Iraq seem to have very different demographic characteristics from the newly popularized archetype of the educated, middle-class terrorist motivated by ideology or grievance alone (Zavis, 2008; Quinn, 2008). Further exploration of the role of charismatic leaders (Weber, 1968) might also help us to better understand the greater appeal of some ideologies over others. Finally, distinguishing leaders from followers should help us to develop more comprehensive theories about terrorism that are context-specific but functionally more useful than what we have today.

Distinguishing Types of Terrorism

The most important task for future researchers, and the one that should be immediately implemented in existing datasets, is the need to distinguish types of terrorism. It is more than likely the case that many disagreements in the literature, as well as some counterintuitive findings, stem from the fact that all terrorism is not the same. And different root causes seem to apply for some types of terrorism (domestic, for example) than for others (such as international). The few datasets that are available lump all types of terrorism together, and both these data as well as case study data are skewed heavily toward long-standing conflicts in which reporting of terrorism has been routinized. One attempt to categorize types of terrorist groups divides them into at least five groups:

1. nationalist separatist
2. religious fundamentalist
3. other religious extremist (for example, millennialist cults)
4. social revolutionaries
5. right-wing extremists.[22]

This variation in types is likely to both stem from and result in various constellations of relevant root causes. A second effort at categorization includes the following:

1. criminal
2. ethno-nationalist
3. religious
4. generic secular
5. right-wing (religious)
6. secular left wing
7. secular right wing
8. single issue

9.   personal/idiosyncratic
10. state-sponsored.[23]

These distinctions are important for understanding critical factors. Breaking out a category for state-sponsored terrorism, for example, dramatically alters the issue of root causes. Post, Ruby and Shaw (2002) developed a framework of causal factors in which some 129 "indicators of risk for terrorism" were identified across four conceptual categories: (1) historical, cultural, and contextual features; (2) key actors affecting the group; (3) the group/organization: characteristics, processes, and structures; and (4) the immediate situation. As expected, certain indicators were more relevant for some groups than for others. Nationalist-separatist groups, for example, required more financial resources to maintain operations than did small cells. Moreover, Abadie (2006) finds that Marxist groups in Western Europe displayed less evidence of root-cause factors than did nationalists, whereas Marxist groups in the developing world displayed more evidence of root-cause factors than did Marxists in Europe. International Islamists displayed less evidence of root-cause factors than did nationalist Muslim groups, but international Islamists displayed more evidence of root-cause factors than did Marxists in Western Europe.

Other examples that suggest the importance of these distinctions include the following. Studies that distinguish between religious and other types of terrorist groups tend to place a greater emphasis on ideology. Some studies distinguish between global and local terrorist groups, and others do not. But determinants of international terrorism, when taken separately, are not informative about determinants of domestic terrorism. This is particularly true with respect to one of the most robust findings in terrorism research—that repression is almost always a necessary condition for terrorism. Although essentially true for domestic terrorism, this is not the case for international terrorism. Economic factors, too, have different relevance depending on whether the type of terrorism is domestic or international.

Distinguishing types of terrorism would also be useful in unpacking the often-conflated categories of political violence literature. Large-scale terrorism that occurs in regions with recent or ongoing hot wars, or in states experiencing foreign occupation, will probably have more in common with findings from the civil war literature than would small, independent, ideologically driven sleeper cells of jihadis in Western Europe. Although most civil war scholars make a clear distinction between their research and terrorism research, the line is much more blurred in the subfield of ethnic conflict, since ethnic terrorism is often a tactically relevant issue.

What is needed is a structured effort to determine which lessons from which subfields are relevant and which are not and to identify what differences account for this lack of relevance between subcategories of political violence. If one agrees with Bjorgo (2005) that terrorism is a (tactical) radicalization of other types of political conflict, then our goal should be to "identify the factors, processes and circumstances that tend to produce such a radicalization into terrorism; as well as to find which factors tend to prevent such conflicts from generating terrorism" (p. 4). This will inevitably require that terrorism be conceptually broken down into more distinct categories. The similarity between nationalist-separatist terrorism (such as in Palestine, Sri Lanka, and Ireland) and the literature on civil wars, revolutions, and ethnic conflict is the most obvious place to start.

Bibliography

Abadie, Alberto, "Poverty, Political Freedom, and the Roots of Terrorism," National Bureau of Economic Research, Inc., Washington, D.C., October 2004. As of December 29, 2008:
http://ideas.repec.org/p/nbr/nberwo/10859.html

————, "Poverty, Political Freedom, and the Roots of Terrorism," *The American Economic Review,* Vol. 96, No. 2, May 2006, pp. 50–56.

Addison, Tony, *Reconstruction from War in Africa: Communities, Entrepreneurs, States,* paper presented at the CSAE Conference: Development Policy in Africa, Centre for the Study of African Economies, University of Oxford, March 29–31, 2001. As of February 6, 2009:
http://www.csae.ox.ac.uk/conferences/2001-DPiA/pdfs/Addison.pdf

Angrist, Joshua, "The Economic Returns to Schooling in the West Bank and Gaza Strip," *American Economic Review,* Vol. 85, No. 5, December 1995, pp. 1065–1087.

Atran, Scott, "Mishandling Suicide Terrorism," *The Washington Quarterly,* Vol. 27, No. 3, 2004, pp. 67–90.

Aust, Stefan, *The Baader-Meinhof Group: The Inside Story of a Phenomenon,* translated by Anthea Bell, Topsfield, Mass.: The Bodley Head/Salem House, 1988.

Auster, Lawrence, "The Search for Moderate Islam," *Front Page Magazine,* January 28, 2005. As of December 29, 2008:
http://www.frontpagemag.com/Articles/Read.aspx?GUID=5F4D7BB5-CA89-4C09-986B-67CF241C2098

Bandura, Albert, *Aggression: A Social Learning Analysis,* New York: Prentice Hall, 1973.

———, *Social Foundations of Thought and Action: A Social Cognitive Theory,* Englewood Cliffs, N.J.: Prentice Hall, 1986.

———, "Mechanisms of Moral Disengagement," in Walter Reich, ed., *Origins of Terrorism: Psychologies, Ideologies, Theologies, States of Mind,* Washington, D.C.: Woodrow Wilson Center Press, (1990) 1998.

Barber, Benjamin, *Jihad vs. McWorld: How Globalism and Tribalism Are Reshaping the World,* New York: Ballantine Books, 1996.

Bergen, Peter, and Swati Pandey, "The Madrassa Scapegoat," *The Washington Quarterly,* Vol. 29, No. 2, Spring 2006, pp. 117–125.

Berrebi, Claude, "Evidence About the Link Between Education, Poverty and Terrorism Among Palestinians," Princeton, N.J.: Princeton University, Industrial Relations Section Working Paper, No. 477, 2003.

———, "Evidence About the Link Between Education, Poverty and Terrorism Among Palestinians," *Peace Economics, Peace Science and Public Policy,* Vol. 13, No. 1, 2007.

———, "The Economics of Terrorism and Counterterrorism: What Matters and Is Rational-Choice Theory Helpful?" in Paul K. Davis and Kim Cragin, eds., *Social Science for Counterterrorism: Putting the Pieces Together,* Santa Monica, Calif.: RAND Corporation, 2009. As of January 20, 2009:
http://www.rand.org/pubs/monographs/MG849/

Bjorgo, Tore, "Introduction," in Tore Bjorgo, ed., *Root Causes of Terrorism: Myths, Reality and Ways Forward,* London: Routledge, 2005.

Blomberg, S. Brock, Gregory D. Hess, and Akila Weerapana, "An Economic Model of Terrorism," *Conflict Management and Peace Science,* Vol. 21, 2004, pp. 17–28.

Bloom, Mia, *Dying to Kill: The Allure of Suicide Terror,* New York: Columbia University Press, 2005.

Bowman, Tom, "U.S. Offers Training, Pay as It Frees Iraqi Detainees," Morning Edition, NPR News, May 16, 2008. As of December 29, 2008: http://www.npr.org/templates/story/story.php?storyId=90506939

Callaway, Rhonda L., and Julie Harrelson-Stephens, "Toward a Theory of Terrorism: Human Security as a Determinant of Terrorism," *Studies in Conflict & Terrorism,* Vol. 29, 2006, pp. 773–796.

Carnegie Commission on Preventing Deadly Conflict, *Preventing Deadly Conflict,* New York: Carnegie Corporation of New York, 1997.

Chai, Sun-Ki, "An Organizational Economics Theory of Antigovernment Violence," *Comparative Politics,* Vol. 26, No. 1, October 1993.

Charrad, Mounira M., *States and Women's Rights: The Making of Postcolonial Tunisia, Algeria, and Morocco,* Los Angeles, Calif.: University of California Press, 2001.

Chua, Amy, *World on Fire: How Exporting Free Market Democracy Breeds Ethnic Hatred and Global Instability,* New York: Doubleday, 2002.

Cohen, Dov, and Richard E. Nisbett, "Self-Protection and the Culture of Honor: Explaining Southern Violence," *Personality & Social Psychology Bulletin,* Vol. 20, No. 5, October 1994, pp. 551–568.

Collier, Paul, and Anke Hoeffler, *On the Incidence of Civil War in Africa,* Washington, D.C.: The World Bank, August 16, 2000.

———, "The Challenge of Reducing the Global Incidence of Civil War," Copenhagen Consensus Challenge Paper, March 26, 2004. Also in Bjørn Lomborg, ed., *Global Crises, Global Solutions,* Cambridge, United Kingdom: Cambridge University Press, 2004. As of March 10, 2008: http://www.copenhagenconsensus.com/Files/Filer/CC/Papers/Conflicts_230404.pdf.

Combs, Cindy C., *Terrorism in the 21st Century,* Upper Saddle River, N.J.: Prentice Hall, 2003.

Cragin, Kim, and Peter Chalk, *Terrorism & Development: Using Social and Economic Development to Inhibit a Resurgence of Terrorism,* Santa Monica, Calif.: RAND Corporation, 2003. As of January 9, 2009: http://www.rand.org/pubs/monograph_reports/MR1630/

Cragin, Kim, Sara Daly, Audra Grant, Kevin O'Brien, Scott Gerwehr, and Susanne Bearne, "Mapping the Terrorist Career Path," Cambridge, United Kingdom: RAND Europe, 2005, unpublished.

Crenshaw, Martha, "The Causes of Terrorism," *Comparative Politics,* Vol. 13, No. 4, 1981, pp. 379–399.

————, "Thoughts on Relating Terrorism to Historical Contexts," in Martha Crenshaw, ed., *Terrorism in Context*, University Park, Pa.: Pennsylvania State University Press, 1995.

————, "The Logic of Terrorism: Terrorist Behavior as a Product of Strategic Choice," in Walter Reich, ed., *Origins of Terrorism: Psychologies, Ideologies, Theologies, States of Mind*, Washington, D.C.: Woodrow Wilson Center Press, (1990) 1998a.

————, "Questions to Be Answered, Research to Be Done, Knowledge to Be Applied," in Walter Reich, ed., *Origins of Terrorism: Psychologies, Ideologies, Theologies, States of Mind*, Washington, D.C.: Woodrow Wilson Center Press, (1990) 1998b.

————, ed., *Terrorism in Context*, 2nd ed., University Park, Pa.: Pennsylvania State University Press, 2001.

————, "The Debate over 'New' vs. 'Old' Terrorism," paper presented at the Annual Meeting of the American Political Science Association, Chicago, Il., August 30–September 2, 2007.

Cronin, Audrey Kurth, "Behind the Curve: Globalization and International Terrorism," *International Security*, Vol. 27, No. 3, 2003, p. 53.

Davis, Paul K., "Representing Social-Science Knowledge Analytically," in Paul K. Davis and Kim Cragin, eds., *Social Science for Counterterrorism: Putting the Pieces Together*, Santa Monica, Calif.: RAND Corporation, 2009. As of January 20, 2009:
http://www.rand.org/pubs/monographs/MG849/

Davis, Paul K., and Brian Michael Jenkins, "A System Approach to Deterring and Influencing Terrorists," *Journal of Conflict Management and Peace Science*, Vol. 21, No. 1, 2004, pp. 3–15.

Davis, Paul K., and Kim Cragin, eds., *Social Science for Counterterrorism: Putting the Pieces Together*, Santa Monica, Calif.: RAND Corporation, 2009. As of January 20, 2009:
http://www.rand.org/pubs/monographs/MG849/

Defense Science Board Summer Study Task Force, "DSB Force Protection Panel Report to DSB," December 1997, p. 8. As of January 24, 2009:
http://www.acq.osd.mil/dsb/reports/trans.pdf

Della Porta, Donatella, "Recruitment Processes in Clandestine Political Organizations: Italian Leftwing Terrorism," in Sidney Tarrow, Bert Klandermans, and Hans Pieter Kriesi, eds., *From Structure to Action*, New York: JAI Press, 1988, pp. 155–169.

————, *Social Movements, Political Violence, & the State*, Cambridge, United Kingdom: Cambridge University Press, 1995.

Doyle, Michael, "Kant, Liberal Legacies, and Foreign Affairs," *Philosophy and Public Affairs,* 1983.

Ehrlich, Paul R., and Jianguo Liu, "Some Roots of Terrorism," *Population and Environment,* Vol. 24, No. 2, November 2002, pp. 183–192.

Eisinger, Peter, "The Conditions of Protest Behavior in American Cities," *American Political Science Review,* Vol. 81, 1973, pp. 11–28.

Elbadawi, Ibrahim, and Nicholas Sambanis, "Why Are There So Many Civil Wars in Africa?" *Journal of African Economies,* Vol. 9, No. 3, 2000. As of December 29, 2008:
http://jae.oupjournals.org/cgi/reprint/9/3/244

Engene, Jan Oskar, *Patterns of Terrorism in Western Europe, 1950–95,* Ph.D. thesis, 1998, Bergen, Norway: University of Bergen.

Eubank, William Lee, and Leonard Weinberg, "Does Democracy Encourage Terrorism?" *Terrorism and Political Violence,* Vol. 6, No. 4, 1994, pp. 417–463.

———, "Terrorism and Democracy: What Recent Events Disclose," *Terrorism and Political Violence,* Vol. 10, No. 1, 1998, pp. 108–118.

———, "Terrorism and Democracy: Perpetrators and Victims," *Terrorism and Political Violence,* Vol. 13, No. 1, 2001, pp. 155–164.

Eyerman, Joe, "Terrorism and Democratic States: Soft Targets or Accessible Systems," *International Interactions,* Vol. 24, No. 2, 1998, pp. 151–170.

Fein, Helen, "More Murder in the Middle: Life Integrity Violations and Democracy in the World," *Human Rights Quarterly,* Vol. 17, No. 1, 1987.

Feldmann, Andreas E., and Maiju Perala, "Reassessing the Causes of Nongovernmental Terrorism in Latin America," *Latin American Politics and Society,* Vol. 46, No. 2, Summer 2004, pp. 101–132.

Fernandez, Roberto, and Doug McAdam, "Social Networks and Social Movements: Multiorganizational Fields and Recruitment to Mississippi Freedom Summer," *Sociological Forum,* No. 3, 1988.

———, "Multiorganizational Fields and Recruitment to Social Movements," in Bert Klandermans, ed., *Organizing for Change: Social Movement Organizations in Europe and the United States,* Greenwich, Conn.: JAI Press, 1989.

Friedman, Thomas, "An Islamic Reformation," *New York Times,* December 4, 2002.

———, "Brave, Young and Muslim," *New York Times,* March 3, 2005. As of December 29, 2008:
http://www.nytimes.com/2005/03/03/opinion/03friedman.htm?_r=1

Galantar, Marc, *Cults: Faith, Healing, and Coercion,* New York: Oxford University Press, 1989.

Gelvin, James L., "Al-Qaeda and Anarchism: A Historian's Reply to Terrorology," *Terrorism and Political Violence,* Vol. 20, No. 4, 2008, pp. 597–600.

Goldstone, Jack, *Revolution and Rebellion in the Early Modern World,* Berkeley, Calif.: University of California Press, 1991.

Goodwin, Jeff, *No Other Way Out: States and Revolutionary Movements, 1945–1991,* Cambridge, United Kingdom: Cambridge University Press, 2001.

———, "Revolutions and Revolutionary Movements," in Thomas Janoski, Robert R. Alford, Alexander M. Hicks, and Mildred A. Schwartz, eds., *The Handbook of Political Sociology,* New York: Cambridge University Press, 2005.

Gupta, Dipak K., *Socio-Economic Costs of Unemployment and Income Inequality: A Cross National Study, 1948–67,* unpublished Ph.D. thesis, Pittsburgh, Pa.: University of Pittsburgh, undated.

———, *The Economics of Political Violence: The Effect of Political Instability on Economic Growth,* New York: Praeger Publishers, 1990.

———, "Accounting for the Waves of International Terrorism," *Perspectives on Terrorism,* Vol. 2, Issue 11, August 2008a, pp. 3–9.

———, *Understanding Terrorism and Political Violence,* London and New York: Routledge, 2008b.

Gurr, Ted Robert, *Why Men Rebel,* Princeton, N.J.: Princeton University Press, 1970.

———, "Terrorism in Democracies: Its Social and Political Biases," in Walter Reich, ed., *Origins of Terrorism: Psychologies, Ideologies, Theologies, States of Mind,* Washington, D.C.: Woodrow Wilson Center Press, (1990) 1998.

———, "Terrorism in Democracies: When It Occurs, Why It Fails," in Charles W. Kegley, Jr., ed., *The New Global Terrorism: Characteristics, Causes, Controls,* Upper Saddle River, N.J.: Prentice Hall, 2003, pp. 202–215.

Gvineria, Gaga, "How Does Terrorism End?" in Paul K. Davis and Kim Cragin, eds., *Social Science for Counterterrorism: Putting the Pieces Together,* Santa Monica, Calif.: RAND Corporation, 2009. As of January 20, 2009: http://www.rand.org/pubs/monographs/MG849/

Hafez, Mohammad, *Why Muslims Rebel: Repression and Resistance in the Islamic World,* Boulder, Colo.: Lynne Rienner, 2003.

Hassan, Nasra, "An Arsenal of Believers," *New Yorker,* November 19, 2001, pp. 36–41.

Helmus, Todd C., "Why and How Some People Become Terrorists," in Paul K. Davis and Kim Cragin, eds., *Social Science for Counterterrorism: Putting the Pieces Together,* Santa Monica, Calif.: RAND Corporation, 2009. As of January 20, 2009:
http://www.rand.org/pubs/monographs/MG849/

Hoffman, Bruce, *Inside Terrorism,* New York: Columbia University Press, 1998.

Hoffman, Bruce, and Gordon H. McCormick, "Terrorism, Signaling, and Suicide Attack," *Studies in Conflict & Terrorism,* Vol. 27, No. 4, 2004, pp. 243–281.

Horgan, John, "Understanding Terrorist Motivation: A Socio-Psychological Perspective," in Magnus Ranstorp, ed., *Mapping Terrorism Research: State of the Art, Gaps and Future Direction,* London and New York: Routledge, 2007.

Horowitz, David, *The Deadly Ethnic Riot,* Berkeley, Calif.: University of California Press, 2003.

Huntington, Samuel P., *The Third Wave,* Norman, Ok.: University of Oklahoma Press, 1991.

Ibrahim, Saad Eddin, "Anatomy of Egypt's Militant Islamic Groups: Methodological Note and Preliminary Findings," *International Journal of Middle East Studies,* Vol. 12, 1980, pp. 423–453.

Ismail, Noor Huda, "The Role of Kinship in Indonesia's Jemaah Islamiya," *Terrorism Monitor,* Vol. 4, No. 11, June 2, 2006.

Jackson, Brian A., "Technology Acquisition by Terrorist Groups: Threat Assessment Informed by Lessons from Private Sector Technology Adoption," *Studies in Conflict & Terrorism,* Vol. 24, No. 3, January 2001, pp. 183–213.

————, "Organizational Decisionmaking by Terrorist Groups," in Paul K. Davis and Kim Cragin, eds., *Social Science for Counterterrorism: Putting the Pieces Together,* Santa Monica, Calif.: RAND Corporation, 2009. As of January 20, 2009:
http://www.rand.org/pubs/monographs/MG849/

Jenkins, Brian Michael, *Terrorism in the 1980s,* Santa Monica, Calif.: RAND Corporation, 1980. As of July 15, 2008:
http://www.rand.org/pubs/papers/P6564/

John Jay and ARTIS Transnational Terrorims Database (JJATT), 2009. As of April 27, 2009:
http://doitapps.jjay.cuny.edu/jjatt

Juergensmeyer, Mark, *Terror in the Mind of God: The Global Rise of Religious Violence,* Berkeley, Calif.: University of California Press, 2001.

Kant, Immanuel, *Zum ewigen Frieden. Ein philosophischer Entwurf,* Bd. VIII, 341–86, 1795. English translation: "Perpetual Peace: A Philosophical Sketch," in *Political Writings,* Cambridge, United Kingdom: Cambridge University Press, 1991, pp. 93–131.

Kaplan, Jeffrey, "The Fifth Wave: The New Tribalism," *Terrorism and Political Violence,* Vol. 19, No. 4, December 2007, pp. 545–570.

Karmon, Ely, "Hizballah and the Anti-Gobalization Movement: A New Coalition?" *PolicyWatch,* Vol. 949, January 27, 2005. As of December 29, 2008: http://www.washingtoninstitute.org/templateC05.php?CID=2244

Karstedt-Henke, Sabine, "Theorien zur Erklärung terroristischer Bewegungen" (Theories for the Explanation of Terrorist Movements), in Erhard Blankenberg, ed., *Politik der inneren Sicherheit* (The Politics of Internal Security), 1980.

Kaye, Dalia Dassa, Frederic M. Wehrey, Audra K. Grant, and Dale Stahl, *More Freedom, Less Terror? Liberalization and Political Violence in the Arab World,* Santa Monica, Calif.: RAND Corporation, 2008. As of February 2, 2008: http://www.rand.org/pubs/monographs/MG772/

Kegley, Charles W., *The New Global Terrorism: Characteristics, Causes, Controls,* Englewood Cliffs, N.J.: Prentice Hall, 2002.

Koopmans, Ruud, "The Dynamics of Protest Waves: West Germany, 1965 to 1989," *American Sociological Review,* Vol. 58, October 1993, pp. 637–658.

―――, "Protest in Time and Space: The Evolution of Waves of Contention," in David A. Snow, Sarah A. Soule, and Hanspeter Kriesi, eds., *The Blackwell Companion to Social Movements,* Oxford, United Kingdom: Blackwell Publishing, 2004, pp. 19–46.

Kristof, Nicholas D., "Behind the Terrorists," *The New York Times,* May 7, 2002.

Krueger, Alan B., and Jitka Maleckova, "Education, Poverty and Terrorism: Is There a Causal Connection?" *Journal of Economic Perspectives,* Vol. 17, No. 4, Fall 2003, pp. 119–144.

Lewis, Bernard, *What Went Wrong: Western Impact and Middle Eastern Response,* New York: Oxford University Press, 2002.

Li, Quan, "Does Democracy Promote or Reduce Transnational Terrorist Incidents?" *Journal of Conflict Resolution,* Vol. 49, No. 2, 2005, pp. 278–297.

Li, Quan, and Drew Schaub, "Economic Globalization and Transnational Terrorism: A Pooled Time-Series Analysis," *Journal of Conflict Resolution,* Vol. 48, No. 2, April 2004, pp. 230–258.

Lia, Brynjar, and Katja H-W Skjolberg, "Causes of Terrorism: An Expanded and Updated Review of the Literature," Kjeller, Norway: FFI 10307, 2004.

Lipsky, Michael, "Protest as a Political Resource," *The American Political Science Review,* Vol. 62, December 1968, pp. 1144–1158.

Maddy-Weitzman, Bruce, "The Islamic Challenge in North Africa," *Middle East Review of International Affairs Journal,* Vol. 1, No. 2, July 1997.

Magouirk, Justin, Scott Atran, and Marc Sageman, "Connecting Terrorist Networks," *Studies in Conflict & Terrorism,* Vol. 31, No. 1, January 2008, pp. 1–16.

Manji, Irshad, *The Trouble with Islam: A Muslim's Call for Reform in Her Faith,* New York: St. Martin's Press, 2004.

Martinez, Luis, *The Algerian Civil War, 1990–1998,* translated by Jonathan Derrick, New York: Columbia University Press, 1998.

Massey, Douglas S., "The Age of Extremes: Concentrated Affluence and Poverty in the Twenty-First Century," *Demography,* Vol. 33, No. 4, 1996, pp. 395–412.

Matthew, Richard, and George Shambaugh, "Sex, Drugs, and Heavy Metal: Transnational Threats and National Vulnerabilities," *Security Dialogue,* Vol. 29, 1998, pp. 163–75.

McAdam, Doug, "Recruitment to High-Risk Activism: The Case of Freedom Summer," *American Journal of Sociology,* Vol. 92, 1986, pp. 64–90.

———, *Freedom Summer,* New York: Oxford University Press, 1988.

———, "'Initiator' and 'Spin-off' Movements: Diffusion Processes in Protest Cycles," in Mark Traugott, ed., *Repertoires and Cycles of Collective Action,* Durham, N.C.: Duke University Press, 1994, pp. 217–239.

———, *Political Process and the Development of Black Insurgency, 1930–1970,* Chicago, Ill.: University of Chicago Press, (1982) 1999.

McCauley, Clark, ed., *Terrorism and Public Policy,* London: Frank Cass, 1991.

Meyer, David S., "Protest and Political Opportunities," *Annual Review of Sociology,* Vol. 30, 2004, pp. 125–45.

Moghadam, Assaf, "Suicide Terrorism, Occupation, and the Globalization of Martyrdom: A Critique of *Dying to Win,*" *Studies in Conflict & Terrorism,* Vol. 29, 2006, pp. 707–729.

Moore, Barrington, Jr., *Social Origins of Dictatorship and Democracy: Lord and Peasant in the Making of the Modern World,* Boston, Mass.: Beacon Press, 1966.

Mousseau, Michael, "Market Civilization and Its Clash with Terror," *International Security,* Vol. 27, No. 3, Winter 2002–2003, pp. 5–29.

Nafziger, Wayne, and Juha Auvinen, "Economic Development, Inequality, War and State Violence," *World Development,* Vol. 30, No. 2, 2002, pp. 153–163.

National Research Council, *Making the Nation Safer: The Role of Science and Technology in Countering Terrorism,* Washington, D.C.: National Academy Press, 2002.

Newman, Edward, "Exploring the 'Root Causes' of Terrorism," *Studies in Conflict & Terrorism,* Vol. 29, 2006, pp. 749–772.

Nisbett, Richard E., "Violence and U.S. Regional Culture," *The American Psychologist,* Vol. 48, No. 4, April 1993, pp. 441–450.

Noricks, Darcy M.E., "Disengagement and Deradicalization: Processes and Programs," in Paul K. Davis and Kim Cragin, eds., *Social Science for Counterterrorism: Putting the Pieces Together,* Santa Monica, Calif.: RAND Corporation, 2009. As of January 20, 2009:
http://www.rand.org/pubs/monographs/MG849/

Pape, Robert A., "The Strategic Logic of Suicide Terrorism," *American Political Science Review,* Vol. 97, August 2003, pp. 343–361.

———, *Dying to Win: The Strategic Logic of Suicide Terrorism,* New York: Random House, 2005.

Paul, Christopher, "How Do Terrorists Generate and Maintain Support?" in Paul K. Davis and Kim Cragin, eds., *Social Science for Counterterrorism: Putting the Pieces Together,* Santa Monica, Calif.: RAND Corporation, 2009. As of January 20, 2009:
http://www.rand.org/pubs/monographs/MG849/

Paxson, Christina, "Comment on Alan Krueger and Jitka Maleckova, 'Education, Poverty, and Terrorism: Is There a Causal Connection?'" Princeton University, Research Program in Development Studies Working Paper No. 207, 2002. As of January 9, 2009:
http://www.princeton.edu/rpds/papers/pdfs/paxson_krueger_comment.pdf

Piazza, James A., "A Supply-Side View of Suicide Terrorism: A Cross-National Study," *The Journal of Politics,* Vol. 70, 2008, pp. 28–39.

Pinker, Steven, *The Blank Slate: Modern Denial of Human Nature,* New York: Viking, 2002.

Piven, Frances Fox, and Richard A. Cloward, *Poor People's Movements: Why They Succeed, How They Fail,* New York: Pantheon Books, 1979.

"Pope: Islamic 'Holy War' Against God's Nature," Newsmax.com, September 12, 2006. As of January 24, 2009:
http://archive.newsmax.com/archives/ic/2006/9/12/123426.shtml?s=ic

Popkin, Samuel, "Political Entrepreneurs and Peasant Movements in Vietnam," in Michael Taylor, ed., *Rationality and Revolution,* Cambridge, United Kingdom: Cambridge University Press, 1987.

Post, Jerrold M., Keven G. Ruby, and Eric D. Shaw, "The Radical Group in Context: 2. Identification of Critical Elements in the Analysis of Risk for Terrorism by Radical Group Type," *Studies in Conflict & Terrorism,* Vol. 25, 2002, pp. 101–126.

"Preachers to the Converted: Reforming Jihadists," *The Economist,* December 13, 2007.

Quinn, Patrick, "Study Gives Info on Foreign Fighters in Iraq," Associated Press, March 17, 2008. As of December 29, 2008: http://www.marinecorpstimes.com/news/2008/03/ap_detaineestudy_031508/

"RAND Worldwide Terrorism Incident Database," undated, Santa Monica, Calif.: RAND Corporation. As of February 5, 2009: http://www.rand.org/ise/projects/terrorismdatabase/

Rapoport, David C., "The Fourth Wave: September 11 in the History of Terrorism," *Current History,* Vol. 100, December 2001, pp. 419–425.

Rapoport, David C., and Leonard Weinberg, eds., *The Democratic Experience and Political Violence,* Portland, Ore.: Frank Cass Publishers: 2001.

Reader, Ian, *Religious Violence in Contemporary Japan: The Case of Aum Shinrikyo,* Honolulu, Hawaii: University of Hawaii Press, 2000.

Reich, Walter, ed., *Origins of Terrorism: Psychologies, Ideologies, Theologies, States of Mind,* Washington, D.C.: Woodrow Wilson Center Press, (1990) 1998.

Reynal-Querol, Marta, "Political Systems, Stability and Civil Wars," *Defence and Peace Economics,* Vol. 13, No. 6, 2002, pp. 465–483.

Rhodes, Richard, "Ideas About the Development of Violent Behavior (Neurobiology over Behavioralism)," *Science,* Vol. 290, No. 5494, November 10, 2000, p. 1093.

Robbins, Thomas, and Dick Anthony, "Sects and Violence: Factors Enhancing the Volatility of Marginal Religious Movements," in Lorne L. Dawson, ed., *Cults in Context,* New Brunswick, Canada: Transaction Publishers, 2004.

Robison, Kristopher K., Edward M. Crenshaw, and Craig J. Jenkins, "Ideologies of Violence: The Social Origins of Islamist and Leftist Transnational Terrorism," *Social Forces,* Vol. 84, No. 4, June 2006, pp. 2009–2026.

Rose, Richard, "Northern Ireland Loyalty Study," Michigan: Inter-University Consortium for Political Research, Study # 7237, 1975.

Rushdie, Salman, "The Right Time for an Islamic Reformation," *The Washington Post,* August 7, 2005, p. B07.

Sageman, Marc, *Understanding Terror Networks,* Philadelphia, Pa.: University of Pennsylvania Press, 2004.

————, *Leaderless Jihad: Terror Networks in the Twenty-First Century,* Philadelphia, Pa.: University of Pennsylvania Press, 2008.

Sandler, Todd, "On the Relationship Between Democracy and Terrorism," *Terrorism and Political Violence,* Vol. 7, No. 4, 1995, pp. 1–9.

Schachter, Jonathan, *The Eye of the Believer: Psychological Influences on Counter-Terrorism Policy-Making,* RAND Dissertation Series, Santa Monica, Calif.: RAND Corporation, 2002. As of December 29, 2008: http://www.rand.org/pubs/rgs_dissertations/RGSD166/

Schmid, Alex P., "Terrorism and Democracy," *Terrorism and Political Violence,* Vol. 4, No. 4, 1992, pp. 14–25.

Schmid, Alex, Albert J. Jongman, et al., *Political Terrorism: A New Guide to Actors, Authors, Concepts, Data Bases, Theories, and Literature*, New Brunswick, Canada: Transactions Books, 1988.

Science, Special Edition on Violence, Vol. 289, No. 5479, July 2000.

Sedgwick, Mark, "Al-Qaeda and the Nature of Religious Terrorism," *Terrorism and Political Violence,* Vol. 16, 2004, pp. 795–814.

————, "Inspiration and the Origins of Global Waves of Terrorism," *Studies in Conflict & Terrorism,* Vol. 30, 2007, pp. 97–112.

Skocpol, Theda, *States and Social Revolutions: A Comparative Analysis of France, Russia and China,* Cambridge, United Kingdom: Cambridge University Press, 1979.

Simon, Jeffrey D., *The Terrorist Trap: America's Experience with Terrorism,* Bloomington, Ind.: Indiana University Press, 1994.

Smelser, Neil J., and Mitchell, Faith, eds., *Discouraging Terrorism: Some Implications of 9/11,* Washington, D.C.: National Academies Press, 2002. As of December 28, 2009: http://www.nap.edu/openbook.php?isbn=0309085306&page=R1

Snow, David A., Louis A. Zurcher, Jr., and Sheldon Ekland-Olson, "Social Networks and Social Movements: Microstructural Approach," *American Sociological Review,* Vol. 45, 1980, p. 792.

Speckhard, Anne, and Khapta Akhmedova, "The New Chechen Jihad: Militant Wahhabism as a Radical Movement and a Source of Suicide Terrorism in Post-War Chechen Society," *Democracy & Security,* Vol. 2, 2006, pp. 1–53.

Sprinzak, Ehud, "Extreme Left Terrorism in a Democracy," in Walter Reich, ed., *Origins of Terrorism: Psychologies, Ideologies, Theologies, States of Mind*, Washington, D.C.: Woodrow Wilson Center Press, 1990.

Stephenson, Matthew, "Rule of Law as a Goal of Development Policy," prepared for the World Bank, undated. As of December 29, 2008:
http://go.worldbank.org/DZETJ85MD0

Stern, Jessica, "Pakistan's Jihad Culture," *Foreign Affairs,* Vol. 79, No. 6, November–December 2000, pp. 115–127.

———, *Terror in the Name of God: Why Religious Militants Kill,* New York: Ecco/Harper Collins Publishers, 2003.

Tarrow, Sidney, *Democracy and Disorder,* Oxford, United Kingdom: Oxford University Press, 1989.

———, "Cycles of Collective Action: Between Moments of Madness and the Repertoire of Contention," in Mark Traugott, ed., *Repertoires and Cycles of Collective Action,* Durham, N.C.: Duke University Press, 1995, pp. 89–116.

———, *Power in Movement: Social Movements, Collective Action and Politics,* Cambridge, United Kingdom: Cambridge University Press, 2nd ed., 1998.

"The ITERATE Data," undated. As of February 5, 2009:
https://www.webdepot.umontreal.ca/Usagers/langlost/MonDepotPublic/recherche/iterate.html

Tilly, Charles, *From Mobilization to Revolution,* Reading, Mass.: Addison-Wesley, 1978.

Travis, Alan, "The Making of an Extremist," *The Guardian,* August 20, 2008a. As of February 2, 2009:
http://www.guardian.co.uk/uk/2008/aug/20/uksecurity.terrorism

———, "MI5 Report Challenges Views on Terrorism in Britain," *The Guardian,* August 21 2008b. As of January 24, 2009:
http://www.guardian.co.uk/uk/2008/aug/20/uksecurity.terrorism1/print

Urdal, Henrik, "A Clash of Generations? Youth Bulges and Political Violence," *International Studies Quarterly,* Vol. 50, 2006, pp. 607–629.

Varshney, Ashutosh, "Ethnic Conflict and Civil Society: India and Beyond," *World Politics,* Vol. 53, No. 3, 2001.

———, *Ethnic Conflict and Civic Life: Hindus and Muslims in India,* 2nd ed., New Haven, Conn: Yale University Press, 2003.

Victoroff, Jeff, "The Mind of the Terrorist: A Review and Critique of Psychological Approaches," *Journal of Conflict Resolution,* Vol. 49, No. 1, 2005.

Wade, Sara Jackson, and Dan Reiter, "Does Democracy Matter: Regime Type and Suicide Terrorism," *Journal of Conflict Resolution,* Vol. 51, No. 2, April 2007, pp. 329–348.

Walton, Norman W, "The Walking City: A History of the Montgomery Boycott," *Negro History Bulletin*, 1956, Vol. 20, pp. 17–20.

Watters, Pat, *Down to Now: Reflections on the Southern Civil Rights Movement*, New York: Pantheon, 1971.

Weber, Max, *On Charisma and Institution Building*, in S. N. Eisenstadt, ed., Heritage of Sociology Series, Chicago, Ill.: University of Chicago Press, 1968.

Weinberg, Leonard, "Turning to Terror: The Conditions Under Which Political Parties Turn to Terrorist Activities," *Comparative Politics,* Vol. 23, No. 4, 1991.

Weinberg, Leonard, and Ami Pedahzur, *Political Parties and Terrorist Groups*, New York: Routledge, 2003.

Weinberg, Leonard, Ami Pedahzur, and Arie Perlinger, *Political Parties and Terrorist Groups*, New York: Routledge, 2008.

Wilkinson, Paul, *Political Terrorism,* London: Macmillan, 1974.

———, *Terrorism Versus Democracy: The Liberal State Response,* London, United Kingdom, and Portland, Ore.: Frank Cass, 2001.

Wisler, Dominique, "Violence Politique et Mouvements Sociaux. Etude Sur les Radicalisations Sociales en Suisse Durant la Période 1969–1990," Ph.D. thesis, University of Geneva, 1993.

Zavis, Alexandra, "A Profile of Iraq's Foreign Insurgents: Interrogations of Fighters in U.S. Custody Yield a Portrait of Young, Lonely Recruits Who Crave Recognition," *Los Angeles Times,* March 17, 2008, p. 3.

Zolberg, Aristide, "Moments of Madness," *Politics & Society*, Vol. 2, No. 2, 1972, pp. 183–207.

Endnotes

[1]   Many other factors contributing to terrorism are described in the companion papers of the larger volume of which this is part (Davis and Cragin, 2009).

[2]   The arguments are that terrorism has not resulted in massive social change and that many theories about revolution are based on socioeconomic and political contexts of a sort much less common in the current day (for example, agricultural revolutions).

[3]   Simon (1994, pp. 4–5) posits a related wave theory—that terrorism can by classified in terms of cycles of tactics. The 1970s cycle emphasized hijackings and attacks on foreign embassies; the 1980s emphasized hostage-taking, suicide truck bombings, and bombs on airplanes. He suggests that the 1990s may be categorized by the use of more sophisticated weapons, such as chemical and biological weapons

and shoulder-fired missiles. Fortunately, that did not come to pass, but his concerns apply today.

4   The Karstedt-Henke citation is in German; her work is also cited in Koopmans (1993, p. 641).

5   Theda Skocpol draws on both Marx and Moore (1966).

6   Karstedt-Henke (1980) refers to this as a "tit-for-tat of violence and counterviolence, which produces terrorist groups" (pp. 213–217).

7   Real-world examples (such as the Weathermen) suggest that repression is not a necessary condition. But the idea that persecution from outside the group has significant effects on the organizational behavior of the group is important, because it often seems to result in an increased tolerance for behaviors that the larger society is more likely to reject. Isolation through persecution is considered a key factor in explaining the likelihood of violent behavior in a cult or sect as negative feedback is removed from the system. See Robbins and Anthony (2004, pp. 354–357) and Galantar (1989).

8   For access to, or information about, the database, see the "RAND Worldwide Terrorism Incident Database," undated.

9   For information about the dataset or access to the dataset, see "The ITERATE Data" (undated), contact Vinyard Software at 2305 Sandburg Street, Dunn Loring, VA 22027, or email Edward W. Mickolus at edwardmickolus@hotmail.com.

10   Cited in Zavis (2008, p. 3); and Quinn (2008).

11   The study is classified but *The Guardian* has reported on it extensively. See Travis (2008a).

12   For a review of the political opportunity literature, see Meyer (2004).

13   See also Varshney (2003, pp. 85–100) for a critique of traditional rational choice theory and a discussion of creative ways to apply rationality to both individual and group actions within the context of ethnic conflict (with its obvious parallels to terrorist group activity).

14   This is particularly true in the social movement literature that focuses on "protest waves." Admittedly, violence is a continuum that runs from destroying private property to taking human life, and, presumably, one end of the continuum is less problematic than the other. Certainly, the punishments for different types of "violent" offenses are distinct.

15   For competing hypotheses on predispositions to violent behavior see Pinker (2002) and *Science* (2000) (for and against).

16   Cited in "Pope: Islamic 'Holy War' Against God's Nature" (2006).

[17] Although the classified 2008 MI-5 study is not publicly available, *The Guardian* newspaper covered its findings in detail. See Travis (2008b).

[18] The Student Nonviolent Coordinating Committee was one of the primary coordinating organizations of the Civil Rights Movement in the 1960s.

[19] Watters (1971), cited in McAdam (1999).

[20] The American political-science tradition also has a small "culture of violence" literature, which explores why the southern and western regions of the United States, initially settled by southerners, are more violent than the rest of the country, measured primarily in terms of homicide rates. Nisbett (1993) and Cohen and Nisbett (1994) argue that southern white males are more violent partly because of a regional ideology that justifies violence to maintain or defend a man's honor. They hypothesize that although the conditions that gave rise to this ideology have dissipated, the culture of violence is sustained "through collective representations emphasizing the importance of honor and through violent self-fulfilling prophecies centering on hypersensitivity to affronts" (p. 566).

[21] For a brief discussion of the rule of law and the pros and cons of using the concept to try and make policy or measure outcomes, see Stephenson (undated).

[22] As developed initially by Schmid, Jongman, et al. (1988) and expanded by Post, Ruby, and Shaw (2002).

[23] This distinction is made in the database maintained by the National Consortium for the Study of Terrorism and Responses to Terrorism (START) at the University of Maryland, a federally funded Department of Homeland Security Center of Excellence.

CHAPTER THREE

Why and How Some People Become Terrorists

Todd C. Helmus

Introduction

Terrorism and the destruction it unleashes are fed and sustained by an ever-willing cadre of new recruits. Saudis, Yemenis, and Jordanians continue to flow into Iraq with the express purpose of achieving martyrdom by participating in operations against coalition forces and Iraqi citizens (Zavis, 2008). Earlier in the decade, Hamas and Fatah were inundated with requests from Palestinians seeking their place in line for suicide operations (Hassan, 2001). Meanwhile, attacks against targets in the United States, Casablanca, London, and Madrid, along with thwarted attacks elsewhere, suggest a more than adequate supply of willing recruits.

Operations designed to kill and capture the leaders and active members of terror organizations and otherwise disrupt terror activities play a critical role in today's fight against terrorism. However, the bountiful number of available new recruits may enable these organizations to fill or even expand depleted ranks. Absent initiatives that effectively disable the organizational capacity for operations, policies will need to also consider approaches that limit the influx of new members.[1]

A counter-recruitment strategy may prove meaningful for several reasons. First, research by Benmelech and Berrebi (2007), which is cited in Berrebi (2009), demonstrates that older and more highly educated suicide bombers, in comparison with the young and less educated, are less likely to be apprehended before detonation, produce more casualties, and are assigned to more important target sets. It is important to note that the age and education of the 148 suicide bombers in their sample differ considerably: Only 18 percent have academic degrees, and their mean age is 21 years, with a standard deviation of 4.7 years (with the most successful bombers averaging close to 26 years). This variance suggests that educated and experienced recruits are in relatively short supply.[2] Counter-radicalization programs may thus lower the number of experienced and educated terrorists and so reduce the success and devastation of suicide attacks.[3] Second, as illustrated by Paul (2009) in his paper on support for terrorism, many of the same factors that motivate terrorist recruitment also motivate broad popular support for terrorists. In theaters of irregular war, these same factors likely motivate the recruitment of counter-government insurgents. Consequently, counter-recruitment strategies can achieve synergy and utility across a broad domain of problem sets that contribute to political violence. Finally, counter-recruitment initiatives may have value in and of themselves. For example, as Berrebi (2009) suggests in his paper, terrorists are indeed sensitive to cost-benefit considerations. Understanding these cost-benefit considerations and effectively targeting those that are malleable should produce advantageous results. For these reasons, this paper reviews the social science research underpinning individual radicalization. The goal is to provide the educational basis on which to derive effective counter-radicalization and counter-recruitment initiatives.

Different disciplines approach the question of radicalization and terrorism in significantly different ways. Consider some examples among many. Psychological approaches examine individual factors that lead to terrorism participation. These can include personality characteristics, mental illness, or previous exposure to traumatic experiences. Social psychology focuses on the motivating role of group dynamics and peer pressure. Models employing rational-choice paradigms evaluate the influential roles of preferences, rewards, and constraints on terrorist behavior. Sociological perspectives focus on the patterns of social relationships, social interactions, and culture. They also support the systematic study of social movements or group-level actions designed to push social change. Political-science approaches examine the overarching role of political environments on individual behavior and often address such factors as occupation by a foreign power and the struggle for liberation. Finally, the study of religion seeks to understand the seemingly influential role of Islam and other religious perspectives as a motivating force (Kimhi and Even, 2004; Hudson, 1999; Moghadam, 2005).

Much of the literature deals with single-discipline approaches to understanding the motivations for terrorism. However, scholars now agree that multidiscipline approaches are required to understand terrorism (Victoroff, 2005; Moghadam, 2005; Hudson, 1999). The next section discusses the many factors believed to be at work simultaneously, drawing as necessary from the various disciplinary studies.

Research on terrorist radicalization and recruitment suffers from several methodological shortfalls that should caution us to not accept and apply the findings with certainty. First, the field of study suffers from a lack of original research. As Andrew Silke notes, "only about 20 percent of research articles provide substantially new knowledge that was previously unavailable to the field" (Silke, 2008, p. 101). Second, much of terrorism research does not use samples of nonterrorist control groups. Studies that look only at terrorist samples themselves are unable to shed light on how these samples truly differ from their nonterrorist counterparts. Other studies evaluate the radicalization phenomenon by relying heavily on secondary analysis of data, such as that available in the press and other open-source reports (Sageman, 2004; Bakker, 2006). Compiling fully complete datasets is nearly impossible, and so missing data points risk skewing study conclusions. Finally, other studies rely on interviews with known terrorists. A key issue here is that factors that propel an individual toward terrorist participation may be fundamentally different from those that maintain that participation. Individuals may thus be inclined to report current attitudes and experiences rather than those experienced at the time of radicalization. As John Horgan notes, "the reason given for involvement [in terrorism] may be a direct reflection of an ideological learning process that comes from being part of the group" (Horgan, 2008, p. 86). Consequently, these personal accounts, as with conclusions based on other forms of research, should be interpreted with caution.

At this point, it should be emphasized that a psychological movement to terrorism is not a discrete choice. In this context, individuals

generally do not make a single and conscious decision to "become" a terrorist. Rather, progression toward violent behavior is gradual (Horgan
and Taylor, 2001; Borum, 2004). Radicalization is, in fact, more like
a process, and within this process, individuals are moved forward by a
host of factors that may include socialization, exposure to rewards, and
other environmental influences.[4] Several papers and reports address
different process theories by tabulating the psychological evolution that
results in terrorist behavior and defining the order in which different
factors exert their influence (Moghaddam, 2005; Taylor and Horgan,
2006; Silber and Blatt, 2007). Such process issues are important but
beyond the scope of this paper. Instead, this paper seeks to simply identify various factors that increase the risk of radicalization.[5]

Looking across the discussions in the various disciplines, a long
list of factors are believed to contribute to radicalization. Subsequent
sections describe these factors under the headings of (1) Radicalizing
Social Groups, (2) Desire for Change, (3) Desire to Respond to Grievance, and (4) Perceived Rewards.

Radicalizing Social Groups

General Observations

Social groups play a critical role in the radicalization process. Basic
research in social science amply demonstrates the influential role that
group interactions have on individual attitudes, beliefs, and commitment to action. The social-psychological processes that influence individuals are many and include processes related to in-group/out-group
biases, conformity, compliance, group think, polarization, and diffusion of responsibility (see Table 3.1 for brief descriptions of these
processes).

Psychiatrist Mark Sageman provides the seminal description
of how these social dynamics influence the radicalization process
(Sageman, 2004). He notes that some Middle Eastern Muslims who
study abroad in Europe become homesick and feel alienated in their
host communities. They seek companionship at their local mosques
and ultimately form small cliques of peers that are centered, in part,

Table 3.1
Relevant Cognitive Biases

Relevant Social-Psychological Processes	Description
In-group/out-group	Groups tend to view themselves positively and view those outside the groups negatively. Terrorist groups that demonize or dehumanize outsiders have a reduced threshold for perpetrating acts of violence against these outsiders (Myers, 2005).
Conformity	Groups provide expectations for individual beliefs and conduct that result in shifting individual attitudes, opinions, and behaviors in favor of group norms (Sherif, 1935).
Compliance	Groups foster increased compliance with group requests and obedience to orders. High group cohesion, isolation from alternative groups, increased cost of defiance, and the degree to which the group satisfies individual needs increase the likelihood and severity of group conformity (Milgram, 1965).
Group think	Groups often engage in excessive efforts to reach agreement or consensus that can result in flawed judgments on the part of individuals (Janis, 1972).
Group polarization	Group interactions polarize individual attitudes to the extreme (Moscovici and Zavalloni, 1969).
Diffusion of responsibility	Responsibility for violence and radical ideology appears spread out over the entire group thus limiting the extent to which individuals assume personal responsibility.

on keeping the Muslim dietary restrictions of halal. Social interactions that take place during these meals and other social settings help push individual members to a radicalized state. In a subsequent analysis, Sageman writes:

> At dinner, they talk about shared interests and traditions and reinforce common values. To conform to conversational courtesy, they stress their commonality and in the process create a micro-culture and develop a collective identity. Over time, they become friends. If their friendship intensifies, they often move in together to save money and further enjoy each other's company. When they are at this stage, they form strong cliques that continue to radicalize over time (2008, p. 68).

The radicalizing role of group interactions is a robust finding in the literature. Edwin Bakker, for example, analyzed the case studies of 242 Europe-based jihadis (a subset of which overlaps with Sageman's sample). Like Sageman (2004), he demonstrated that networks of friends or relatives were instrumental in the radicalization process, with such networks preexisting radicalization in over 35 percent of his sample and operating independently of formal recruitment tactics (Bakker, 2006). In other analyses by Thomas Hegghammer, friends and relatives were influential in motivating many Saudis to enter Afghan training camps. He specifically states, "Group dynamics such as peer pressure and intra-group affection seem to have been crucial in the process" (Heghammer, 2006, p. 50).[6] Such processes have similarly been implicated in samples of Saudi militants in Iraq (Hegghammer, 2007), Southeast Asian militants belonging to Jemaah Islamiyyah, and Filipino militant groups (Cragin et al., 2006).

The social movement literature confirms these findings. In a seminal study on recruitment into the 1960s civil rights project Freedom Summer, black political participation in the American South was in part associated with links to individuals already involved in the campaign (McAdam, 1986). More broadly, Snow, Zurcher, and Eckland-Olson (1980) identified ten quantitative studies that address the recruitment process into religious organizations. In eight of these studies, relatives or acquaintances helped recruit between 59 and 100 percent of study participants. Donatella della Porta's study on left-wing terror groups in Italy reached similar conclusions: that linkages with close friends and kin were influential in recruitment (della Porta, 1996). Observes della Porta and her colleague, Mario Diani, in a more recent analysis: "Available evidence suggests that the more costly and dangerous the collective action, the stronger and more numerous the ties required for individuals to participate" (della Porta and Diani, 2006, p. 117).

Exposure to the radicalizing effect of peer and social groups can occur in any number of ways and settings. Overt and top-down organizational recruitment efforts that routinely harness the power of social milieus are but one example. Informal peer or family groups also influence individual radicalization. In addition, a growing sense of alienation among susceptible youth may give these social groups an increasing degree of power and allure. Organizational recruitment efforts, informal peer and family groups, and alienation are addressed in greater detail below.

Terrorist Recruitment

Group socialization processes are inherently active in organizational recruitment and indoctrination efforts. Recruiters from Saudi Arabia's al-Qaeda, for example, held informal gatherings in private homes or at religious summer camps. Potential candidates were invited to smaller gatherings or one-on-one conversations where motivation and qualifications were assessed and potential participants selected (Hegghammer, 2006). Gatherings at private homes or mosques were also used to target Saudi recruits for Iraq (Hegghammer, 2007) and served as a venue where a "harmless discussion about Islam" turned to the U.S. war in Iraq and U.S.-committed atrocities (Zavis, 2008). Jemaah Islamiyyah is also known for its systematic recruitment campaigns in Indonesia, Malaysia, and Singapore.[7] Recruiters for the Liberation Tigers of Tamil (LTTE) have targeted efforts at Tamil schools and the Kurdistan Workers Party has sent recruiters to cultural centers and summer camps (Pedahzur, 2005). Enlistment officers for Hamas would seek out bombers among universities, social clubs, schools, and mosques (Pedahzur, 2005). The teacher-student instruction and interactions illustrated here certainly point to group dynamic effects that help facilitate ideology. Interactions among students that motivate conformity, in-group processes, and group think dynamics will likely prove equally critical.

Other recruitment efforts are more informal. An analysis of European recruitment by Petter Nesser demonstrates that many recruiters look to personally indoctrinate those in their immediate circle of friends (Nesser, 2006a). Such recruiters are active, for example, in the Algerian Armed Islamic Group and the Salafist Group for Preaching and Combat (identified by French acronyms GIA and GSPC, respectively) (Lia and Kjøk, 2001). However, little is known about the actual interactions that take place between recruiters and their subjects (Nesser, 2006a).

Bottom-Up Peer Groups

Several peer groups play a critical role in fostering bottom-up radicalization. Sageman's (2004) analysis suggests that numerous recruits made their connections with al-Qaeda and other organizations only after radicalizing among peer groups. Alternatively, the Hofstad Group attack in the Netherlands was conducted by a cell with limited affiliation with al-Qaeda. This potentially emerging trend in terrorism has been referred to as the decentralization of al-Qaeda or growth of Salafism as a social movement (Kirby, 2007). Radical Salafi ideology provides the overarching inspiration for radicalization and subsequent attacks, although al-Qaeda as an organization may exert limited command and control. In Bakker's (2006) analysis of 242 European jihadis, although networks of friends or relatives facilitated involvement in terrorism, there were generally no formal ties with global Salafi networks. These bottom-up processes may further interact with organizational recruitment efforts by increasing individual susceptibility to a recruiter's advances.

These groups meet or are otherwise exposed to social influence at many venues. Examples of these venues include religious settings that advocate violence, radicalized families, and prisons. The Internet also provides a virtual meeting ground for radicalization. The following paragraphs describe several of these in more detail.

Religious Settings Advocating Violence. Marc Sageman notes that 50 percent of his sample attended only 12 Islamist institutions. The individuals that formed the core of the Madrid terrorist cell came to know each other while attending Madrid's M-30 mosque (Sageman, 2004). In addition, spiritual leaders not infrequently use their roles as authoritative religious sources to indoctrinate radical ideology. These religious leaders often play a critical role in cultivating radicalization through sermons and the leadership of prayer meetings. In one reported case that may be representative of many others, a 24-year-old Saudi student was motivated to join the Iraqi resistance in part through the forceful rhetoric of a local cleric who was sympathetic to al-Qaeda (Obaid and Cordesman, 2005).

Family. Sustained exposure to and interaction with radicalized parents and siblings are also suggested as aids to radicalization (Cragin et al., 2006). Twenty percent of Edwin Bakker's sample were related to the terrorist cell or organization through kinship. One example is the Benchellali family, which had six members arrested on terrorism charges (Bakker, 2006). Bakker suggests that these family ties played a critical role in radicalization and recruitment. Militant family ties also predominate in Southeast Asia. In Indonesia, surveyed terrorism experts argue that many recruits come from families already associated with Jemaah Islamiyyah or the Darul Islam movement (Cragin et al., 2006). In the Philippines, membership in MILF (Moro Islamic Liberation Front) and ASG (Abu Sayyaf Group) is frequently the result of family, tribal, or clan introductions (Cragin et al., 2006).

Prisons. Prisons are seen as a growing breeding ground for radicals. In 1994, the French police intensified operations against GIA networks, resulting in a burgeoning population of incarcerated militants. The prisons provided these radicals a captive audience with which relationships could be developed and used to inculcate radical ideology (Lia and Kjøk, 2001). Radical Islamic proselytism in French prisons has recently increased and is a significant security concern (Siegel, 2006).

Internet. The Internet increasingly functions as a virtual meeting ground for radicalization. Marc Sageman argues that, in 2004, connectivity within radicalizing social groups increasingly went through cyberspace. He notes the following examples:

> People involved in the Crevice case spanned two continents and kept in touch via the Internet. The Madrid bombers were inspired by a document posted on the Global Islamic Media Front Website in December 2003. The Hofstad Group in the Netherlands interacted through dedicated forums and chat rooms and inspired other young Muslims to join them physically after making contact with them on the forums. The April 2005 Cairo Khan al-Khalili bombing was aided by the Internet, with the perpetrators downloading bomb-making instructions from jihadi websites. . . . The people who tried to plant bombs on trains in Germany in the summer of 2006 met in an Internet forum (2008, pp. 109–110).

Several characteristics of the Internet may help account for this trend. First, the World Wide Web provides a venue where radicals can post and share training material, ideological manifestos, radicalized e-magazines, and videos depicting attacks against U.S. forces. These sources provide a key information and ideological source to would-be recruits. In addition, such Web-based technologies as blogs, wikis, chat rooms, social networking, and video-sharing sites increasingly create an environment where groups of like-minded individuals can interact and develop mutually supportive relationships. The Internet may thus act as a meeting ground and social milieu no less than a mosque or prison setting. Group psychological processes thus operate in these environments that facilitate radicalization in ways similar to those for live interaction. The Internet also offers a newly egalitarian environment where, for example, all participants in a chat room carry nearly equal weight and influence (Sageman, 2008).[8]

As addressed subsequently in this paper, the Internet functions as a key propaganda outlet for terrorist organizations. Its users can incite anger at grievances committed against Muslims, facilitate Salafi ideology, and heap praise on martyred militants. The Internet may also function as a means by which individuals gain access to terror organizations. Al-Qaeda in Iraq used the Internet to facilitate connection between interested recruits and community gatekeepers (Hegghammer, 2007). Iraqi insurgents and their sympathizers are said to have monitored users of their Internet sites and then contacted candidates who seemed the most willing to participate in the campaign (Curiel, 2005). Written instructions were also provided on how to join the Iraqi jihad (Hegghammer, 2007). These instances may be the exception, however, as few other examples of the Internet as gatekeeper exist.

A word of caution should be noted, however. In a review of the social movement literature, della Porta and Diani argue that the "jury is still out" with respect to whether the World Wide Web facilitates organizational activism by reinforcing links established in the real world or whether they can effectively establish brand new links from scratch. However, they do observe that "virtual networks operate at their best when they are backed by real social linkages in specifically localized communities" (della Porta and Diani, 2006, p. 133).

Alienation

Feelings of social alienation may contribute to the allure and power of radicalizing social groups. As noted above, Marc Sageman argues that Middle East Muslims who lived in European Diaspora communities felt homesick, lonely, marginalized, and otherwise excluded from society, and second-generation Muslims felt subjected to discrimination and exclusion from society. This alienation ultimately drove them to local mosques to seek Muslim companionship (Sageman, 2004). From 1993 to 2003, 86 percent of 212 suspected and convicted terrorists were Muslim immigrants (Leiken, 2004). Similar evidence of social alienation is documented by members of the cell that perpetrated the 2006 London attacks (Kirby, 2007), by Afghan veterans who returned to Saudi Arabia and ultimately joined al-Qaeda's outlet in Saudi Arabia (Hegghammer, 2006), and by the radical Islamist group al-Muhajiroun (Wiktotwicz, 2004).

One means by which alienation contributes to radicalization is through the promotion of cognitive openings. Drawing on the social movement literature, Wiktotwicz (2004) argues that an individual must be willing to be exposed to the radical views of terror organizations or peers. Socialization that gradually opens an individual's mind to the radical message is one way to create a cognitive opening. Another is through a crisis that "shakes certainty in previously accepted beliefs and renders an individual more receptive to the possibility of alternative views and perspectives" (Wiktotwicz, 2004, p. 7). Perceptions of social alienation play a critical role in facilitating these cognitive openings, leading individuals to reorient themselves to a radical Islamist perspective. Political repression and personal crises (such as death in the family) can also facilitate cognitive openings. In Europe-based Muslim communities, this alienation likely results from widespread perceptions of social, economic, and political discrimination (see the section below titled "Discrimination").

Desire for Change

Some terrorist recruits may be motivated out of a desire to institute changes in their environment. These changes are often related to the objectives of the terrorist organizations or to movements they seek to join. There is some reason to suspect that organizational goals do contribute to radicalization. For example, organizational goals filter down to ideological arguments that undergird recruitment and propaganda. Recruiters in European terror cells, for example, seek to "convince and socialize young Muslims in their social surroundings" into accepting the tenets of Jihad (Nesser, 2006a, p. 20). Kruglanski and Golec (2004) likewise suggest that many terrorists trust in and ultimately incorporate ideological manifestos of terror organizations and their leaders. Many terrorists' own words provide evidence of the motivating role of these organizational objectives, although it should be noted that the attitudes these words express may be adopted after radicalization (Crenshaw, 1981; Horgan, 2005).

However, other lines of evidence counter the notion that organizational goals influence individual radicalization. Max Abrahms, in his 2008 paper titled "What Terrorists Really Want: Terrorist Motives and Counterterrorism Strategy," argues that terror-organizational goals are rarely stable and consistent. He notes that even al-Qaeda's goals shifted frequently in the late 1990s, from waging defensive jihad against the Soviets in Afghanistan, to fighting local conflicts in the Philippines and Bosnia, to targeting the "far enemy." To this end, he notes that al-Qaeda's members have criticized the organization for inconsistent messages. Other evidence cited by Abrahms suggests that many interviewed terrorists, and even their leaders, are unaware of the organization's political or religious objectives. Such is the case with members of the Irish Republican Army (IRA) (White, 1992) and al-Qaeda (Richardson, 2006). Abrahms also cites organizational research that suggests that personal inducements are more important in motivating participation than a confluence of personal and organization motivations. Abrahms's primary conclusion is that the social bonds that form among members of a terrorist organization are far more influential as a motivating force than is ideological commitment (see the section titled

"Perceived Rewards/Friendship"). It is also possible that terrorists learn to adopt organizational goals as a consequence of their participation. With this limitation understood, the following sections enumerate desired changes, which relate to political, religious, and legal goals.

Political Change: Desire for an Independent State and to Sow Anarchy

Some terrorists seek to use their actions to motivate broad political changes. Some terrorist organizations seek to overthrow capitalist social and economic systems and are exemplified by such 1970s terrorist groups as the Red Army Faction in Germany and the Red Brigades in Italy (Post, Ruby, and Shaw, 2002). Donatella della Porta's research into the Italian and German movements shows that many would-be militants had previous and extensive experience in legal social movements of the radical left. She argues that "One essential fact about the activists of the left-wing underground organizations in Italy and Germany is that they all had political motivations" (della Porta, 1996, p. 166).

Other terrorist organizations and their members are motivated by a desire for an independent state or to remove a perceived occupier from power. Robert Pape (2003) argues, from an organizational and strategic perspective, that suicide terrorism is often a response to perceived occupation and is designed to coerce governments into making territorial concessions. His analysis suggests that the tactics have proven successful in such areas as Lebanon, Palestinian territories, and Sri Lanka. Others argue that modern terrorism is more than just a response to occupation, as is evident in the many attacks that have plagued Western Europe (Moghadam, 2006). Regardless, concerns of occupation appear to play a motivating role in at least a large subset of terror operations.

Such overt strategic objectives certainly inspire the organizations themselves to use the tactic and identify new recruits. The objectives also appear to motivate individual recruits. For example, Wells's and Horowitz's (2007) review of biographical data on Palestinian suicide terrorists and surveys of terrorism experts suggest an important role for political motivations in Palestinian terrorist recruitment. Similarly,

Farhad Khosrokhavar and David Macey observe that, "martyrdom in Iran, Algeria and Palestine obeys an internal logic born of the frustrated ambition to have a nation whose existence has been denied" (Khosrokhavar and Macey, 2005, p. 109). Individual quotes from terrorists further attest to this political motivation.[9]

Religious Changes: Caliphate and Millennialism

The desire to institute religion-related changes constitutes another motivational set. For example, Islamist terrorists frequently seek the goal of a worldwide and united Muslim community where a Muslim caliphate and Sharia law or Islamic law reign supreme.[10] Analyses of jihadi ideological manifestos and propaganda confirm that a united worldwide community of Muslims, or umma, is indeed a key priority. McCants and Brachman observe that

> Jihadis will fight until every country in the Middle East is ruled only by Islamic law. Once they are in power, the punishments of the Qur'an (such as cutting off the hand of a thief) will be implemented immediately. Not even Saudi Arabia has it right; the Taliban state was the only state that was closest to their vision (2006, pp. 9–10).[11]

A worldwide caliphate motivates at least some Muslim participation in terrorism. The New York Police Department's assessment of radicalization in the West argues that religious and political ideology, part of which is related to the caliphate, has worked to inspire "all or nearly all of the homegrown groups" they analyzed to include the Madrid 2004 bombers, the Hofstad Group, and London's 7/7 bombers (Silber and Blatt, 2007). Desire for an Islamic state also appears to have motivated members of Jemaah Islamiyyah (Gunaratna, 2005).[12]

Single-Issue Change: Environmental Rights and Anti-Abortion

Some individuals are motivated by desire to institute change on a single and focused issue. Two examples in this regard focus on environmental rights and anti-abortion militant movements. Environmental rights movements seek to protect the ecological environment and promote animal rights. These groups include the Animal Liberation Front (ALF), the Earth Liberation Front (ELF), and Stop Huntingdon Animal Cruelty (SHAC). The actions of these groups range from property vandalism and arson to violent assaults and murder. For example, SHAC has specifically targeted executives of Huntingdon Life Sciences, a British drug-testing facility that uses animals to test drugs for safety before they are tested on people. These militant groups receive at least tacit support from more radical elements of People for the Ethical Treatment of Animals (PETA), whose member Bruce Friedrich is quoted as providing the following rational for eco-terrorism:

> If we really believe that animals have the same right to be free from pain and suffering at our hands, then of course we're going to be blowing things up and smashing windows (Southern Poverty Law Center, 2002).

Anti-abortion violence is another form of single-issue terrorism. Anti-abortion militants, such as Eric Rudolph, Juames Kopp, and Paul Hill, have launched arson and bombing attacks against abortion clinics and have murdered or attempted to murder clinic staff. Many of their militant actions are focused on intimidating abortion providers. The Army of God is a radical anti-abortion group that permits the use of violence to stop abortion and justifies this violence on theological grounds.[13] The group makes the following statement on its Web site:

> We the undersigned, declare the justice of taking all godly action necessary, including the use of force, to defend innocent human life (born and unborn). We proclaim that whatever force is legitimate to defend the life of a born child is legitimate to defend the life of an unborn child (Army of God, 2008).

Discrimination

Perceived social, economic, and political discrimination can play a critical role in the radicalization process. As noted above, perceived discrimination can facilitate a sense of alienation that may drive participation in radical milieus. Discrimination and other perceived injustices that are perpetrated by governing authorities may promote a sense that those regimes should be removed.

Many Muslims in Europe believe that they are subject to broad forms of social, economic, and political discrimination. For example, a European survey confirms that Muslims across Europe perceive widespread negative attitudes toward their religion and at times experience verbal and physical attacks (International Helsinki Federation for Human Rights, 2005). Muslims are also underperforming in British secondary education, are badly underrepresented in institutions of higher education (Silke, 2008, pp. 112–113), and suffer from disproportionately high unemployment rates (Office for National Statistics, 2004).[14] Low political representation is also problematic. In the United Kingdom, Muslims represent 3 percent of the population but have only 0.3 percent of the country's Parliament members and 0.9 percent of district councilors. Consequently 70 percent of British Muslims feel politically underrepresented. Similar problems are reflected across Europe (Greif, 2007).[15]

Such factors have been directly attributed to radicals seeking political independence. In Northern Ireland, three factors cited as helping to drive political conflict include economic deprivation, educational underperformance, and insufficient political representation (O'Leary, 2007). In the Philippines, surveyed experts attribute individual predilections for radicalism to several historical discrimination factors: economic neglect by the Filipino government, dispossession of ancestral Muslim lands by Christians, and attempts to forcibly assimilate Muslim communities into wider Catholic Philippine polity (Cragin et al., 2006).

Desire to Respond to Grievance

A desire to respond to some perceived grievance appears to motivate radicalization in a subset of individuals. These perceived grievances can be inflicted on either the individual personally (to include the individual's family and friends) or they can be inflicted on a larger collective group with whom the individual closely identifies. In the former case, seeking revenge against the alleged perpetrator is likely an intermediary motivation. In the latter case, individuals may be motivated out of a desire to "defend" the collective. Some elaboration of grievance issues is warranted.

Personal Grievance: Revenge

The desire to exact revenge against instigators of injustice and abuse is often cited as a powerful motivational force for terrorism. As Martha Crenshaw observed, "If there is a single common emotion that drives the individual to become a terrorist, it is vengeance on behalf of comrades or even the constituency the terrorist aspires to represent" (Crenshaw, 1981, p. 394).

Numerous examples from the literature highlight the motivational role of revenge. Palestinian terrorists frequently cite revenge as a key motivator for martyrdom operations (Argo, 2004; Soibelman, 2004). Vengeance has also been cited as influential in the Chechen conflict (Speckhard and Ahkmedova, 2006). In explaining motivation for becoming a member of ETA (Euskadi Ta Askatasuna, which is Basque for "Basque Homeland and Freedom"), one terrorist stated:

> Yes, I think there was an important factor which was revenge. What I mean is that I was coming from a family who had suffered political reprisals so the fact that I had the opportunity to take revenge. . . . I do consider that I was obliged to take revenge. . . . I consciously did it, being aware of the fact that I was going to hurt [someone] and with great satisfaction for doing so (Alonso, 2006, p. 195).[16]

Personal Attacks Directed at Self or Loved Ones. The experience of personal attacks directed at individuals or their loved ones underpins the revenge motivation. Many who enter terrorism report personal abuse at the hands of governing authorities or an occupying force. Psychiatrist Anna Speckhard and Khapta Ahkmedova interviewed 34 family members of martyred Chechen terrorists. They found that virtually all of the deceased terrorists "had personally witnessed the death and beatings of close family members or experienced torture themselves" (Speckhard and Ahkmedova, 2006, p. 455). Likewise, Palestinian bombers have frequently lost a friend or relative to the actions of Israeli soldiers.[17]

Mali Soibelman's interviews with five failed suicide bombers suggest that all had negative experiences with the Israeli military (2004, pp. 175–190). Individual examples are also prominent in Southeast Asia, where many terrorists suffered abuse at the hands of Filipino and Indonesian governments (Cragin et al., 2006). Protestant vigilante violence against Catholics and repression by British troops are also argued to have prompted the Provisional IRA to terrorist retaliation and increased the legitimacy of their efforts within local the population (Alonso, 2006). However, it is important to note that many civilians in these and other locales have experienced such abuses, but only a small percentage of them ultimately resort to terrorist actions.

Post-Traumatic Stress Disorder. Traumatic experiences frequently result in the development of the psychological condition known as post-traumatic stress disorder (PTSD). At least one study suggests that the development of PTSD in trauma survivors increases their propensity to seek revenge. Speckhard's and Ahkmedova's analysis, for example, showed that many of the Chechen terrorists who were exposed to abuse developed psychological symptoms indicative of PTSD to include foreshortened life and survivor guilt. They argue that

> A trauma victim who really does not expect to live a long life and who feels he does not deserve to have survived when others did not is logically and much more easily than an individual with a normal history able to surrender his life for a cause (Speckhard and Ahkmedova, 2006, pp. 459–460).

In another study that was cited by Speckhard and Ahkmedova (2006) but otherwise unavailable for review, Ahkmedova (2003) reportedly analyzed 653 traumatized Chechans and found that 39 percent of them had a desire for revenge. In addition, with increased traumatization, "revenge became both sufficient and acceptable" (Speckhard and Ahkmedova, 2006, p. 467). The relationship between PTSD and a revenge motive for terrorism has yet to be assessed in other zones of conflict, such as Iraq or the Palestinian territories.

Collective Grievance: Duty to Defend

"Collective grievance" refers to perceived problems experienced by members of an identifiable social group. Often, individuals increasingly identify with a social group whom they see as subject to unjust policies or actions. In the case of Islamic extremism, radicals increasingly identify with their Muslim heritage and the broader Muslim community. They may see discriminatory practices unjustly affecting this group or become incensed at abusive acts perpetrated against faraway Muslim "kin." The desire to address these collective grievances may prove a motivating force in the radicalization process.

A need to defend the perceived collective may underpin a desire to respond to collective grievances. Petter Nesser evaluated motivational patterns that appear to have precipitated terrorist attacks in Europe. He noted that key operatives and leaders regularly cited the occupation of Palestine, French support for the Algerian regime, Russian military operations in Chechnya, the Iraq war, and perceived European discrimination and persecution of Muslims (Nesser, 2006b). He states, "The doctrine and idea of global defensive *jihad* against aggressors attacking Islam and Muslims stands out as the single most important motivational factor at the group level" (p. 327).

A major factor implicated in radicalization relates to perceived "attacks on the collective," such as the anger incited by exposure to atrocities committed against Muslims living in faraway lands. The Internet is replete with depictions of anti-Muslim violence from conflicts in Kashmir, Bosnia, Chechnya, the Palestinian territories, Iraq, and Afghanistan. They are usually worst-case stories that are intended to inflame public opinion. They are available to anyone with Internet access. Many terrorist and radical organizations also distribute DVD and cassette videos (Speckhard, 2006; Silke, 2008). Numerous studies have implicated exposure to videos of foreign Muslim conflicts in the radicalization of European and Middle East–based militants (Obaid and Cordesman, 2005; Hegghammer, 2006, 2007; Sageman, 2008).

Research suggests that mere exposure to death-related imagery can increase support for martyrdom operations. The "mortality salience" effect refers to a series of research findings that suggest that exposure to such imagery has a number of pronounced effects, including increased pride in one's country, religion, and race. A team of researchers found that Iranian college students who were reminded of death were more likely than other students to voice support for martyrdom attacks and were more likely to state that they would take part in those attacks themselves (Pyszczynski et al., 2006).

Identity

Identification with the broader Muslim community plays a critical role in facilitating the concept of a collective grievance. Andrew Silke (2008) notes that many terrorist recruits report that they perceived a strong connection to the worldwide community of Muslims before they decided to participate in terrorism. The connection brings with it a sense of responsibility for helping Muslims they had never even met. This observation is backed up by Olivier Roy, who argues that the sense of community within the Islamic Diaspora more than rivals that of the local secular community. The uprooted and alienated condition of many Muslims leads to a process in which they reassess what Islam means for them (Roy, 2004).

Recent research bears out these observations. Humayun Ansari and colleagues assessed the interconnecting roles of national, ethnic, and religious identity using attitudinal questionnaires and in-depth interviews in a sample of British Muslims. Results demonstrated an identity hierarchy, whereby religious identity superseded ethnic identity, which was itself greater than national identity. It is important to note that Muslim identity was positively correlated with the perceived importance of Jihad and martyrdom, whereas British identity negatively predicted these attitudes (Ansari et al., 2005). A British government survey found that British Muslims, more than any other religious group, tended to rate their faith as their primary identity. This effect was most pronounced in young people ages 16–24 (Attwood et al., 2003).

Perceived Rewards

Behaviorism is a branch of psychology that postulates the simple notion that consequences influence behavior. Consequences that increase the likelihood of a behavior are called rewards, whereas consequences that decrease the likelihood of behavior are called punishments. The role that consequences play in motivating individual behavior is a well established phenomenon in behavioral research (Skinner, 1974; Higgins, 1997). Rational choice is a related paradigm that asserts that individuals choose the best action according to stable preference functions and constraints facing them and accordingly respond to incentives (see Berrebi, 2009). With both behaviorism and rational-choice theories, it is highly likely that both real and expected consequences motivate terrorist participation. Several of these motivating consequences are listed below. They include religious rewards, social status, financial rewards, friendship, and excitement.

Religious Rewards

The most prominently identified reward for participation in suicide terrorism is martyrdom. The perceived benefits associated with a martyr's afterlife include forgiveness of the martyr's sins, access to heaven and communion with God, the ability to guarantee access to paradise for 70 relatives or friends, and the belief that the martyr will be greeted in heaven to enjoy the sexual pleasure of 72 virgins (Soibelman, 2004). The concept of martyrdom and its heavenly rewards rests on ideological notions that the act of terrorism fulfils a divinely inspired imperative to act.

In numerous analyses, evidence suggests that suicide bombers and other terrorists believe that heavenly rewards await their final act. In Palestinian areas, suicide bombings are frequently referred to as "martyrdom operations." Analyses of last wills and testaments and videotaped statements demonstrate the fervent belief that suicide bombers have in the martyrdom concept. Motivations for martyrdom also appear prominent among Saudi militants recruited for operations in Iraq. Although political motivations were also present, the investigators noted that militants deeply aspired martyrdom and this factor may have been more important than achieving political objectives (Hegg-hammer, 2007).

Religion also serves as a motivating factor, in that participation in terrorism is seen as fulfilling a divine mandate. For example, many Singapore-based Jemaah Islamiyyah members saw membership in that group as a "'no fuss' path to heaven'" and they "believed they could not go wrong, as [the group's] leaders had quoted from holy texts" (Gunaratna, 2005).

Social Status

The role of increased social status for militants appears to be important in motivating recruitment. In Palestinian-controlled areas, popular support for terrorist attacks against civilians reached 70 percent in the summer of 2001. With this support comes high public veneration of those who participate in militant organizations and who conduct suicide bombings; they are viewed as "heroic soldiers" who "participate in a great struggle" (Soibelman, 2004). Ritualized pomp and circumstance greet suicide bombers as they prepare for their missions. Social applause can reach its peak after a suicide bomber's death, where posters, Web sites, and public exhibits pay homage to the martyr (Hafez, 2006). Mourning ceremonies provide further veneration of the dead.

Incarcerated Palestinian operators attest to the importance of increased social status. Militants interviewed by Post, Sprinzak, and Denny (2003) observed that recruits were treated with great respect and that a member of Hamas or Fatah was more highly regarded than someone unaffiliated. In another study, an incarcerated bomber claimed, "It made me feel good about myself. I got respect and justice to the family. There was a lot of excitement and everybody looked at me with honour to the shaheed" (Soibelman, 2004, p. 184).

The motivating role of social status may be less pronounced in other locales. Popular support for terrorist actions does not appear present in Chechnya, where even family members of suicide bombers were reluctant to express support for terrorist actions (Speckhard and Ahkmedova, 2006). However, popular support for terrorism is still high in the Middle East.[18] Nonetheless, the greater need for operational secrecy in some of these locales may limit the extent to which recruits can openly receive praise from their community.

Financial Rewards

Some rewards for terrorist participation are financial. In the early days of al-Qaeda, many members were given regular salaries (Wright, 2006). Financial rewards have bolstered recruitment by the Abu Sayyaf Group, where money from kidnappings has led to an increase in the number of recruits, who "spanned the spectrum from out of work farmers to opportunistic youths looking to make 'a fast buck'" (Cragin et al., 2006).

The families of terrorists killed in action have also received financial rewards. Reports suggest that the families of suicide bombers in Palestinian-controlled areas receive a one-time financial payment that can range from $1,000 to $5,000, with other reports estimating payments as high as $15,000 (Soibelman, 2004). Other families receive compensation for homes destroyed by Israeli forces (Post, Sprinzak, and Denny, 2003).

The effect of these rewards on recruitment is unclear. Financial incentives would certainly hold the greatest lure during times of economic downturn or among the poor, but there is no evidence that the pace of terrorist actions against Israel wax or wane with Palestinian economic indicators, and Palestinian suicide bombers do not generally come from impoverished families (Krueger and Maleková, 2003).

Friendship

Bonds of friendship and community frequently form among members of a close-knit terrorist organization or operational cell. As noted above, these close relationships and the intergroup dynamics they foster are often fundamental to the radicalization process itself (Crenshaw, 1981). As individuals become increasingly isolated from the broader community, they must rely even more on the group interrelationships (Kirby, 2007). It seems self-evident that such relationships serve to bond individual participants to one another. Interviews with former IRA and ETA members attest to the powerful role of friendship. Former IRA members cite the closeness among group members and the "sense of shared risk and common purpose" as something they miss about their old life as active members (Silke, 2008). An ETA member observed the following: "This is great. I mean great in the sense that at that time the support, the warmth around the organization was . . . , well, it was like signing for Athletic [Football Club] or something like that" (Alonso, 2006, p. 195).

Excitement

Finally, life as a terrorist promises excitement. Recruitment videos reportedly showcase this aspect of terrorism, portraying training with various weapons along with footage of actual operations (Silke, 2008). Reports suggest that many Saudis who went to Afghanistan for jihad in the 1990s did so out of a sense "adventure" (Hegghammer, 2007). Interviews with former IRA members further attest to this factor:

> Actually the motivation [was that] I was young. When you are young there is an excitement to it. You are seeing guns, you had only ever seen them on the TV or in the comics, "somebody has given me a gun, this is great" (Alonso, 2006).

Psychological research with nonterrorist subjects strongly suggests that exciting activities are perceived as rewarding by a significant subset of the general population. However, direct evidence for the role of excitement as a motivating factor in terrorism is limited, as no studies have identified any kind of unique "sensation seeking" characteristic among recruits.[19]

Relationships and Hierarchies

The factors described above have numerous interrelationships, as suggested by Figure 3.1, which shows a potential framework for these relationships. Individuals radicalize through heterogeneous pathways, with life histories, group experiences, and motivations all affecting radicalized individuals in different ways. Consequently, not all factors are necessary conditions for radicalization. However, the literature and basic social- science research suggest that other features in the environment are

Figure 3.1
Hypothesized Relationships Among Factors Implicated in Radicalization

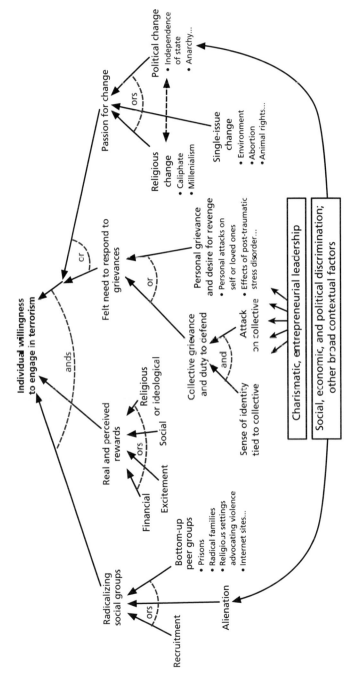

necessary for the radicalization process. These necessary and unnecessary conditions combine to make and/or distinctions that are enumerated in detail below.

The first-order factors in the figure are Group Socialization Processes, Expected Rewards, Passion for Change (even a "felt need for change"), and Felt Need to Respond to Grievance. *Abundant evidence suggests that socialization processes are a necessary precondition for radicalization.* Few individuals radicalize in social isolation. Group processes assure individuals that their chosen path is a correct one and provide the socially motivated courage to carry out attacks and the means by which to dehumanize selected targets. Group processes also stir and solidify ideologically based motivations for action and, as such, are interlinked to motivational desires for change and a desire to respond to grievances. Another factor that is probably necessary is the presence of perceived or real rewards for terror participation. Friendships solidified in the terror cell or organization, the social status that is derived from membership, and the heavenly gains of martyrdom are three such rewards. Group processes and rewards ultimately combine with one of two key motivational factors: a passion for change or felt need to respond to grievance. These two factors form the ideological basis for terrorism and constitute overt reasons for terror action. Neither is in itself a necessary factor, but at least one is likely required. Consequently, individuals may become radicalized when the following conditions exist: group processes *and* expected/real rewards *and* (passion for change *or* felt need to respond to grievance).

Group processes ultimately take place in one of two ways. They are either implemented through top-down recruitment strategies initiated by a terror organization or cell, or they result from bottom-up processes, whereby peer bonds and other social influences spiral individuals toward radicalization. Within the bottom-up trajectory, groups of individuals meet and interact in many different settings to include prisons, radical families, religious settings advocating violence, and the Internet. Groups within both the top-down and bottom-up settings may hold particularly strong sway because of widespread perceptions of social and religious alienation. Feelings of alienation commonly attributed to Muslim communities throughout Europe and the Middle East may draw individuals to settings where they can meet and identify with like-minded individuals and groups. This alienation is likely fed by perceptions of social, economic, and political discrimination. The relationships between these different factors can be summarized in the following way: alienation *and* (recruitment *or* bottom-up peer groups); prisons *or* radical families *or* religious settings advocating violence *or* the Internet.

The desire to implement change may be an ideologically based motivating factor for terrorism. Three broad-based motivational influences feed into this: political change *or* religious change *or* single-issue change. Politically motivated change may be related to a desire for political independence or an effort to sow anarchy. Religious change can include a desire to bring about a Muslim caliphate/Sharia law or millennialist-related change. Examples of single-issue changes relate to environmental rights and abortion. It is also important to note that although these motivations are listed separately, they likely cross-feed each other. For example, religion-motivated changes can feed into politically motivated conflicts. This is the case in the Palestinian territories, where religious motivations have intermixed with nationalist ones.

The desire to respond to perceived grievances constitutes another motivational set. Grievances perceived by a collective and fostered through identification with that collective are a likely driving force for initiating a collective defense. Alternatively, a desire for revenge is frequently the result of personal attacks directed at individuals and their friends and families. Some evidence suggests that the personal traumatization, which is often manifested in PTSD, helps facilitate motivations for revenge.

Perceived rewards (which may or may not be real) that stem from terrorist involvement are a presumed motivating factor. Five separate rewards have been illustrated in this paper: financial, excitement, friendship, religious, and social status. It is difficult to summarize these factors with an and/or discussion. These rewards can operate with differing degrees of intensity (for example, close and personal relationships engendered in a terrorist group will prove more rewarding than relatively weak relationships), and they can combine in numbers to increase the reward value (for example, terrorist involvement that is rewarded by friendships, social status, and perceived martyrdom may prove more influential than any single reward).

Finally, note at the bottom of Figure 3.1 that some factors may influence a number of items in the tree. These include contextual factors, such as discrimination and the influence of a terrorist organization's leaders, who may be quite charismatic and entrepreneurial (Gupta, 2008).

Possible Implications for Policy

Although this is not a policy paper, its review of the steps toward individual radicalization suggests some implications for U.S. and international policy options. Perhaps they merely reinforce familiar points, but collecting them may be useful:

- Since organizational recruitment and bottom-up socialization processes are critical to the radicalization process, they merit attention from international police and intelligence agencies. This can include monitoring of Islamic mosques supportive of militant ideology, as well as stronger steps against troublesome imams.
- Since social relationships not only help drive the radicalization process but can also reward participation in radical organizations, attacking the social bonds within terrorist organizations may prove to be an important counterterrorism tactic (Abrahms, 2008, p. 104).[20]
- Since alienation helps feed collective grievances and enhances the lure of radical milieus, and for other reasons, strides should be taken to integrate Muslim populations who otherwise reside in separated diaspora communities. Limiting Muslim-directed hostility from host populations will play a key role in these efforts. Increasing Muslim participation in governance may also have benefits.
- Since Muslim perceptions of both collective and personal attack appear to be key drivers of radicalization, the United States and its

allies should be cautious about military interventions and, when those are necessary, they should take considerable pains to limit civilian casualties and the human rights abuses that war often entails.

- Since militants use a sophisticated propaganda campaign that increasingly exploits the World Wide Web, a systematic countercampaign is needed on the Web. However, since tainted U.S. credibility among disaffected Muslims will limit the effectiveness of a U.S.-based message campaign, other mechanisms are needed. Success will likely depend on the Web-based activities of moderate Muslims.

Bibliography

Abadie, Alberto, "Poverty, Political Freedom, and the Roots of Terrorism," Cambridge, Mass.: National Bureau of Economic Research, Working Paper 10859, October 2004. As of March 10, 2008:
http://www.nber.org/papers/w10859

Abrahms, Max, "What Terrorists Really Want: Terrorism Motives and Counterterrorism Strategy," *International Security*, Vol. 32, No. 4, Spring 2008, pp. 78–105.

Ahkmedova, K., "Fanaticism and Revenge Idea of Civilians Who Had PTSD," *Social and Clinical Psychiatry*, Vol. 12, No. 3, 2003, pp. 24–32.

Al-Shishani, Murad, "The Salafi-Jihadist Movement in Iraq: Recruitment Methods and Arab Volunteers," *Terrorism Monitor*, Vol. 3, No. 23, December 2, 2005.

Alonso, Rogelio, "Individual Motivations for Joining Terrorist Organizations: A Comparative Qualitative Study on Members of ETA and IRA," in Jeff Victoroff, ed., *Social and Psychological Factors in the Genesis of Terrorism*, Amsterdam: IOS Press, 2006, pp. 187–202.

Ansari, Humayun, Marco Cinnirella, M. Brooke Rogers, Kate Miriam Loewenthal, and Christopher Alan Lewis, "Perceptions of Martyrdom and Terrorism Amongst British Muslims," *Proceedings of the British Psychological Society Seminar Series Aspects of Terrorism and Martyrdom*, eCommunity: International Journal of Mental Health and Addiction, 2005.

Argo, Nichole, "Understanding and Defusing Human Bombs: The Palestinian Case and the Pursuit of a Martyrdom Complex," paper presented to the Panel on "Challenges and Opportunities in Combating Transnational Terrorism," International Studies Association, 2004. As of January 2, 2009: http://www.allacademic.com//meta/p_mla_apa_research_citation/0/7/3/8/7/pages73877/p73877-1.php

Army of God, Web page, 2008. As of January 2, 2009: http://www.armyofgod.com/defense2.html

Atran, Scott, and Marc Sageman, "Global Network Terrorism: Comparative Anatomy and Evolution," NSC briefing, Washington, D.C.: White House, April 28, 2006.

Attwood, C., G. Singh, D. Prime, and R. Creasey, *2001 Home Office Citizenship Survey: People, Families and Communities,* Home Office Research Study 270, London: Home Office, 2003.

Bakker, Edwin, "Jihadi Terrorists in Europe," Security Paper No. 2, Clingendael: Netherlands Institute of International Relations, December 2006.

Bar, Shmuel, "The Religious Sources of Islamic Terrorism," *Policy Review,* No. 125, 2004.

Barber, Brian K., "Political Violence, Social Integration, and Youth Functioning: Palestinian Youth from the Intifada," *Journal of Community Psychology,* Vol. 29, No. 3, 2001, pp. 259–280. As of March 12, 2008: http://www2.uni-jena.de/svw/igc/studies/ss03/barber_2001.pdf

Barber, Brian K., and Joseph A. Olson, "Adolescents' Willingness to Engage in Political Conflict: Lessons from the Gaza Strip," in Jeff Victoroff, ed., *Social and Psychological Factors in the Genesis of Terrorism,* Amsterdam: IOS Press, 2006, pp. 203–226.

Benmelech, Efraim, and Claude Berrebi, "Human Capital and the Productivity of Suicide Bombers," *Journal of Economic Perspectives,* Vol. 21, No. 3, 2007, pp. 223–238.

Berrebi, Claude, "The Economics of Terrorism and Counterterrorism: What Matters and Is Rational-Choice Theory Helpful?" in Paul K. Davis and Kim Cragin, eds., *Social Science for Counterterrorism: Putting the Pieces Together,* Santa Monica, Calif.: RAND Corporation, 2009. As of January 20, 2009: http://www.rand.org/pubs/monographs/MG849/

Blackwell, Aaron D., "Terrorism, Heroism, and Altruism: Kin Selection and Socio-Religious Cost-Benefit Scaling in Palestinian Suicide Attack," presentation in Anthropology department colloquium series, Eugene, Ore.: University of Oregon, April 2005. As of December 22, 2008: http://www.uoregon.edu/~ablackwe/Documents/terrorism%20heroism%20altruism%20-%20hbes%20poster.pdf

Bokhari, Laila, "'Paths to Jihad—Faces of Terrorism:' Interviews with Radical Islamist Movements in Pakistan," in Laila Bokhari, Thomas Hegghammer, Brynjar Lia, Petter Nesser, and Truls H. Tonnessen, eds., *Paths to Global Jihad: Radicalization and Recruitment to Terror Networks,* Proceedings from an FFI Seminar, Oslo, Norway, March 25, 2006.

Borum, Randy, "Understanding the Terrorist Mindset," *Crime & Justice International,* Vol. 19, No. 77, November/December 2003. As of December 23, 2008:
http://www.cjcenter.org/cjcenter/publications/cji/archives/cji.php?id=680.

————, *Psychology of Terrorism,* Tampa, Fla.: University of South Florida, 2004.

Boyns, David, and James David Ballard, "Developing a Sociological Theory for the Empirical Understanding of Terrorism," *The American Sociologist,* Vol. 35, No. 2, June 2004, pp. 5–25.

Clevenger, Marie, and Christopher Birkbeck, "The Causes of Delinquency," Institute for Social Research, University of New Mexico, Working Paper No. 2, January 1996.

Cragin, Kim, "Understanding Terrorist Ideology: Testimony Presented to the Senate Select Committee on Intelligence, June 12, 2007," Santa Monica, Calif.: RAND Corporation, 2007. As of December 15, 2008:
http://www.rand.org/pubs/testimonies/CT283/

Cragin, Kim, Peter Chalk, Audra Grant, Todd C. Helmus, Donald Temple, and Matt Wheeler, "Curbing Militant Recruitment in Southeast Asia," Santa Monica, Calif.: RAND Corporation, unpublished research, 2006.

Crenshaw, Martha, "The Causes of Terrorism," *Comparative Politics,* Vol. 13, No. 4, July 1981, pp. 379–399.

————, "The Psychology of Terrorism: An Agenda for the 21st Century," *Political Psychology,* Vol. 21, No. 2, 2000.

————, "Have Motivations for Terrorism Changed?" in Jeff Victoroff, ed., *Social and Psychological Factors in the Genesis of Terrorism,* Amsterdam: IOS Press, 2006, pp. 51–56.

Curiel, Jonathan "Terror.Com," *San Francisco Chronicle,* July 10, 2005.

Della Porta, Donatella, *Social Movements, Political Violence, and the State: A Comparative Analysis of Italy and Germany,* Cambridge and New York: Cambridge University Press, 1996.

Della Porta, Donatella, and Mario Diani, *Social Movements: An Introduction,* 2nd ed., Malden, Oxford, and Carlton: Blackwell Publishing, 2006.

Faria, João Ricardo, and Daniel G. Arce M., "Terror Support and Recruitment," *Defence Peace and Economics,* Vol. 16, No. 4, August 2005, pp. 263–273.

Flaherty, Lois T., "Youth Ideology, and Terrorism," *Adolescent Psychiatry,* Vol. 27, 2003, pp. 29–58. As of March 10, 2008:
http://findarticles.com/p/articles/mi_qa3882/is_200301/ai_n9209836

Forster, Peter M., "An Introduction to the Psychology of Terror," 2008. As of January 2, 2009:
http://web.mac.com/petermforster/Site/Psychology_of_terrorism.html

Greer, Tammy, Mitchell Berman, Valerie Varan, Lori Bobrycki, and Sheree Watson, "We Are a Religious People; We Are a Vengeful People," *Journal for the Scientific Study of Religion,* Vol. 44, No. 1, 2005, pp. 45–56.

Greif, Adi, *Double Alienation and Muslim Youth in Europe,* United States Institute of Peace, August 21, 2007. As of May 9, 2008:
http://www.usip.org/pubs/usipeace_briefings/2007/0821_muslim_youth.html

Gunaratna, Rohan, "Ideology in Terrorism and Counter Terrorism: Lessons from Combating Al Qaeda and Al Jemaah Al Islamiyah in Southeast Asia," in Ehsan Ahrari, Fariborz Mokhtari, Richard L. Russell, et al., eds., *Countering Terrorist Ideologies Discussion Papers,* Swindon: Advanced Research and Assessment Group, September 2005, pp. 1–29. As of January 2, 2009:
http://www.isn.ethz.ch/isn/Digital-Library/Publications/Detail/?ots591=0C54E3B3-1E9C-BE1E-2C24-A6A8C7060233&lng=en&id=43967

Gupta, Dipak, K., "Tyranny of Data: Going Beyond Theories," in Jeff Victoroff, ed., *Social and Psychological Factors in the Genesis of Terrorism,* Amsterdam: IOS Press, 2006, pp. 37–50.

———, *Understanding Terrorism and Political Violence: The Life Cycle of Birth, Growth, Transformation, and Demise,* New York: Routledge, 2008.

Güss, C. Dominik, Teresa Tuason, and Vanessa B. Teixeira, "A Cultural-Psychological Theory of Contemporary Islamic Martyrdom," *Journal for the Theory of Social Behavior,* Vol. 37, No. 4, 2007, pp. 415–445.

Hafez, Mohammed M., "Rationality, Culture, and Structure in the Making of Suicide Bombers: A Preliminary Theoretical Synthesis and Illustrative Case Study," *Studies in Conflict & Terrorism,* Vol. 29, 2006, pp. 165–185.

Hassan, Nasra, "An Arsenal of Believers," *The New Yorker,* November 19, 2001, pp. 36–41.

Hegghammer, Thomas, "Terrorist Recruitment and Radicalization in Saudi Arabia," *Middle East Policy,* Vol. 13, No. 4, 2006, pp. 39–60. As of March 25, 2008:
http://findarticles.com/p/articles/mi_qa5400/is_200612/ai_n21403836

———, *Saudi Militants in Iraq: Backgrounds and Recruitment Patterns,* Norwegian Defence Research Establishment (FFI), February 5, 2007. As of March 25, 2008:
http://rapporter.ffi.no/rapporter/2006/03875.pdf

Higgins, S. T., "The Influence of Alternative Reinforcers on Cocaine Use and Abuse: A Brief Review," *Pharmacology, Biochemistry, and Behavior,* Vol. 57, No. 3, July 1997, pp. 419–426.

Horgan, John, *The Psychology of Terrorism,* New York: Routledge, 2005.

————, "From Profiles to Pathways and Roots to Routes: Perspectives from Psychology on Radicalization into Terrorism," *The Annals of the American Academy of Political and Social Science,* Vol. 618, 2008, pp. 80–93.

Horgan, John, and Max Taylor, "The Making of a Terrorist," *Jane's Intelligence Review,* December 1, 2001.

Hudson, Rex A., *The Sociology and Psychology of Terrorism: Who Becomes a Terrorist and Why?* Washington, D.C.: Federal Reserve Division, Library of Congress, September 1999. As of May 17, 2006:
http://www.fas.org/irp/threat/frd.html

International Helsinki Federation for Human Rights, *Intolerance and Discrimination Against Muslims, Developments Since September 11, 2005,* undated. As of January 2, 2009:
http://www.ihf-hr.org/viewbinary/viewdocument.php?doc_id=6237

Ismail, Noor Huda, "Schooled for Jihad; They Turned to Terrorism. I Wanted to Know Why," *The Washington Post,* June 26, 2005, p. B.01.

Janis, I. L., *Victims of Groupthink: A Psychological Study of Foreign Policy Decisions and Fiascoes,* Boston, Mass.: Houghton Mifflin Company, 1972.

Jordan, Javier, and Nicola Horsburgh, "Mapping Jihadist Terrorism in Spain," *Studies in Conflict & Terrorism,* Vol. 28, No. 3, 2005, pp. 169–191.

Khosrokhavar, Farhad, and David Macey, *Suicide Bombers: Allah's New Martyrs,* London: Pluto Press, 2005.

Kimhi, Shaul, and Shemuel Even, "Who Are the Palestinian Suicide Bombers?" *Terrorism and Political Violence,* Vol. 16, No. 4, 2004, pp. 815–840.

Kirby, Aidan, "The London Bombers as 'Self-Starters': A Case Study in Indigenous Radicalization and the Emergence of Autonomous Cliques," *Studies in Conflict & Terrorism,* Vol. 30, No. 5, 2007, pp. 415–428.

Krueger, Alan B., and Jitka Maleková, "Education, Poverty and Terrorism: Is There a Causal Connection?" *Journal of Economic Perspectives,* Vol. 17, No. 4, Fall 2003, pp. 119–144.

Kruglanski, Arie W., "Inside the Terrorist Mind," paper presented to the National Academy of Science, Washington, D.C, April 29, 2002. As of May 17, 2006:
http://www.wam.umd.edu/~hannahk/ULTIMATE%2029%20APRIL%20TALK.doc

————, "The Psychology of Terrorism: 'Syndrome' Versus 'Tool' Perspectives," in Jeff Victoroff, ed., *Social and Psychological Factors in the Genesis of Terrorism,* Amsterdam: IOS Press, 2006, pp. 61–73.

Kruglanski, Arie W., and Agnieszka Golec, "Individual Motivations, The Group Process and Organizational Strategies in Suicide Terrorism," in E. M. Meyersson Milgrom, ed., *Suicide Missions and the Market for Martyrs, A Multidisciplinary Approach,* Princeton, N.J.: Princeton University Press, 2004.

Larsson, J. P., "The Role of Religious Ideology in Modern Terrorist Recruitment," in James J.F. Forest, ed., *The Making of a Terrorist,* Vol. 1, Westport, Conn.: Praeger Security International, 2005.

Leiken, Robert S., *Bearers of Global Jihad? Immigration and National Security after 9/11,* Washington, D.C.: The Nixon Center, 2004. As of January 2008:
http://www.nixoncenter.org/publications/monographs/Leiken_Bearers_of_Global_Jihad.pdf

————, "Europe's Angry Muslims," *Foreign Affairs,* Vol. 84, No. 4, July–August 2005, pp. 120–135.

Lia, Brynjar, and Åshild Kjøk, *Islamist Insurgencies, Diasporic Support Networks, and Their Host States: The Case of the Algerian GIA in Europe 1993–2000,* Kjeller, Norway: Forsvarets Forskningsinstitutt, FFI/RAPPORT-2001/03789, August 8, 2001. As of March 10, 2008:
http://rapporter.ffi.no/rapporter/2001/03789.pdf

MacKerrow, Edward P., "Understanding Why: Dissecting Radical Islamist Terrorism with Agent-Based Simulation," *Los Alamos Science,* No. 28, 2003, pp. 184–191.

Mazarr, Michael J., "The Psychological Sources of Islamic Terrorism," *Policy Review,* June–July 2004.

McAdam, D., "Recruitment to High-Risk Activism: The Case of Freedom Summer," *The American Journal of Sociology,* Vol. 92, No. 1, 1986, pp. 64–90.

McCants, William, and Jarret Brachman, *Militant Ideology Atlas: Executive Report,* West Point, N.Y.: Combating Terrorism Center, November 2006. As of March 10, 2008:
http://www.ctc.usma.edu/atlas/Atlas-ExecutiveReport.pdf

Merari, Ariel, "Terrorism as a Strategy of Struggle: Past and Future," *Terrorism and Political Violence,* Vol. 11, No. 4, 1999, pp. 52–65.

Middle East Media Research Institute, "The Iraqi Al-Qa'ida Organization: A Self-Portrait," *MEMRI Special Dispatch Series,* No. 884, March 24, 2008. As of May 9, 2008:
http://memri.org/bin/articles.cgi?Page=archives&Area=sd&ID=SP88405

Milgram, Stanley, "Some Conditions of Obedience and Disobedience to Authority," *Human Relations,* Vol. 18, No. 1, 1965.

Moghadam, Assaf, "The Roots of Suicide Terrorism: A Multi-Causal Approach," Austin, Tex.: Harrington Workshop on the Root Causes of Suicide Terrorism, April 12, 2005.

———, "Suicide Terrorism, Occupation, and the Globalization of Martyrdom: A Critique of Dying to Win," *Studies in Conflict & Terrorism,* Vol. 29, No. 8, 2006, pp. 707–729.

Moghaddam, Fathali M., "Cultural Preconditions for Potential Terrorist Groups: Terrorism and Societal Change," in Fathali M. Moghaddam and Anthony J. Marsella, eds., *Understanding Terrorism,* Washington, D.C.: American Psychological Association, 2004.

———, "The Staircase to Terrorism: A Psychological Exploration," *American Psychologist,* Vol. 60, No. 2, 2005, pp. 161–169.

Moscovici, S., and M. Zavalloni, "The Group as a Polarizer of Attitudes," *Journal of Personality and Social Psychology,* Vol. 12, 1969.

Myers, David G., Social Psychology, New York: McGraw-Hill, 2005.

National Consortium for the Study of Terrorism and Responses to Terrorism, "Support for the Caliphate and Radical Mobilization," research brief, January 2008. As of May 9, 2008:
http://www.start.umd.edu/start/publications/research_briefs/20080131_Caliphate_and_Radicalization.pdf

Nesser, Petter, *Jihad in Europe: A Survey of the Motivations for Sunni Islamist Terrorism in Post-Millennium Europe,* Kjeller, Norway: Norwegian Defence Research Establishment, FFI/RAPPORT-2004/01146, 2004. As of December 7, 2007:
http://www.investigativeproject.org/documents/testimony/35.pdf

———, "Jihad in Europe; Recruitment for Terrorist Cells in Europe," in Laila Bokhari, Thomas Hegghammer, Brynjar Lia, Petter Nesser, and Truls H. Tonnessen, eds., *Paths to Global Jihad: Radicalization and Recruitment to Terror Networks,* Proceedings from an FFI Seminar, Oslo, Norway, March 25, 2006a.

———, "Jihadism in Western Europe After the Invasion of Iraq: Tracing Motivational Influences from the Iraq War on Jihadist Terrorism in Western Europe," *Studies in Conflict & Terrorism,* Vol. 29, No. 4, 2006b, pp. 323–342.

Obaid, Nawaf, and Anthony Cordesman, *Saudi Militants in Iraq: Assessment and Kingdom's Response,* Washington, D.C.: Center for Strategic and International Studies, September 19, 2005. As of March 25, 2008:
http://www.csis.org/media/csis/pubs/050919_saudimiltantsiraq.pdf

Office for National Statistics (UK), *Labour Market: Muslim Unemployment Rate Highest,* 2004. As of January 2, 2009:
http://www.statistics.gov.uk/cci/nugget.asp?id=979

O'Leary, Brendan, "The IRA: Looking Back; Mission Accomplished?" in Marianne Heiberg, Brendan O'Leary, and John Tirman, eds., *Terror, Insurgency, and the State,* Philadelphia, Pa.: University of Pennsylvania Press, 2007.

Pape, Robert A., "The Strategic Logic of Suicide Terrorism," *American Political Science Review,* Vol. 97, No. 3, August 2003.

Paul, Christopher, "How Do Terrorists Generate and Maintain Support?" in Paul K. Davis and Kim Cragin, eds., *Social Science for Counterterrorism: Putting the Pieces Together,* Santa Monica, Calif.: RAND Corporation, 2009. As of January 20, 2009:
http://www.rand.org/pubs/monographs/MG849/

Paz, Reuven, "Arab Volunteers Killed in Iraq: An Analysis," *PRISM Series of Global Jihad,* No. 1/3, March 2005. As of January 2, 2008:
http://www.imra.org.il/story.php3?id=24396

Pedahzur, A., *Suicide Terrorism,* Cambridge, United Kingdom: Polity Press, 2005.

Pedahzur A., A. Perliger, and L. Weinberg, "Altruism and Fatalism: The Characteristics of Palestinian Suicide Terrorists," *Deviant Behavior,* Vol. 24, No. 4, July–August 2003, pp. 405–423.

Pew Research Center, "Islamic Extremism: Common Concern for Muslim and Western Publics," *Pew Global Attitudes Project,* July 14, 2005. As of April 3, 2008:
http://pewglobal.org/reports/display.php?ReportID=248

———, "The Great Divide: How Westerners and Muslims View Each Other: Europe's Muslims More Moderate," *Pew Global Attitudes Project,* July 22, 2006. As of January 2, 2009:
http://pewglobal.org/reports/display.php?ReportID=253

Post, Jerrold M., "When Hatred Is Bred in the Bone: Psycho-Cultural Foundations of Contemporary Terrorism," *Political Psychology,* Vol. 26, No. 4, August 2005, pp. 615–636.

Post, Jerrold M., Keven G. Ruby, and Eric D. Shaw, "The Radical Group in Context: 2. Identification of Critical Elements in the Analysis of Risk for Terrorism by Radical Group Type," *Studies in Conflict & Terrorism,* Vol. 25, 2002, pp. 101–126.

Post, Jerrold M., Ehud Sprinzak, and Laurita M. Denny, "The Terrorists in Their Own Words: Interviews with 35 Incarcerated Middle Eastern Terrorists," *Terrorism and Political Violence,* Vol. 15, No. 1, Spring 2003, pp. 171–184.

Pyszczynski, Tom, "Crusades and Jihads: An Existential Psychological Perspective on the Psychology of Terrorism and Political Extremism," in Jeff Victoroff, ed., *Social and Psychological Factors in the Genesis of Terrorism,* Amsterdam: IOS Press, 2006, pp. 85–96.

Pyszczynski, Tom, Abdolhossein Abdollahi, Sheldon Solomon, Jeff Greenberg, Florette Cohen, and David Weise, "Mortality Salience, Martyrdom, and Military Might: The Great Satan Versus the Axis of Evil," *Personality and Social Psychology Bulletin,* Vol. 32, No. 4, April 2006, pp. 525–536.

Rabasa, Angel, "Moderate and Radical Islam: Testimony Presented Before the House Armed Services Committee Defense Review Terrorism and Radical Islam Gap Panel on November 3, 2005," Santa Monica, Calif.: RAND Corporation, 2005. As of December 23, 2008:
http://www.rand.org/pubs/testimonies/CT251/

Republic of Singapore, Ministry of Home Affairs, *The Jemaah Islamiyah Arrests and the Threat of Terrorism,* White Paper, January 2003.

Richardson, Louis, *What Terrorists Want: Understanding the Enemy, Containing the Threat,* New York: Random House, 2006.

Roy, Olivier, *Globalized Islam: The Search for a New Ummah* (CERI Series in Comparative Politics and International Studies), New York: Columbia University Press, 2004.

———, "The Ideology of Terror," *International Herald Tribune,* July 23, 2005. As of March 10, 2008:
http://www.iht.com/articles/2005/07/22/opinion/edroy.php

Ryan, Johnny, "The Four P-Words of Militant Islamist Radicalization and Recruitment: Persecution, Precedent, Piety, and Perseverance," *Studies in Conflict & Terrorism,* Vol. 30. No. 11, 2007, pp. 985–1011.

Sageman, Marc, *Understanding Terror Networks,* Philadelphia, Pa.: University of Pennsylvania Press, 2004.

———, *Leaderless Jihad: Terror Networks in the Twenty-First Century,* Philadelphia, Pa.: University of Pennsylvania Press, 2008.

Scheuer, Michael, "Reinforcing the Mujahideen: Origins of Jihadi Manpower," *Terrorism Focus,* Vol. 3, No. 18, May 9, 2006.

Schweitzer, Yoram, "Palestinian Female Suicide Bombers: Reality vs. Myth," in Yoram Schweitzer, ed., *Female Suicide Bombers: Dying for Equality?* Tel Aviv, Israel: Jaffee Center for Strategic Studies, Tel Aviv University, Memorandum No. 84, August 2006, pp. 26–41. As of January 2, 2009:
http://www.isn.ethz.ch/isn/Digital-Library/Publications/Detail/?lng=en&id=91112

Sherif, M., "A Study of Some Social Factors in Perception," *Archives of Psychology,* Vol. 27, No. 187, 1935.

Siegel, Pascale Combelles, "Radical Islam and the French Muslim Prison Population," *Terrorism Monitor,* Vol. 4, No. 15, July 27, 2006, pp. 1–3.

Silber, Mitchell D., and Arvin Blatt, *Radicalization in the West: The Homegrown Threat,* The New York City Police Department, 2007. As of March 25, 2008: http://sethgodin.typepad.com/seths_blog/files/NYPD_Report-Radicalization_in_ the_West.pdf

Silke, Andrew, "The Devil You Know: Continuing Problems with Research on Terrorism," *Terrorism and Political Violence,* Vol. 13, No. 4, Winter 2001, pp. 1–14.

———, "Holy Warriors: Exploring the Psychological Processes of Jihadi Radicalization," *European Journal of Criminology,* Vol. 5, No. 1, 2008, pp. 99–123.

Skinner, B. F., *About Behaviorism,* New York: Vintage Books, 1974.

Snow, D. A., L. A. Zurcher, Jr., and S. Ekland-Olson, "Social Networks and Social Movements: A Microstructural Aproach to Differential Recruitment," *American Sociological Review,* Vol. 45, No. 5, 1980, pp. 787–801.

Soibelman, Mali, "Palestinian Suicide Bombers," *Journal of Investigative Psychology and Offender Profiling,* Vol. 1, 2004, pp. 175–190.

Southern Poverty Law Center, "From Push to Shove: Radical Environmental and Animal-Rights Groups Have Always Drawn the Line at Targeting Humans. Not Anymore," *Intelligence Report,* Fall 2002. As of January 2, 2009: http://www.splcenter.org/intel/intelreport/article.jsp?pid=91

Speckhard, Anne, "Sacred Terror: Insights into the Psychology of Religiously Motivated Terrorism," *Radicalism: Christianity, Islam and Judaism Between Constructive Activism and Destructive Fanaticism,* Antwerp, Belgium: USCIA, 2006. As of March 5, 2009: http://www.uwmc.uwc.edu/alumni/news_items/speckhard/sacred_terror%20.pdf

Speckhard, Anne, and Khapta Ahkmedova, "The Making of a Martyr: Chechen Suicide Terrorism," *Studies in Conflict & Terrorism,* Vol. 29, No. 5, 2006, pp. 429–492.

Taarnby, Michael, *Motivational Parameters in Islamic Terrorism,* University of Aarhus, Denmark: Centre for Cultural Research, 2002.

Taylor M., and J. Horgan, "A Conceptual Framework for Addressing Psychological Process in the Development of the Terrorist," *Terrorism and Political Violence,* Vol. 18, 2006, pp. 585–601.

"The U.S. Army's Limited-War Mission and Social Science Research, March 26, 27, 28, 1962," symposium proceedings, Washington, D.C.: The American University, 1965.

Tudge, C., "Natural Born Killers," *New Scientist,* Vol. 174, 2002, pp. 36–39.

Victoroff, Jeff, "The Mind of the Terrorist: A Review and Critique of Psychological Approaches," *Journal of Conflict Resolution*, Vol. 49, No. 1, February 2005.

———, "Executive Summary," in Jeff Victoroff, ed., *Social and Psychological Factors in the Genesis of Terrorism*, Amsterdam: IOS Press, 2006, pp. 455–461.

Wells, Linton III, and Barry M. Horowitz, "Implementation of a Methodology for the Prioritizing of Suicide Attacker Recruitment Preferences," *Journal of Homeland Security and Emergency Management*, Vol. 4, No. 2, 2007, pp. 1–16.

Wiktotwicz, Quintan, *Joining the Cause: Al-Muhajiroun and Radical Islam*, Memphis, Tenn.: Rhodes College, Department of International Studies, 2004. As of March 31, 2008: http://www.cis.yale.edu/polisci/info/conferences/Islamic%20 Radicalism/papers/wiktorowicz-paper.pdf

White, Robert W., "Political Violence by the Nonaggrieved," in Donatella della Porta, ed., *International Social Movement Research*, Vol. 4, Greenwich, Conn.: Jai Press, 1992, p. 92.

Wright, Lawrence, *The Looming Tower: Al-Qaeda and the Road to 9/11*, New York: Vintage Books, 2006.

Zavis, Alexandrea, "Foreign Fighters in Iraq Seek Recognition, U.S. Says," *Los Angeles Times*, March 17, 2008. As of April 1, 2008: http://www.latimes.com/news/nationworld/iraq/complete/la-fg-iraq17mar17,1,6951742.story

Endnotes

[1] One challenge with this line of reasoning, however, is the argument that terrorism is actually driven by demand from the terrorist organizations themselves, which is in turn aided by the need to operationalize only a small number of recruits. The arguments in this paper attempt to address this line of reasoning.

[2] If well-educated and experienced recruits were readily available, terror organizations would systematically rely on them for more of their operations.

[3] This paper makes frequent use of the terms "radical" and "radicalization." In this context, they mean "terrorist" and "the process of becoming a terrorist." In another context, the same words would mean something more benign, such as "strong proponent of a view" and "the process of coming to accept that strong proponency."

[4] The process does not have a single outcome, that of being a terrorist. Individuals can begin the cycle but escape its grasp at any point in time.

[5] What is referred to as "radicalization" in this paper can be regarded as psychologically very similar to the process by which otherwise unremarkable teenagers are indoctrinated into military service. Factors such as socialization, recruitment, and

duty to defend the collective are likely all influential in this regard. Needless to say, this does not suggest a moral equivalency.

[6]   Interviews with incarcerated Middle East terrorists reveal a similar pattern where convicted terrorists cited peer influence as a major reason for joining the group and "Membership was described as being associated with a fusion of the young adult's individual identity with the group's collective identity and goals" (Post, Sprinzak, and Denny, 2003).

[7]   According to a government white paper report, the Singapore paper published by Jemaah Islamiyyah invited prospective recruits to attend otherwise innocuous religious classes. The desire for religious instruction and Muslim fellowship initially motivated many to attend. Teachers would then insert references to Jihad and Muslim suffering. Students who expressed additional interest in the Muslim plight were identified and submitted to further screening and indoctrination and were invited to join the organization. A series of tactics were then implemented to solidify in-group bonding. These included the use of secrecy, the idea that group outsiders were infidels, and escalation of commitment to include pledging public allegiance to the organization and psychological contracting. See "The Jemaah Islamiyah Arrests and the Threat of Terrorism" (2003).

[8]   Twenty-year-old British Muslim Younis Tsouli took up the Web name Irhabi007 and became a highly influential voice in al-Qaeda chat rooms. He soon became a public relations mouthpiece for Abu Mus'ab al-Zarqawi of al-Qaeda in Iraq and proved prolific in efforts to post Jihadi videos, to disseminate terror training materials, and to raise funds (Sageman, 2008).

[9]   A Fatah member was quoted as saying, "I belong to the generation of occupation. My family are refugees from the 1967 war. The war and my refugee status were the seminal events that formed my political consciousness, and provided the incentive for doing all I could to help regain our legitimate rights in our occupied country" (Post, 2005, p. 622).

[10]   This aspect of jihadi-Salafi ideology was in part inspired by Egyptian author and Islamist Sayyid Qutb, who was the leading intellectual of the Muslim Brotherhood in the 1950s and 1960s. He argued that militant jihad was necessary to overthrow non-Islamic governments and institute a "pure" Islamic society (Silber and Blatt, 2007).

[11]   Abu Maysara, who is believed to be a commander of al-Qaeda in Iraq, listed the following as one of al-Qaeda's key organizational goals: "To re-establish the Rightly-Guided Caliphate in accordance with the Prophet's example, because [according to the tradition] 'whoever dies without having sworn allegiance to a Muslim ruler dies as an unbeliever'" (as transcribed by the Middle East Media Research Institute, 2008).

[12]   In addition, survey data suggest strong although not necessarily violent support for a caliphate among a majority of Muslims. Authors of the survey report argue

that "For some Muslims, the imagery of an Islam reflective of the golden era of Muhammad is a religious value worthy of pursuit in terms of life goals, finances, and personal sacrifice 'in the cause of Allah'" (National Consortium for the Study of Terrorism and Responses to Terrorism, 2008).

[13] The group partly justifies its militant stance using the following Biblical texts: "Whoso sheddeth man's blood, by man shall his blood be shed: for in the image of God made he man" (Genesis 9:6) and "So ye shall not pollute the land wherein ye are: for blood it defileth the land: and the land cannot be cleansed of the blood that is shed therein, but by the blood of him that shed it" (Numbers 35:33).

[14] In one experimental study, despite submitting identical resumes, Arab-named applicants for French jobs received five times fewer callbacks than applicants with western names (Greif, 2007).

[15] One factor presumed to contribute to limited political representation is the fact that immigrant Muslims may lack knowledge on how to participate in the political process (Greif, 2007)

[16] Several psychological studies highlight the powerful role of revenge. Studies show that individuals engaged in a cooperative task are frequently willing to sacrifice their own rewards to exact revenge on an opponent caught cheating (Tudge, 2002). Other studies suggest that young males and religious individuals are particularly prone to attitudes of vengeance. See Silke (2008, p. 105) for a more detailed review of the psychology of vengeance and its relationship to terrorism.

[17] Data collected by the National Security Studies Center of the University of Haifa, as cited in Moghadam (2006).

[18] Popular support for violence against civilian targets is 29 percent for Jordan, 28 percent for Egypt, 14 percent for Pakistan and 16% for French and Spanish Muslims (Pew Global Attitudes Project, 2006).

[19] For a review on the hypothesized role of sensation seeking in radicalization, see Victoroff (2005).

[20] Abrahms notes, for example, that commuting the prison sentences of Italian Red Brigade members in exchange for actionable intelligence against former comrades has helped breed resentment among movement members. Government programs that seek to infiltrate radical online chat rooms may also have a similar effect. The intent, where possible, should be to sow confusion and mistrust among radical social networks.

CHAPTER FOUR

How Do Terrorists Generate and Maintain Support?

Christopher Paul

Introduction

Objectives

How do terrorists generate support initially, and how do they maintain it over time? What benefits do terrorists or insurgents draw from such support, and how critical are these benefits?

A starting point may be to remember Mao Tse-tung's (1937) admonition that "the guerrilla must move amongst the people as a fish swims in the sea." The suggestion in our context would be that terrorists and insurgents desperately need popular support, and that disconnecting them from that support is a potentially highly effective approach to combating them. Although the proposition is quite plausible intuitively and is part of a near-consensus view, it is not as straightforward as it might seem. There are disagreements on the matter and the empirical base is fairly thin.[1] Nonetheless, it appears to be an important proposition and what follows surveys much of the relevant literature. It then seeks to structure the information in a coherent way and to draw some conclusions for counterterrorism.

Disciplinary Approaches to Studying Support for Terrorism

Where support for terrorism is addressed in the literature, the focus is typically limited to state support, financial support, or expressions of support as captured in public opinion polls. More often, support is *assumed*, as authors address other questions. As a result, the survey in this paper includes insights from studies that touch on support for terrorism only tangentially.

Support for terrorism (or, more generally, for insurgency) is treated differently and to different depths in the various social-science disciplines. For example, the topic is minimally treated in sociology but with slightly greater frequency in economics, where much of the relevant work is focused on finance (see, for example, Basile, 2004; Levitt, 2007) or public opinion (see Jaeger et al., 2008, for example). The most useful insights about support come from studies that cross traditional disciplinary boundaries. Case studies (be it from historians, political scientists, or interdisciplinary scholars) are the best sources for empirical work on factors and relationships. However, most such are aimed primarily at other topics. They offer little unifying theory.

The theory that exists is not tied to specific cases and typically stems from the work of policy analysts and military theorists (primarily those associated with counterinsurgency; see, for example, SHARP, 2006; Kilcullen, undated; Metz and Millen, 2004).

On the empirical side, an area of social science that does regularly and explicitly focus on support is public-opinion research (for example, Myers, 2004; Fair and Shepherd, 2006; Khashan, 2003; PCPSR, 2001; Abdallah, 2003; Haddad, 2006; Haddad and Khashan, 2002; Pew Research Center, 2005; and Hayes and McAllister, 2005). However, as discussed below, integrating this work is complicated by the many different denotations of "support."

A broader set of materials is available when one approaches questions of support obliquely, drawing on criminological research on organized crime and street gangs (for example, Makarenko, 2002; Roth and Sever, 2007; Kleemans and Bunt, 1999; Cottino, 1999; Shulte-Bocholt, 2006; and Jankowski, 1991) as well as on sociology and political science on social movements (such as Tilly, 1979). Even less directly related (but with relatively concrete findings) are contributions from behavioral science on community, identity, and baseline cultural and social processes, such as the development of trust (for example, Kenny, 2007; Welch, Sikkink, and Loveland 2007; Farrell, 2003; and Putnam, 2001).

What Do We Mean by "Support"?

The variety of different approaches (tangential or otherwise) to the analysis of terrorist or insurgent support available in the social sciences begs one very important question: What do we mean by "support"? In the existing literature, "support" is used to discuss two overlapping but distinct concepts, as illustrated in Figure 4.1.

The first denotation is support in the form of feelings or expressions of sympathy; the second is actual material support or other direct or indirect aid or abetment. Here, I will refer to the former explicitly as "sympathetic of" and the latter as "supporting."

Of course, one could assume that those who are supporting are also sympathetic. In many cases, this is a completely reasonable assumption. However, imagine an environment in which citizens were concerned that the state's security apparatus had penetrated the polling entity. Under those circumstances, a group's most ardent financial supporters might decline to answer a pollster or might offer a "cover" answer to prevent suspicion from adhering to them.

Figure 4.1
Discriminating Between "Sympathetic of" and "Supporting"

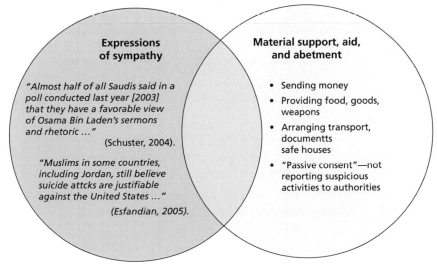

Expressions of sympathy

"Almost half of all Saudis said in a poll conducted last year [2003] that they have a favorable view of Osama Bin Laden's sermons and rhetoric ..."
(Schuster, 2004).

"Muslims in some countries, including Jordan, still believe suicide attcks are justifiable against the United States ..."
(Esfandian, 2005).

Material support, aid, and abetment

- Sending money
- Providing food, goods, weapons
- Arranging transport, documentts safe houses
- "Passive consent"—not reporting suspicious activities to authorities

It is even less straightforward to infer from polls alone how much those sympathetic of terrorism are actually supporting it materially. Expressions of support can stem from a sense of frustration, can be a political statement, or can indicate genuine sympathy at a lower level while not translating to active material support. Gerges (2005) notes, "The critical question is not whether Muslims sympathize with bin Laden's rhetoric of victimhood but if they are ready to shed blood to support it" (p. 233). His findings suggest the answer is "not really." Existing RAND research (Cragin et al., unpublished) finds different levels of involvement in terrorist groups: "Participating in militant group activity is a direct form of activity, while endorsing militant activity is much more of an indirect form" (p. 84). Providing material support presumably falls between endorsement and active participation on the radicalization scale.

Research on public opinion and communications reports significant differences between public and private expressions of opinions, perhaps analogous to the difference between an anonymous polled sympathy and detectable active support (Boyle et al., 2006; Scheufele and Eveland, 2001). In many cases, polled expressions of support may provide very good second-order approximations of the likelihood of certain kinds of direct material support or may be evidence of changing trends in proclivity to offer support. In other cases, they may not.

Polled support is also difficult to connect to support needs. Even if only a small proportion of a population expresses sympathy for a terrorist group, the ardent support of that small fraction may be sufficient to meet the group's needs. (See the discussion of need for support, below.)

Expressed support is undoubtedly an important indicator of material support, but the details of the connection between the two are not wholly clear. This important distinction has a prominent place in the discussion in the subsection below on need for and sources of support. Throughout this paper, when I use the term "support," I intend to denote "supporting" as in providing material support, aid, or abetment.

Relationships to Topics in Other Papers

The kinds of factors that lead individuals to express sympathy or to offer material support are certainly similar to the radicalization processes that lead individuals to decide to become terrorists. Where such processes are in play, see the extended discussion in Helmus (2009) on radicalization. The rise of terrorism includes efforts to garner support; this makes the discussion of root causes in Noricks (2009) relevant here. Finally, reducing popular support is often hypothesized as a way to defeat terrorism and is explored in greater detail in Gvineria's paper on how terrorism ends.

This paper first enumerates factors that contribute to determining groups' needs for support and the sources of that support and then lists factors motivating decisions to offer support. With the factors in place, the discussion then turns to their strength and consensus in the social sciences, the relationships among the factors, and the policy implications of these findings.

Support Factors

This section discusses the factors identified in the social sciences as relevant or contributing to support for terrorist or insurgent groups. In the first subsection, I present factors found to determine terrorist support needs and the sources that have traditionally met those needs; in the second subsection, I present factors contributing to *decisions* to support terrorists or insurgents.

Types of Support Needed and Sources for That Support

Kinds of Needs. From sociologists Boyns and Ballard (2004), we learn that "a basic tenet of theories of collective action is that social movements and organizations are dependent upon the availability of resources." Metz and Millen (2004) identify five categories of resource needs for insurgents:

- manpower (to include people with special knowledge or skill)
- funding

- materiel
- sanctuary
- intelligence.

To that I would add a sixth:

- tolerance of activities.

Most of the requirements identified by other scholars fit easily under one of these umbrella categories.[2] The exception is "sanctuary," since the term is used in slightly different ways by different scholars. Also called "havens," sanctuaries are spaces safe from harassment and surveillance that foster oppositional culture and group solidarity (Fantasia and Hirsch, 1995; Haussler, Russel, and Baylouny, 2005). Scholars disagree on the amount of space and freedom of action needed to qualify a safe place as a sanctuary. One strand of thought demands a fairly high threshold as an ideal type (Lia and Kjøk, 2001): "a secure base within which an insurgent group is able to organize the politico-military infrastructure needed to support its activities." Depending on the usage, sanctuaries or havens can refer to a country outside the conflict country, regions or cities within the conflict country, specific neighborhoods, or areas as small as single safe houses (although this last example of a micro-sanctuary is extreme and arguably a corruption of the concept). Havens must provide some kind of relative security, tolerance for political-military mobilization, and at least some military-support-related activities (Lia and Kjøk, 2001). Traditionally, sanctuaries have been found in the hinterlands of countries adjacent to the country of conflict. In the contemporary era, globalization and the free flow of information make it possible for havens to be either within the country of conflict or globally quite removed; and urban environs are much more attractive than rural ones (Metz and Millen, 2004).

One notable category of resources that is not mentioned by Metz and Millen and does not fit easily within their five categories is passive support or tolerance, either from the local population or from a state. Passive support is similar to a sanctuary or havens, in that it allows groups to operate with the resources they already have without interruption. In fact, several scholars suggest that some havens require little more than the passive tolerance of the host state (see Byman, 2005; Forrest et al., 2006; Lia and Kjøk, 2001; Byman et al., 2001; Haussler, Russel, and Baylouny, 2005).

Although terrorists or insurgents want the active support of the population within which they operate, for certain segments of the populations or for certain kind of insurgent groups, passive acceptance may be adequate (Haussler, Russel, and Baylouny, 2005). One can easily imagine operations for which a group needs to avoid having anyone notify the forces of their preparations but needs very little support beyond that. Social processes of silencing or subtle coercion can lead to passive acceptance (Flanigan, 2006), as can more active coercion or intimidation.

Factors Determining the Magnitude of Needs. Terrorist groups have differing levels of need for the six resources identified (Metz and Millen, 2004; Haussler, Russel, and Baylouny, 2005). The size of the group, the group's goals, the nature of operations undertaken, and the extent to which the group is overt or covert all contribute to the need in each category. Here the literature's reductionist tendencies are revealed, with most authors pointing toward two "ideal types" of organization: the typical insurgent or guerilla organization and the typical transnational terrorist organization. The former is typed to be large, seeks the overthrow and replacement of the current government through military means, undertakes a wide range of quasi-military operations (up to and including force-on-force conventional attacks), and is largely overt; the latter is held to be small, seeks often underspecified goals, engages in infrequent but symbolically painful attacks, and is almost wholly covert (Turk, 2004). After reviewing this literature, I conclude that there is no reason for naïve acceptance of either of these ideal types. In practice, analyzing the connection between a specific identified group's characteristics and its resource needs is not very difficult and it seems more fruitful to do so than to lose context-dependent subtleties.

Of the identified factors contributing to magnitudes of support needs, overtness or covertness is the most contentious. Some assert that wholly clandestine groups are markedly different from other groups and that a core difference is having few support needs, including very little or no need for passive support (see Rodriguez, 2005, or Tsvetovat and Carley, 2005). Others assert that wholly clandestine groups are very resource-dependent, as their concealment prevents them from doing much in the way of self-supply (Boyns and Ballard, 2004). Under certain circumstances, both could be correct.

Sources. Groups use four basic methods to meet their varied resource needs: self-supply, looting, purchasing, and relying on an external source (Vinci, 2006). "External source" usually, but not always, refers to what in this paper I call "support." A group that requires a resource of any kind from an external source usually requires what in this paper I call support, but—especially in an open society—many needs could be met by purchasing goods or services from people who have no inkling that they are selling to terrorists. The literature identifies a host of possible external sources. These include

- communities (for example, "the population")
- states (Levitt, 2007; Lia and Skjolberg, 2004; Gerges, 2005; Richardson, 2006; Byman et al., 2001)
- diasporas (Levitt, 2007; Smith, 2007; Richardson, 2006; Byman et al., 2001)
- charities and nongovernmental organizations (NGOs) (Lia and Skjolberg, 2004; Basile, 2004; Mascini, 2006; Smith, 2007)
- organized criminal groups (Lia and Skjolberg, 2004; Metz and Millen, 2004)
- other insurgent or terrorist groups (Middle East Newsline, 2005).

Other possibilities exist, of course, such as the lone-wolf forgers of documents and merchants who sell to everyone without knowing anything much about the buyers.

Some of these sources can provide support passively, some are coerced into providing support, and others need not even be aware that they are supporting terrorists or insurgents. This can happen in a state because of its sympathy with the terrorist or insurgent cause persuading it to look the other way, a lack of capability to act against hosted terrorists, or self-imposed restrictions on surveillance and repression within its borders (Lia and Kjøk, 2001). Charities are another mechanism for providing support through ignorance. Many donors may not know that a charity is supporting (or is a front for) a terrorist or insurgent group. This is especially true if the charity is well known for other noncontroversial and beneficial activities.[3]

Individual Decisions to Support

The following discussion explores the factors that cause individuals to support terrorist or insurgent groups. I divide motives for support into three analytical bins: contextual factors, factors based on social or cultural processes, and motives resulting directly from actions taken by the terrorist or insurgent group. These three bins are not exclusive; that is, some could be in multiple or different bins, depending on the circumstances (for example, some factors rely on social processes such as kinship ties that terrorists actively leverage in some cases). I use these bins for sorting and presentational purposes only and do not rely on them for analytical use.

In the existing social-science literature, popular support is too often assumed and the processes by which it is generated and maintained are not often problematized. This is particularly so when specific cases are considered and an insurgency claims to represent a group and appears to have that group's expressed support. In these cases, it is frequently implied that there are "obvious" (if unstated) reasons for support. Fortunately, some studies are explicit about motives for support, and there is a substantial body of research in the social sciences more broadly contributing useful explanations for why people do certain things.

Note that motives for support and motives for individual radicalization have substantial overlap. See Helmus's paper in this monograph for a full discussion of the latter.

Motivations for Support in Response to Context. It is not surprising that many attribute support motives (at least in part) to aspects of oppression. I identify six broad topics in existing social-science research that discuss factors contributing to support:

- humiliation, intolerable frustration, alienation, and hatred
- repression and occupation

- lack of regime legitimacy, lack of opportunity for political expression, and lack of political freedom
- desire for resistance/action by proxy/"public good" (including self-defense)
- social movements (including ideology)
- grievances.

Each is discussed in greater detail below.

Humiliation, Intolerable Frustration, Alienation, and Hatred. Humiliation is often invoked as part of an explanation for support. Richardson (2006) describes the humiliation experienced by Palestinians as they pass through Israeli checkpoints and by Irish Catholics during protestant Orange Order marches. Many scholars invoke Muslim humiliation among Muslims at the decline of traditional Muslim societies in the face of Western advancement as a route to increased sympathy for terrorism, if not offering a direct connection to support (SHARP, 2006).

Frustration (or even "intolerable frustration") creates an opening for terrorist or insurgent groups to gain support (Khashan, 2003). Frustration can result from injustice, economic woes (Clutterbuck, 1995), a repressive political environment (Turk, 2004), or lack of voice (Boyns and Ballard, 2004), and so connects to several factors listed below. Metz and Millen (2004, p. 6) go so far as to name frustration as a basic precondition for insurgency.

Although not offered as stand-alone factors for support generation, alienation and hatred both receive regular mention in discussions of how terrorist sympathy evolves. See Argo (2006) regarding alienation and Hicks (2007) regarding hatred.

Repression and Occupation. The presence of an occupying power or living under a repressive regime is regularly reported as contributing to the popular support of terrorist or insurgent groups (see Metz and Millen, 2004; Richardson, 2006; Pape, 2006, for example). If an individual finds government or occupiers oppressive, that individual is more likely to contribute to those who oppose that government or occupation. Logic of this kind underlies many support factors; the face validity of such logic explains the frequency with which support is simply assumed.

Repression, especially disproportionate response, tends to increase support for terrorists and insurgents (Marks, 2004; Haussler, Russel, and Baylouny, 2005; Libicki et al., 2007). Violence is polarizing, and tit-for-tat reprisals increase the likelihood of further violence and support for the same (Argo, 2006). Efforts to sever the connection between an insurgency and the population through repressive means can backfire (Richardson, 2006).

Lack of Regime Legitimacy, Lack of Opportunity for Political Expression, and Lack of Political Freedom. Independent of repression, the lack of a way to voice a desire to see grievances redressed can lead to the endorsement and support of those who will do so through terrorism or insurgency. As Beckett (2005) asserts, "Above all, however, insurgency remains invariably a competition in government and in perceptions of legitimacy" (p. 2). Metz and Millen (2004) offer "the belief that this [the grievance or frustration] cannot be ameliorated through the existing political system" as the "most basic precondition for insurgency" (p. 6). Krueger (2007) indicates that "the importance of guaranteeing civil liberties has been underemphasized as a means of prosecuting the war on terrorism and the war in Iraq" (p. 87).

Lack of legitimacy and lack of opportunity for political expression are commonly offered as contributors to support for terrorism, although more frequently by assumption than empiricism. An exception is recent RAND research (Cragin et al., unpublished), which reports results from focus groups in the Philippines, Indonesia, and Malaysia. They confirm that the absence of political channels for the expression of concerns is indeed important in motivating sympathetic opinions of terrorism.

Also, Pauly and Redding (2007) find that legitimacy of the government is inversely correlated with legitimacy accorded a terrorist or insurgent group.

Desire for Resistance in Support of the Public Good. Some researchers are more explicit about the connection between pressure by an occupying power, repressive government, or other sources of intolerable frustration and support for terrorist or insurgent groups. Recognizing that individuals experiencing frustration or repression and wishing to act can do so in many ways, decisions to support are argued to follow a more fully specific logic. Support becomes a means of resistance with less personal cost and risk. Such logic appears most prevalently in economic or econometric research. Shulte-Bocholt (2006) draws on enterprise theory to explain the persistence of organized crime or terrorist groups even after their leaders are killed; there is a "demand" for such "services" that others will step in and provide. Similarly, Fair and Shepherd (2006) note a body of research arguing that terrorism can be considered a "public good." The logic of this argument suggests that terrorists provide "resistance," and that anyone who also wants resistance benefits from that provision (and, by extension, by supporting its continued provision).

Desire for resistance by proxy encompasses and includes a variety of self-defense arguments. Many evoke a perception of defenders versus aggressors on the part of those who are or those who support terrorists. See Argo (2006) and Richardson (2006) for examples of communities supporting terrorists because they (the community) are "victims" in need of protection.

Social Movements. Several theoretical tangents can be drawn from social-movement literature on motives for supporting terrorists or insurgents. For example, Boyle et al. (2006) note that social movements must attract participants outside their activist hardcore. This is no doubt also the case for terrorist and insurgent movements. They also argue that framing group activities in collectivist terms might contribute to a group's ability to recruit supporters. Other classic social movement strategies and processes also surely contribute.

The presence of social movements includes the existence of relevant ideologies. Whether developed explicitly by a terrorist or insurgent organization or simply leveraged by them, ideology is argued to make an important contribution to support decisions. Juergensmeyer (argues that "ideologies of validation" are critical for terrorist groups to generate and maintain support. These ideologies are proposed to build on other factors identified: a sense that communities are already under attack, a broader sense of threat or of humiliation and response to those dangers or affronts (Juergensmeyer, 2003, p. 2; SHARP, 2006).

Grievances. A final factor often invoked is the catch-all term "grievances." The presence of grievance is certainly exacerbated if there is no opportunity to give voice to it. Cragin et al. (unpublished, p. 98) report on focus groups attributing increases in militant activity and support to "global and domestic injustice towards Muslims." Beckett (2005) asserts some sense of grievance as foundational for insurgency and for support of insurgency.

Not everyone is so convinced about the independent effect of grievance; Marks (2004) asserts that successful insurgencies base support on both grievance and ideology, and movements lacking either are likely to struggle or be forced to transform. Tilly (1979) agrees with the general point, noting that grievances of some sort are almost always present, so successful explanations of contentious politics need some further ingredient.

Support Motivations Stemming from Cultural or Social Processes. Individuals also make decisions (or nondecisions; see Bachrach and Baratz, 1963) to support terrorists or insurgents based in whole or in part on cultural or social processes. I found the following social or cultural processes offered as factors motivating such support:

- identity processes
- kinship or fictive kinship ties, including tribal motivations
- cultural and social obligations
- revenge
- normative acceptability of violence
- cost-benefit calculations
- misperception and self-deception.

Each is discussed briefly below.

Identity Processes. The literature on identity in the social sciences is truly massive. It is quintessentially human to divide people and places into categories and to separate the world into "us" and "them" to make sense of a complex world (Fiske, 1993; Nelson, 2002; Duckitt, 1992). Many scholars of terrorism note the connections between contextual factors and identify processes as part of a causal chain (de la Roche, 2001, for example). Gerstenfeld (2002) connects context and hatred/intergroup bias. Cragin et al. (unpublished) note that Muslim perceptions of their own status in a country contribute to their views about Islamic militancy as a solution. Richardson (2006) also makes an explicit connection between context, identity, and support, indicating that many who want revenge on behalf of those they identify with join or actively contribute to terrorist organizations.

One seminar at the Summer Hard Problem Program (2006) noted broad agreement in the social sciences that perception of an external threat is the most reliable source of in-group cohesion and associated idealization of in-group values and support for in-group leaders. This creates a more explicit causal pathway for some of the contextual factors identified above to lead to support.

Countries in which an ethnic or cultural group constitutes a small minority see increases in the tightness of identity ties, especially when the group lives in a limited geographic area or enclave (Guild, 2005). When a terrorist group arises out of such a community, community support is more likely (Haahr, 2006).

Identity is complex, and when individuals strongly experience multiple identities (such as religious, tribal, political party, or nationalist), it becomes more difficult to mobilize a single identity for support for terrorism. Cragin et al. (unpublished) find a striking difference between strength of identity with Islam in countries they consider that are Muslim minority (strong identity) and countries that are Muslim majority (less strong).

Identity provides a reinforcing feedback loop for terrorism in two directions. A successful terrorist attack is a symbolic act inspiring solidarity among those sharing interests with the terrorists and promotes unification of identity among the victims (Boyns and Ballard, 2004). Identity can work against terrorist organizations, too, if the terrorist group is defined as being outsiders or others by the relevant population.

Kinship or Fictive Kinship Ties, Including That of Tribes. An oft-invoked form of identity in the literature on terrorism is kinship or fictive kinship ties. Such ties follow similar processes to identity more broadly but are much more easily mobilized (attachment to kin is much more constant than political or ethnic identity across contexts). These factors often intermix; Haussler, Russel, and Baylouny (2005) find that group structures are based locally, on either kinship or other forms of communal solidarity (identity).

Some terrorist organizations use kinship as an explicit tie as they grow. Jemaah Islamiya insiders regularly marry sisters and daughters of other Jemaah Islamiyyah members and supporters (Ismail, 2006). This increases in-group solidarity and spreads kinship throughout the network. The technique, of course, was also practiced by the royal families of Europe and is probably as old as history.

Fictive kinship networks can be as powerful as actual blood or marriage ties. In cultures that have a strong sense of community belonging, members of terrorist groups from that community continue to be treated as kin. For example, Levitt (2007) finds that Hezbollah receives the unambiguous support of close-knit Shia communities in Beirut, where the community is tied together though family relations and shared neighborhood experiences reaching back into childhood.

Speckhard and Ahkmdova (2006) find that a "sense of 'fictive kin' is also commonly created in terror groups that make use of Islamic-based ideologies building on common religious practices of considering a worldwide 'brotherhood' of believers" (p. 448). These fictive kinship ties extend to peripheral members and supporters and appear not just in Islamic terrorist or insurgent organizations but in those of other religions and even nonreligion-based groups. Such brotherhoods are not the exclusive provenance of Islamists, either; consider the ties formed within Elks Lodges, Lions Clubs, or the Masonic Order in contemporary America.

Clan or tribal connection is a form of fictive kinship tie frequently invoked in research on terrorism and insurgency (see McCallister, 2005, for example). McCallister (2005) notes tribal ties not only as a motive but as a connection to a ready made support network: "Consider the fact that tribal society already has at its disposal affiliated social, economic, and military networks easily adapted to war-fighting. The ways in which the insurgents are exploiting the tribal network does not represent an evolved form of insurgency but the expression of inherent cultural and social customs" (p. 3).

Tribal values are central to connecting tribes to support. Tribes usually place a premium on in-group solidarity, personal and group honor, manliness, loyalty, hospitality, and pride (Eisenstadt, 2007). Most of these values can be leveraged by terrorist or insurgent groups with tribal ties for various forms of support. As Brown (2007) notes: "First and foremost, tribes will protect their own. Individuals willing to provide information about insurgents or criminals would do so about members of other tribes, but never about members of their own" (p. 29). Hospitality, particularly common as a tribal value, is also particularly open to exploitation by terrorists or insurgents (Chatty, 1986).

Ronfeldt (2005) suggests that the types of ties inherent in al-Qaeda's network can be effectively characterized in tribal terms. He further argues that a tribal paradigm is useful for framing counterterrorism efforts.

Cultural and Social Obligations. Any of a number of cultural or social obligations, including an obligation to provide hospitality from whatever source, can lead to certain kinds of support for terrorists or insurgents from individuals who are otherwise opposed to those groups' goals.

Although a full discussion of the various types of social obligations identified in social science is beyond the scope of this effort, a baseline can be provided. In a review of social-science's potential contribution to terrorism modeling, Resnyansky (2007) references classic sociologist Max Weber's four "ideal types" of social action: traditional, affectional, value-rational, and instrumental. Cultural and social obligations are most likely to fall into the value-rational category (action is rational in pursuit of the value, which itself may not be rational), or traditional category (action is as described by tradition, independent of the other logics of the situation).

Social forces that could be argued to play a role here include hospitality (already mentioned), answerability (Kenny, 2007), trust (Welch, Sikkink, and Loveland, 2007; Farrell, 2003), or norms of reciprocity and trustworthiness in the form of "social capital" (Putnam, 2001). In terrorism studies, an explicit connection has been proposed between popular support and the social processes of honor and shame (SHARP, 2006). Another factor addressed in terrorism studies is the "spiral of silence" theory, where perceptions of others' opinions influence an individual's willingness to speak up when they hold opinions that differ from those of the perceived majority (Boyle et al., 2006).

Both of these specific examples represent only sample processes picked out from a large range of possible cultural or social obligations.

Some terrorism scholars recognize the multifaceted nature of contributing social or cultural obligations; Richardson (2006) describes what she calls a "complicit surround" (p. 49), where individuals in a community are exposed to a host of assumptions, conditions, and obligations that make supporting or joining a terrorist organization appear either natural or unavoidable (see also Stern, 2003).

Revenge. As a motive, revenge connects quite logically with desire for resistance by proxy, as discussed above. Because revenge itself is a complex psychological process mentioned regularly in the terrorism literature (see Richardson, 2006, for example), I include it as a separate factor.

Normative Acceptability of Violence. Part of Richardson's (2006) "complicit surround" includes a community acceptance of violent means. Scholars studying the growth of acceptance of suicide bombing among Palestinians mention long-term learning and socialization processes that gradually develop the acceptance of violence in a community (Kelley, 2001; Khashan, 2003). Hayes and McAllister (2005) find widespread latent public support for the use of paramilitary violence as a political tool in Northern Ireland. Alonso and Rey (2007) suggest that Moroccans reject violence against civilians and do not support suicide terror in their country. Others highlight acceptance of violence as a preexisting societal or cultural fact that lowers barriers to participation in terrorist or insurgent movements, either as a supporter or as an insurgent (see Shulte-Bocholt, 2006, for example). From an economist's perspective, a social acceptance of violence lowers the entry cost for participation in terrorist activities; sociologically, acceptance of violence lowers barriers to identifying with terrorist actions and thus with terrorists themselves.

Cost-Benefit Calculations. To the extent that humans are rational actors (and Weber's ideal types, discussed above, allow for rational behavior in several types of action), several factors could contribute to calculated decisions to support terrorism or insurgency. These could include a host of monetary or nonmonetary incentives, including a desire to earn recognition within an identity group for support of its champions. The "bandwagon" effect, where large segments of a population throw their support behind the side they believe will win, can also play a role (Metz and Millen, 2004; Byman et al., 2001).

Axelrod (1984), in his study of cooperation over time, suggests that prospects for cooperation are reduced as the probability of future interaction decreases. Counterinsurgency scholars have found this to play into the hands of insurgents when occupation or peacekeeping forces must at some point leave (Beckett, 2005, for example). As the old Afghan saying suggests about the Taliban perspective on patience, "the west may have all the watches, but we have all the time" (Allen, 2006). Fear of future consequences can contribute to decisions to support.

Misperception and Self-Deception. Although not mentioned very often in the terrorism literature, misperception or self-deception can play a role in support decisions. Especially if a terrorist or insurgent organization is providing critical services or engaging in successful propaganda (both discussed below), it might be fairly easy to convince oneself that "they aren't that bad." Cragin et al. (unpublished) found several Filipino and Indonesian focus-group respondents with different levels of awareness of and endorsement of militant groups in their region.

Support Motivated by the Direct Activities of the Terrorist or Insurgent Group. The final category of factors that I use consists of factors that are the direct result of action by the terrorist or insurgent group. These factors include

- intimidation
- propaganda (including propaganda by deed and mobilization efforts)
- provision of social services
- identification with the group (as separate from identity as a process, and including ideology, shared goals, and legitimacy)
- excessive civilian casualties or other unacceptable group behavior
- corruption or penetration of the state.

Each is discussed in greater detail below.

Intimidation. Intimidation or coercion can be extremely effective at generating certain kinds of support, particularly passive support. Where "collaborators" are punished, average citizens are much more likely to keep their heads down and chose to ignore evidence of terrorist or insurgent activity.

Many case studies in terrorism and insurgency reveal the role of intimidation and coercion. Clutterbuck (1995) reports the importance of judicial intimidation in Peru and Colombia, where judges were often offered the "choice between silver and lead" (p. 87). Beckett (2005) reports that the FLN [*Front de Libération Nationale;* National Liberation Front] in Algeria ensured popular support through the use of terror and intimidation. Trinquier's (1985) account concurs; he further argues that intimidation is an extremely effective strategy for controlling the populace and highlights the difficulty faced by the forces of order when trying to ensure the safety of the population.

Looney (2005) highlights the utility of kidnapping both as a form of intimidation and as a way to secure funds.

Propaganda. Propaganda is the pejorative term for information-based influence efforts. Although the dictionary definition is neutral on the matter, the usual informal meaning of "propaganda" includes the suggestion of untruthful or highly selective and misleading information. Information in U.S. psychological operations (PSYOPs) is typically intended to be truthful but "relevant" for influence. However that may be, the tendency is to refer to the enemy's influence operations as propaganda and to those of friendly operations as strategic communications, PSYOPs, or information operations. Current U.S. military doctrine contains advice on facing the often rather difficult challenges posed by terrorist and insurgent PSYOPs (see the discussions in Paul, 2008, or Helmus, Paul, and Glenn, 2007). The paper by Egner in this monograph discusses implications of social science for strategic communications.

Like advertising, propaganda works. Propaganda can be used to mobilize or leverage many of the contextual or cultural and social factors identified above and is part of why separating factors directly influenced by terrorist or insurgent groups from these other organizational bins can be so difficult. Insurgent or terrorist propaganda, disseminated on TV, through the Internet, or in printed form, can call attention to salient aspects of the context (repression, lack of voice), can attempt to mobilize social processes (invoke identities, call for revenge, suggest a means/benefit calculation), or can advertise the effectiveness of the group in serving as a proxy actor for resistance. Propaganda is not just information; it includes "propaganda by deed," where actions taken by the terrorist or insurgent group can have influence. Traditionally, "propaganda by deed" refers to violent action against political enemies to inspire or otherwise catalyze an audience. A broader interpretation includes actions that provoke disproportionate responses from authorities and good deeds (discussed under provision of social services, below) done by the organization.

Examples of carefully orchestrated propaganda campaigns abound in the terrorism and insurgency literature. See, for example, Schleifer's (2006) discussion of Hezbollah PSYOP or Metz's and Millen's (2004) discussion of the topic.

Propaganda can be also used as a tool for mobilization (in the tradition of social movements; see Marks, 2004, and Haussler, Russel, and Baylouny, 2005). Propaganda is an important element in strategic communications considerations.

Provision of Social Services. In addition to offering would-be supporters a way to strike out against "enemies," terrorist or insurgent organizations can make other contributions to exchange relationships. Large, public organizations often offer area communities a variety of basic services, especially over areas they control or would like to (see the discussion in Sinai, 2007). This has been found to be very effective in generating positive opinions, endorsement, and support (Helmus, Paul, and Glenn, 2007; Bloom, 2007).

Protection and access to resources are classic elements of the gang/community exchange, and the logic continues to apply when insurgents are viewed as "3rd generation gangs" (Haussler, Russel, and Maylouny, 2005).

Flanigan (2006) notes that Hezbollah has used provision of services to build a dedicated and indebted constituency, and that organizations in Turkey, Egypt, and Algeria have followed similar approaches with some success. Flanigan (2006) further notes that areas where a terrorist or insurgent group is the *only* provider of such services contain the highest levels of support for such organizations; Bloom (2007) concurs.

Identification with the Group. As noted above, shared identity is a powerful social process that can lead to support for terrorist or insurgent groups. Where such identity is not actually shared or is not sufficiently salient, it can be mobilized by insurgents (through such means as propaganda, discussed above). This factor includes the promotion of shared ideology, shared goals, and steps taken to increase the legitimacy of the group (Metz and Millen, 2004; Richardson, 2006).

Excessive Civilian Casualties or Other Unacceptable Group Behavior. Tied to group legitimacy and related to normative acceptability of violence, the extent and character of civilian casualties from terror attacks (or other potentially "unacceptable" group behavior) can contribute to decisions to support. Historically, when groups have committed (or have been perceived as committing) particularly atrocious attacks, there have been backlashes in sympathy and presumably in material support as well (see Murphy, 2005; Bloom, 2007; or Levinson, 2008, for example). The effect of this factor differs considerably by context; some groups depend much more on general popular support than others do, and some populations are much more tolerant of civilian casualties (Jaeger et al., 2008, show little relationship between polled support by Palestinians and Israeli casualties, for example). Bloom (2007) ties the use of suicide bombing explicitly to public approval for such acts and explicitly connects withholding of material support as expressions of disapproval. Scholars differ on the effect of this factor. Suicide terrorism in Morocco in 2003 met massive public condemnation and street demonstrations opposed to the acts (Alonso and Rey, 2007).

"Unacceptable" group behavior need have nothing to do with violence. Although perhaps deserving of a broader heading, anything that alienates the population from the group or works to sever existing (or potential) identity ties will decrease prospects for support. As Taarnby (2007) notes regarding foreign Mujaheddin in Bosnia, "Their complete disregard for the local culture alienated the Mujaheddin from their only source of support" (p. 169).

Corruption or Penetration of the State. A final factor stemming directly from insurgent or terrorist group action is the corruption or penetration of the state. Both Haussler, Russel, and Baylouny (2005) and Looney (2005) report that penetration of the state by insurgent groups in Iraq has allowed those groups to siphon resources directly from the state into their own coffers or to various patronage groups in exchange for other kinds of support. Clutterbuck (1995) compares Peruvian military-officer wages with the bribes available from narcotics-funded Shining Path guerillas and concludes that it is no surprise that many officers were on the take and not prosecuting operations against the guerillas to full effect.

Weakening the state can also generate certain kinds of support. Bibes (2001) discusses the symbiotic relationship between terrorism and organized crime, whereby both benefit from actions that weaken the state and thus increase passive tolerance of both types of organizations' activities.

Cautions, Consensus, and Disagreements

Cautions

As suggested in the paper's introduction, existing social science cannot yet fully explain support for terrorist or insurgent groups. Interestingly, Jongman (2007) includes "determinants of popular support for terrorist organizations" (p. 279) on his list of research desiderata in the field of terrorism. On the plus side, historical accounts of individual cases (typically in narrative form) offer a very plausible discussion of motivations. Further, all of the factors listed in the previous sections can be found in or extrapolated from existing theories of support.

The empirical weaknesses are due in part to the fact that decisions to support terrorist groups are often assumed and are seldom a researcher's focus of inquiry; such analysis of support as exists often relies on intuition or broad application of general social-science processes. Existing theories of support have not been subjected to particularly rigorous assessment. In short, although many possible and reasonable motives for terrorist or insurgent support have been identified, the current empirical base does not allow us to say with confidence which factors or combinations of factors are either causally absolutely necessary or absolutely sufficient to engender support of terrorists or insurgents across diverse situations.

However, just because we cannot discern "absolutely" necessary or sufficient factors or collections of factors does not mean that we cannot identify factors that are *likely* to be more or less important, to appear more or less frequently, and to be amenable to policy influence. Indeed, there can be a considerable agreement on these items among scholars.

Consensus

Let us now turn to points of consensus within the social sciences regarding the identified factors. There is broad consensus (although only limited empirical support) for the view that popular support is critical for most terrorist or insurgent groups. Some (Mascini, 2006; SHARP, 2006) moderate this strong claim and show that sympathizer support is essential for some activities but not for others, observing that some groups have been able to sustain themselves with the support of very small fractions of a population. This dissenting view is limited in application, however, to a fairly narrow class of terrorist or insurgent group.

There is also broad consensus about the resource needs of terrorist groups (although the necessity of passive support, especially for covert groups, is debated) and the sources by which those needs can be met.

The state of research and degree of consensus are more ambiguous for other factors. Many are uncontested but not well substantiated. Even where there is consensus and significant empirical confirmation, the level of detail at which these processes are understood is often modest and most of the evidence in this area of research is observational. Only when substantial existing bodies of research outside terrorism studies are leveraged (that for social movements, for example) has the research been structured to explicitly test relevant theoretical propositions.

Disagreements

I highlight three substantive disagreements:

- Some authors doubt the centrality and criticality of popular support in general.
- Some reject the importance of passive support, at least in some situations.
- Some disagree about the effect of excessive civilian casualties on a population's willingness to support terrorism or insurgency and its importance.

The three disagreements have some similarities. One group of researchers holds that certain terrorist groups are sufficiently covert that they do not require the support of the population; this extends to not needing passive support from the population. This also means that the consequences of a population withdrawing its support because of reactions to excessive levels of civilian casualties should not be exaggerated, in their view.

My own conclusion is that the dissents are merely cautions: There exist circumstances in which these relatively broad conclusions do not apply.

One reason for there being so few disagreements is that there are so few strong claims; if anyone claimed to have identified a universal necessary or sufficient causal factor, someone would undoubtedly disagree. Since factors are offered as contributors or as important in a specific case, and since all have some face validity, they are not sufficiently provocative, by and large, to inspire great dissent.

Making Sense of the Myriad of Factors

In an effort to better connect the various factor identified and to posit relationships between them, I have developed hypothesized causal path diagrams.

I begin by arranging factors that contribute to determining a terrorist or insurgent group's support needs and the sources for that support. As will be discussed below, identifying the needs of a specific group and the sources through which those needs are being met should be the logical first step in any effort to deny support for a group. See Figure 4.2.

Figure 4.2 illustrates the relationships between groups' needs and the sources for meeting those needs. The output of the function implied by Figure 4.2 is at the top, "adequacy of support received." Moving down the figure one level suggests that adequacy outcomes depend on self-provided support, support obtained from external sources, and actual support needs. The two lower tiers of the figure explore types and sources of external support. Here, the figure closely follows the discussion in sections above. The bottom tier of the figure indicates that active support comes from a variety of sources and can potentially meet needs in all six categories, whereas passive support likely consists of only intelligence, sanctuary, or tolerance of activities.

Once important sources of support are identified, attention can be turned to the providers of that support. Figure 4.3 proposes

Figure 4.2
Relationships Between Needs for and Sources of Support

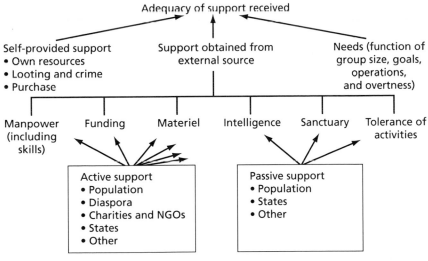

Figure 4.3
Relationships Between Factors Contributing to Strength of Support

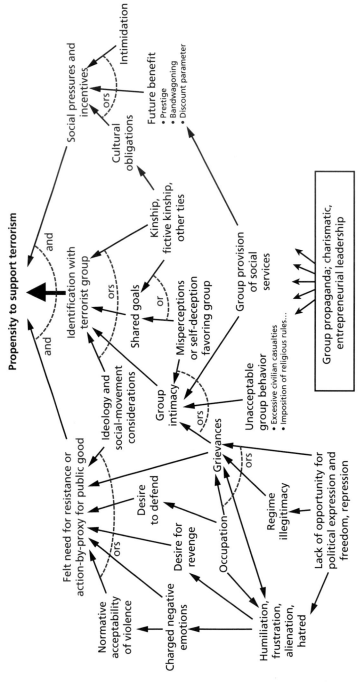

relationships between factors that contribute to decisions to support. If factors are removed or diminished, the likelihood of positive decisions to support diminish as well.

In Figure 4.3, I break support motives into three core chains. At the highest level, individuals decide to support a terrorist organization in part because of (1) a desire (indeed, a "felt need") to contribute to resistance or action for a common good, (2) identification with the group, or (3) social pressure and incentives (either positive or negative). Below these core motivating logics sit all of the various factors mentioned above. Each factor contributes to one or more other factors or logics and accumulates to a notional likelihood of offering support at the top of the figure. Allowing social pressure to include both positive and negative pressure allows me to include both active and passive support motives in the model without specifying separate processes for each. Note that Figure 4.3 presents contributing pathways; from a causal perspective, all arrows should be viewed as partial influence pathways rather than as strictly causal, and all arrows should be viewed as "and/or," indicating multiple possible paths operating simultaneously or, in certain circumstances, one path operating strongly and sufficiently. None is strictly necessary and any could be sufficient in extremis. (See the discussion above regarding the state of social-science research with regard to criticality and confirmatory evidence for these factors.)

Also note that each factor can exist sui generis and does not depend on factors with arrows leading to it; for example, "normative acceptability of violence" (in the top left of the figure) can occur as a result of its own independent processes, but can be accelerated by charged negative emotions. Similarly, identification with the group (center, toward the top) can be preexisting but can be strengthened and increased in likelihood by the group's legitimacy, the extent to which goals are shared, and other forms of identity ties.

The complex chains of relationships and the multiplicity of influence arrows partially reflects the complexity of the situation. The situation is actually more complex than depicted; not every posed relationship arrow is shown, only the major ones; further, propaganda (at the bottom of the figure) can be used to affect or leverage so many of the other factors that the figure would be rendered incomprehensible were I to draw in those arrows. Propaganda is also a special case, in that it can be considered to point to arrows between factors as well as to factors themselves; that is, propaganda can affect the relationships between factors. For example, effective propaganda could magnify the effect of repression, causing deeper grievances or greater desire to seek revenge; propaganda could mobilize kinship or tribal ties, increasing the likelihood of identifying with the terrorist group.

With those caveats in place, Figure 4.3 does give a reasonable picture of primary relationships between proposed factors and tenable influence paths for support motives. Policy options for leveraging these relationships to attack support are discussed below.

Table 4.1 offers alternative evaluating factors contributing to support. For instances in which a terrorist or insurgent groups needs the support provided by a community or population, it asks whether the various factors are reasonably likely to be in play, important, and amenable to the influence of strategy and policy. Each question is answered in the affirmative by an X in Table 4.1. A "slash" indicates a "half-check" and indicates that the factor is either less likely, less important (or important in only some of the situations in which it occurs), or less easy to influence through policy.

Factors that are likely to be present, are important when present, and are amenable to influence via policy are the factors that provisionally should be considered high priority, in the abstract.

When facing an actual situation instead of an abstraction, presence, importance, and amenability to influence may all differ, depending on conditions on the ground, and a quick review of factors for which those boxes can be checked in an actual case will lead to the identification of factors critical there.

Implications for Strategy and Policy

Given that the vast majority of terrorist and insurgent groups depend on external support for at least some of their resource needs, attacking support continues to be a potentially effective policy lever. Krueger

Table 4.1
Likelihood of Factors' Being Operative, Important, and Mutable, by Strategy

Factor	Reasonably Likely	Important	Amenable to Policy Influence
Desire for resistance or action by proxy ro the public good	X	X	/
Charged negative emotions, including humiliation, intolerable frustration, alienation, or hatred	X	/	/
Revenge, defense, or other grievances	X	/	X
Normative acceptability of violence		X	
Identification with the group	X	X	
Legitimacy of group, including civilian casualties or other unacceptable behavior	X	X	/
Shared goals	/	/	X
Social movements and ideology	X	/	
Kinship, fictive kinship, or other identity ties	X	X	
Social pressure or influence	/	X	/
Cultural obligations	/	/	
Net positive incentives (cost-benefit calculus)[a]	/	/	X
Group propaganda	X	X	/
Provision of social services		X	X
Intimidation	X	X	/

[a] An economist would almost assuredly assume that net positive incentives are fully present (X) wherever terrorists receive support.

(2007) suggests a focus on policy that decreases "demand for pursuing grievances through terrorist tactics" (p. 50) and degrades terrorist organizations' financial or technical capabilities. Interrupting popular support has the potential to do both. This section briefly discusses the policy implications of the findings of this paper.

The discussion of factors determining support needs and sources produces an important revelation: Support is *not* "one size fits all." This leads to perhaps the most important policy implication of this paper: *Ascertain the specifics of the case.* Before developing and implementing a generic effort to undermine support, identify the type of group (size, goals, nature of operations, and covertness), the extent of each group's support needs (manpower, funding, materiel, intelligence, sanctuary, and tolerance of activities), and how it is meeting these needs. With a particular type and source of support in mind, it becomes much easier to specify actions to reduce support motives than in the generic case. To put it differently, although ideal cases have a long and valued role in academic studies, applying the lessons of social science is another matter. It is not that all the details matter but rather that different details matter in different cases.

The value of case specifics continues when considering specific support motivating factors. In the generic case, and with the current state of the art in social-science understanding, it is impossible to assert the importance of one factor or set of factors over another. In a specific case, the presence, importance, and amenability to influence of certain factors will be more clear, although it may also change within days or weeks as the result of events. This leads to the second policy implication: Once you know the specifics of your case, *focus on factors that matter and that can be changed.*

Once the specifics of a group and its support relationships are understood, Figure 4.3 should help in the establishment of targets and goals. With a support source in mind, work from the top down. Identify the primary motive for support (in the top rank of the hierarchy) and then identify which supporting logics for that motive are in place.

Bibliography

Abdallah, Abdel Mahdi, "Causes of Anti-Americanism in the Arab World: A Socio-Political Perspective," *Middle East Review of International Affairs*, Vol. 7, No. 4, December 2003.

Allen, Nick, "US Forces Feel the Heat in Afghanistan's 'Forgotten War,'" *Monsters and Critics.com*, December 18, 2006.

Alonso, Rogelio, and Marcos Garcia Rey, "The Evolution of Jihadist Terrorism in Morocco," *Terrorism and Political Violence*, Vol. 19, No. 4, 2007, pp. 571–592.

Argo, Nicole, "The Role of Social Context in Terrorist Attacks," *The Chronicle of Higher Education*, January 13, 2006.

Axelrod, Robert, *The Evolution of Cooperation*, New York: HarperCollins, 1984.

Bachrach, Peter, and Morton J. Baratz, "Decisions and Non-Decisions," *American Political Science Review*, Vol. 57, No. 3, September 1963, pp. 632–642.

Basile, Mark, "Going to the Source: Why Al Qaeda's Financial Network Is Likely to Withstand the Current War on Terrorist Financing," *Studies in Conflict & Terrorism*, Vol. 27, No. 3, May 2004, pp. 169–185.

Beckett, Ian F.W., *Insurgency in Iraq: An Historical Perspective*, Carlisle, Pa.: Strategic Studies Institute, U.S. Army War College, January 2005.

Bibes, Patricia, "Transnational Organized Crime and Terrorism: Colombia, a Case Study," *Journal of Contemporary Criminal Justice*, Vol. 17, No. 3, August 2001, pp. 243–258.

Bloom, Mia, *Dying to Kill: The Allure of Suicide Terror*, New York: Columbia University Press, 2007.

Boyle, Michael P., Mike Schmierbach, Cory L. Armstrong, et al., "Expressive Responses to News Stories about Extremist Groups: A Framing Experiment," *Journal of Communication*, Vol. 56, No. 2, 2006, pp. 271–288.

Boyns, David, and James David Ballard, "Developing a Sociological Theory for the Empirical Understanding of Terrorism," *The American Sociologist*, Summer 2004, pp. 5–25.

Brown, Lieutenant Colonel Ross A., "Commander's Assessment: South Baghdad," *Military Review*, January–February 2007.

Byman, Daniel L., *Confronting Passive Sponsors of Terrorism*, Saban Center for Middle East Policy, Analysis Paper No. 4, Washington, D.C.: Brookings Institution, February 2005.

Byman, Daniel, Peter Chalk, Bruce Hoffman, William Grey Rosenau, and David Brannan, *Trends in Outside Support for Insurgent Movements*, Santa Monica, Calif.: RAND Corporation, 2001. As of December 21, 2008: http://www.rand.org/pubs/monograph_reports/MR1405/

Chatty, Dawn, *From Camel to Truck: The Bedouin in the Modern World*, New York: Vantage Press, 1986.

Clutterbuck, Richard, "Peru: Cocaine, Terrorism and Corruption," *International Relations*, Vol. 12, No. 5, 1995, pp. 77–92.

Coleman, James "Social Capital in the Creation of Human Capital," *American Journal of Sociology*, Vol. 94 Supplement, 1988, pp. S95–S120.

Cottino, Amedeo, "Sicilian Cultures of Violence: The Interconnections Between Organized Crime and Local Society," *Crime, Law and Social Change*, Vol. 32, No. 2, 1999, pp. 103–113.

Cragin, Kim, Peter Chalk, Audra Grant, Todd C. Helmus, Donald Temple, and Matt Wheeler, "Curbing Militant Recruitment in Southeast Asia: Factors That Influence Individual Motivation and Popular Support for Violence," Santa Monica, Calif.: RAND Corporation, unpublished.

De la Roche, Roberta Senechal, "Why Is Collective Violence Collective?" *Sociological Theory*, Vol. 19, No. 2, July 2001, pp. 126–144.

Duckitt, J., "Psychology and Prejudice: A Historical Analysis and Integrative Framework," *American Psychologist*, Vol. 47, No. 10, October 1992, pp. 1182–1193.

Eisenstadt, Lieutenant Colonel Michael, U.S. Army Reserve, "Iraq: Tribal Engagement Lessons Learned," *Military Review*, September–October 2007.

Esfandiari, Golnaz, "World: Poll Finds Muslim Approval of Terrorism Declines," *Radio Free Europe/Radio Liberty*, July 15, 2005. As of March 13, 2008: http://www.rferl.org/content/article/1059961.html

Fair, C. Christine, and Bryan Shepherd, "Who Supports Terrorism? Evidence from Fourteen Muslim Countries," *Studies in Conflict & Terrorism*, Vol. 29, 2006, pp. 51–74.

Fantasia, Rick, and Eric L. Hirsch, "Culture in Rebellion: The Appropriation and Transformation of the Veil in the Algerian Revolution," in Hank Johnston and Bert Klandermans, eds., *Social Movements and Culture*, Minneapolis, Minn.: University of Minnesota Press, 1995.

Faria, João Ricardo, and Daniel G. Arce, "Terror Support and Recruitment," *Defence and Peace Economics*, Vol. 16, No. 4, 2005, pp. 263–273.

Farrell, Henry, "Trust Institutions, and Institutional Change: Industrial Districts and the Social Capital Hypothesis," *Politics & Society*, Vol. 31, No. 4, December 2003, pp. 537–566.

Fiske, S. T., "Social Cognition and Social Perception," *Annual Review of Psychology*, Vol. 44, 1993, pp. 155–194.

Flanigan, Shawn Teresa, "Charity as Resistance: Connections Between Charity, Contentious Politics, and Terror," *Studies in Conflict & Terrorism*, Vol. 29, 2006, pp. 641–655.

Forrest, James J.F., Thomas A. Bengtson, Jr., Hilada Rosa Martinez, Nathan Gonzalez, and Bridget C. Nee, *Terrorism and Counterterrorism: An Annotated Bibliography*, Vol. 2, West Point, N.Y.: Combating Terrorism Center, September 11, 2006.

Gerges, Fawaz A., *The Far Enemy: Why JIHAD Went GLOBAL*, Cambridge, U.K.: Cambridge University Press, 2005.

Gerstenfeld, Phyllis B., "A Time to Hate: Situational Antecedents of Intergroup Bias," *Analyses of Social and Public Policy*, January 2002, pp. 61–67.

Guild, Elspeth, *International Migration and Security: Immigrants as an Asset or Threat?* London: Routledge, June 17, 2005.

Haahr, Kathryn, "Emerging Terrorist Trends in Spain's Moroccan Communities," *Terrorism Monitor*, Vol. 4, No. 9, May 4, 2006.

Haddad, Simon, "The Origins of Popular Support for Lebanon's Hezbollah," *Studies in Conflict & Terrorism*, Vol. 29, 2006, pp. 21–34.

Haddad, Simon, and Hilal Khashan, "Islam and Terrorism: Lebanese Muslim Views on September 11," *The Journal of Conflict Resolution*, Vol. 46, No. 6, December 2002, pp. 812–828.

bin Hassan, Muhammad Haniff, "Key Considerations in Counterideological Work Against Terrorist Ideology," *Studies in Conflict & Terrorism*, Vol. 29, 2006, pp. 531–558.

Haussler, Nicholas I., James Russel, and Anne Marie Baylouny, *Third Generation Gangs Revisited: The Iraq Insurgency*, Monterey, Calif.: Naval Postgraduate School, Thesis, September 2005.

Hayes, Bernadette C., and Ian McAllister, "Public Support for Political Violence and Paramilitarism in Northern Ireland and the Republic of Ireland," *Terrorism and Political Violence*, Vol. 17, 2005, pp. 599–617.

Helmus, Todd C., "Why and How Some People Become Terrorists," in Paul K. Davis and Kim Cragin, eds., *Social Science for Counterterrorism: Putting the Pieces Together,* Santa Monica, Calif.: RAND Corporation, 2009. As of January 20, 2009:
http://www.rand.org/pubs/monographs/MG849/

Helmus, Todd C., Christopher Paul, and Russell W. Glenn, *Enlisting Madison Avenue: The Marketing Approach to Earning Popular Support in Theaters of Operation*, Santa Monica, Calif.: RAND Corporation, 2007. As of December 21, 2008:
http://www.rand.org/pubs/monographs/MG607/

Hicks, Tessa, "Humanizing the Other in 'Us and Them,'" *Peace Review: A Journal of Social Justice*, Vol. 18, 2007, pp. 499–506.

Hodges, Donald C., ed., *Philosophy of the Urban Guerrilla: The Revolutionary Writings of Abraham Guillen*, New York: William Morrow, 1973.

Ismail, Noor Huda, "The Role of Kinship in Indonesia's Jemaah Islamiya," *Terrorism Monitor*, Vol. 4, No. 11, June 2, 2006.

Jaeger, David A., Esteban F. Klor, Sami H. Miaari, and M. Daniele Paserman, "The Struggle for Palestinian Hearts and Minds: Violence and Public Opinion in the Second Intifada," Working Paper 72, Williamsburg, Va.: Department of Economics, College of William and Mary, 2008.

Jankowski, Martin Sanchez, *Islands in the Street: Gangs in American Urban Society*, Berkeley, Calif.: University of California Press, 1991.

Jongman, Berto, "Research Desiderata in the Field of Terrorism," in Magnus Ranstorp, ed., *Mapping Terrorism Research: State of the Art, Gaps, and Future Direction*, London: Routledge, 2007, pp. 255–291.

Juergensmeyer, M., *Terror in the Mind of God: The Global Rise of Religious Violence*, 3rd ed., Berkeley, Calif.: University of California Press, 2003.

Kelley, J., "Wired for Death: Ignoring Islam's Mainstream Message of Peace, Israel's Most Bitter Enemies Embrace the Ultimate Weapon in Modem Warfare—The Human Bomb," *Reader's Digest*, Vol. 159, No. 954, 2001, pp. 78–81.

Kenny, Robert Wade, "The Good, the Bad, and the Social: On Living as an Answerable Agent," *Sociological Theory*, Vol. 25, No. 3, September 2007, pp. 268–291.

Khashan, Hilal, "Collective Palestinian Frustration and Suicide Bombings," *Third World Quarterly*, Vol. 24, No. 6, December 2003, pp. 1049–1067.

Kilcullen, David, "Twenty-Eight Articles: Fundamentals of Company-Level Counterinsurgency," undated.

Kleemans, Edward R., and Henk G. van de Bunt, "The Social Embeddedness of Organized Crime," *Transnational Organized Crime*, Vol. 5, No. 1, 1999, pp. 19–36.

Krueger, Alan B., *What Makes a Terrorist: Economics and the Roots of Terrorism*, Princeton, N.J.: Princeton University Press, 2007.

Levinson, Charles, "Al-Qaeda Targets Hearts, Minds: New Tactics Seek to Raise Local Image," *USA Today*, February 7, 2008.

Levitt, Matthew, "Hezbollah Finances: Funding the Party of God," *Terrorism Financing and State Responses: A Comparative Perspective*, Monterey, Calif.: Center for Homeland Defense and Security, Naval Postgraduate School, March 2007.

Lia, Brynjar, and Åshild Kjøk, *Islamist Insurgencies, Diasporic Support Networks, and Their Host States: The Case of the Algerian GIA in Europe 1993–2000*, Kjeller, Norway: Norwegian Defence Research Establishment, FFI Rapport-2001/03789, 2001.

Lia, Brynjar, and Katja Skjolberg, *Causes of Terrorism: An Expanded and Updated Review of the Literature*, Kjeller, Norway: Norwegian Defence Research Establishment, FFI/Rapport, 2004.

Libicki, Martin C., David C. Gompert, David R. Frelinger, and Raymond Smith, *Byting Back—Regaining Information Superiority Against 21st-Century Insurgents: RAND Counterinsurgency Study—Volume 1*, Santa Monica, Calif.: RAND Corporation, 2007. As of December 21, 2008: http://www.rand.org/pubs/monographs/MG595.1/

Looney, Robert, "The Business of Insurgency: The Expansion of Iraq's Shadow Economy," *The National Interest*, Fall 2005, pp. 1–6.

Makarenko, Tamara, "On the Border of Crime and Insurgency," *Jane's Intelligence Review*, Vol. 14, No. 1, January 2002, pp. 33–35.

Manwaring, Max G., *Shadows of Things Past and Images of the Future: Lessons for the Insurgencies in Our Midst*, Carlisle, Pa.: Strategic Studies Institute, U.S. Army War College, November 2004.

Marks, Thomas A., "Ideology of Insurgency: New Ethnic Focus or Old Cold War Distortions?" *Small Wars & Insurgencies*, Vol. 15, No. 1, Spring 2004, pp. 107–128.

Mascini, Peter, "Can the Violent *Jihad* Do Without Sympathizers?" *Studies in Conflict & Terrorism*, Vol. 29, 2006, pp. 343–357.

Masoud, Tarek E., "The Arabs and Islam: The Troubled Search for Legitimacy," *Daedalus*, Vol. 128, No. 2, Spring 1999, pp. 127–145.

McCallister, William S., "The Iraq Insurgency: Anatomy of a Tribal Rebellion," *First Monday*, Vol. 10, No. 3, March 2005.

Metz, Steven, and Raymond Millen, *Insurgency and Counterinsurgency in the 21st Century: Reconceptualizing Threat and Response*, Carlisle, Pa.: Strategic Studies Institute, U.S. Army War College, November 2004.

Middle East Newsline, "U.S. Finds Rat Line from N. Africa to Iraq," July 7, 2005.

Murphy, Dan, "Terror Shifts Muslim Views," *Christian Science Monitor*, July 26, 2005.

Myers, Steven Lee, "From Dismal Chechnya, Women Turn to Bombs," *The New York Times*, September 10, 2004.

Nelson, T. D., *The Psychology of Prejudice*, Boston, Mass.: Allyn and Bacon, 2002.

Palestinian Center for Policy & Survey Research (PCPSR), PSR-Survey Research Unit: Public Opinion Poll #3, December 19–24, 2001. As of January 3, 2008: http://www.pcpsr.org/survey/polls/2001/p3a.html

Pape, Robert, *Dying to Win: The Strategic Logic of Suicide Terrorism*, New York: Random House, 2006.

Paul, Christopher, *Information Operations—Doctrine and Practice: A Reference Handbook*, Westport, Conn.: Praeger Security International, 2008.

Pauly, Robert J., and Robert W. Redding, "Denying Terrorists Sanctuary Through Civil Military Operations," in James J.F. Forest, ed., *Countering Terrorism and Insurgency in the 21st Century: International Perspectives, Volume 1: Strategic and Tactical Considerations*, Westport, Conn.: Praeger Security International, 2007, pp. 273–297.

PCPSR–*See* Palestinian Center for Policy & Survey Research.

Pew Research Center, *Support for Terror Wanes Among Muslim Publics: Islamic Extremism: Common Concern for Muslim and Western Publics*, 17-Nation Pew Global Attitudes Survey, July 14, 2005.

Putnam, Robert D., *Bowling Alone: The Collapse and Revival of American Community*, New York: Simon & Schuster, 2001.

Resnyansky, L., "Integration of Social Sciences in Terrorism Modeling: Issues, Problems and Recommendations," *Command and Control Division Defence Science and Technology Organisation*, Edinburgh, South Australia, DSTO-TR-1955, February 2007.

Richardson, Louise, *What Terrorists Want*, New York: Random House, 2006.

Rodriguez, Jose A., *The March 11th Terrorist Network: In Its Weakness Lies Its Strength*, Barcelona, Spain: Department of Sociology and Analysis of Organizations, Working Papers EPP-LEA:03, December 2005.

Ronfeldt, David, "Al Qaeda and Its Affiliates: A Global Tribe Waging Segmental Warfare?" *First Monday*, 2005.

Roth, Mitchell P., and Murat Sever, "The Kurdish Workers Party (PKK) as Criminal Syndicate: Funding Terrorism Through Organized Crime, a Case Study," *Studies in Conflict & Terrorism*, Vol. 30, 2007, pp. 901–920.

Scheufele, D. A., and W. P. Eveland, "Perceptions of 'Public Opinion' and 'Public' Opinion Expression," *International Journal of Public Opinion Research*, Vol. 13, 2001, pp. 25–44.

Schleifer, Ron, "Psychological Operations: A New Variation on an Age Old Art: Hezbollah Versus Israel," *Studies in Conflict & Terrorism*, Vol. 29, 2006, pp. 1–19.

Schuster, Henry, "Poll of Saudis Shows Wide Support for bin Laden's Views," *CNN.com*, June 8, 2004. As of March 13, 2008:
http://www.cnn.com/2004/WORLD/meast/06/08/poll.binladen/index.html

SHARP—*See* Summer Hard Problem Program.

Shulte-Bocholt, Alfredo, *The Politics of Organized Crime and the Organized Crime of Politics: A Study in Criminal Power*, Lanham, Md.: Rowman & Littlefield Publishers, 2006.

Sinai, Joshua, "New Trends in Terrorism Studies: Strengths and Weaknesses," in Magnus Ranstorp, ed., *Mapping Terrorism Research: State of the Art, Gaps, and Future Direction*, London: Routledge, 2007, pp. 31–50.

Smith, Paul J., "Terrorism Finance: Global Responses to the Terrorism Money Trail," in James J.F. Forest, ed., *Countering Terrorism and Insurgency in the 21st Century: International Perspectives, Volume 2: Combating the Sources and Facilitators*, Westport, Conn.: Praeger Security International, 2007, pp. 142–162.

Speckhard, Anne, and Khapta Ahkmedova, "The Making of a Martyr: Chechen Suicide Terrorism," *Studies in Conflict & Terrorism*, Vol. 29, 2006, pp. 429–492.

Stern, Jessica, *Terror in the Name of God: Why Religious Militants Kill*, New York: HarperCollins, 2003.

Summer Hard Problem Program (SHARP), Director of National Intelligence, *White Papers*, 2006.

Taarnby, Michael, "Understanding Recruitment of Islamist Terrorists in Europe," in Magnus Ranstorp, ed., *Mapping Terrorism Research: State of the Art, Gaps, and Future Direction*, London: Routledge, 2007, pp. 164–186.

Tilly, Charles, "Social Movements and National Politics," CRSO Working Paper #197, Ann Arbor, Mich.: Center for Research on Social Organization, University of Michigan, 1979.

Trinquier, Roger, "Modern Warfare: A French View of Counterinsurgency," Fort Leavenworth, Kan.: Combined Arms Research Library, U.S. Army Command and General Staff College, January 1985. As of January 14, 2008: http://www-cgsc.army.mil/carl/resources/csi/trinquier/trinquier.asp

Tse-tung, Mao (trans. Samuel B. Griffith II), *On Guerrilla Warfare*, Chicago, Ill.: University of Illinois Press, 1937 (2000).

Tsvetovat, Maksim and Kathleen M. Carley, "Structural Knowledge and Success of Anti-Terrorist Activity: The Downside of Structural Equivalence," *Journal of Social Structure*, Vol. 6, 2005.

Turk, Austin T., "Sociology of Terrorism," *Annual Review of Sociology*, Vol. 30, 2004, pp. 271–86.

Vinci, Anthony, "The 'Problems of Mobilization' and the Analysis of Armed Groups," *Parameters*, Spring 2006, p. 51.

von Lampe, Klaus, "The Interdisciplinary Dimensions of the Study of Organized Crime," *Trends in Organized Crime*, Vol. 9, No. 3, 2006, pp. 77–95.

Welch, Michael R., David Sikkink, and Matthew T. Loveland, "The Radius of Trust: Religion, Social Embeddedness and Trust in Strangers," *Social Forces*, Vol. 86, No. 1, September 2007, pp. 23–24.

Endnotes

[1] Nonetheless, there is consensus around the assertion that popular support is a critical terrorist or insurgent center of gravity. See, for example, Hodges (1973); Trinquier (1985); bin Hassan (2006); Manwaring (2004); Haussler, Russel, and Baylouny (2005); Kilcullen (undated); Libicki, Gompert, Frelinger, and Smith (2007); Bloom (2007); Sinai (2007); and Beckett (2005). This consensus is not perfect. Some argue that some terrorist groups require very little or almost no popular support (e.g., Mascini, 2006).

[2] Vinci (2006) identifies three needs: "[P]eople who will fight. It needs the means of force, including weapons and the basics of survival. Finally, it needs the ability to exercise direction" (p. 51). Without much effort, these can be fit into manpower, materiel, and intelligence. Regarding manpower requirements, see Chapter 3. Funding requirements and resources are broadly discussed; see Mascini (2006), for example. Intelligence is less broadly discussed as a requirement, but no one would dispute its importance. See Haussler, Russel, and Baylouny (2005) for a good discussion of the importance of intelligence to insurgencies.

[3] See Basile (2004); for a more general point about the suborning of organizational resources to other purposes in groups like social movements, see Coleman (1988).

CHAPTER FIVE

The Economics of Terrorism and Counterterrorism: What Matters and Is Rational-Choice Theory Helpful?

Claude Berrebi

Introduction

What is the relationship between terrorism and such potential root causes as poverty, education, religion, and mental health? Is it useful to discuss cause-effect relationships in terms of a rational-choice model? The questions are related in the following way. First, many have sought to explain terrorism in terms of various structural factors such as those mentioned, without reference to issues of choice. In this case, the factors are thought of as preconditions; the imagery is then of the form "Because of such-and-such powerful factors, people are driven to or drawn into terrorism." The empirical evidence has tended to disconfirm such approaches, as decisively as one finds in social science. An alternative approach is to explain terrorism as the result of what individuals and groups perceive (whether or not correctly) as rational choices. Accordingly, it is not so much that terrorists are victims of some external pressures, but rather that they are acting in sensible ways given their preferences and surrounding state of the world (whether perceived or accurate). Evidence on this is still being sorted out. It seems clear that simple-minded rational-choice models (such as those that limit considerations solely to monetary reward benefits and costs) do not work well. However, I shall argue that more-sophisticated rational-choice models appear to have substantial explanatory power. If this is true, such models should be useful in assessing alternative counterterrorism strategies.

Relationships Between Terrorism and Postulated Root Causes

Conventional Wisdom

Although terrorism experts have been changing their minds on this over the last five years or so, it is probably still conventional wisdom that people become terrorists because of some combination of economic conditions, educational attainment, religious zealotry, or mental illness. They lack the knowledge or ability to make reasoned decisions, or they are in such desperate circumstances as to seek extreme measures.

Sometimes, the arguments favoring such views are intuitive, so-called "common-sense" notions. At other times, they are based on, for example, the logic suggested by the traditional economic theory of crime (Becker, 1976),[1] by the economic theory of suicide (Hamermesh and Soss, 1974),[2] or a theory of the economics of religious sects (Berman, 2000, 2003). In all of these, the common denominator is that the terrorists possess relatively inferior marketable alternatives, and therefore their opportunity costs are low. I shall not discuss these arguments in any length because a large body of empirical work tends to disconfirm the underlying common denominator—both at the individual and organizational levels:

- Terrorists are *not* particularly poor, ignorant, mentally ill, or religious. Their most notable characteristic is normalcy.

In what follows, I summarize the evidence for this conclusion.[3] Although the evidence I present is of the form favored by economists studying these issues, the conclusions are, perhaps, not what might be expected as they point to social, behavioral, and political factors as being most important.

The following sections deal with what the economist-lens literature has to say about our knowledge regarding (1) poverty and education, (2) religion, and (3) mental health. After that, I return to the issue of rational-choice explanations.

Poverty and Education

At the beginning of the decade, there was widespread belief that poverty and education were root causes of terrorism. However, evidence emerged to the contrary—informal evidence, inconclusive scientific evidence, and then increasingly definitive evidence.

Informal Evidence. Anecdotal evidence came to contradict the conventional wisdom about terrorists being predominantly poor and ignorant. Of course, there was the well-known example of Osama bin Laden, a man of impressive wealth and a fine education. Nonetheless, in an article in the *New York Times* on the characteristics of the September 11, 2001, terrorist hijackers, Jodi Wilgoren (2001) reported:

> They were adults with education and skill . . . [who] spent years studying and training in the United States, collecting valuable commercial skills and facing many opportunities to change their minds. . . . [T]hey were not reckless young men facing dire economic conditions and dim prospects but men as old as 41 enjoying middle-class lives.

In the same year, an intriguing publication by Hassan (2001) also suggested that economic incentives probably cannot explain terrorist activity. In an article summarizing interviews of nearly 250 terrorists and associates (including failed suicide bombers, families of deceased bombers, and those who trained and prepared the bombers for their missions), she reported:

> None of them were uneducated, desperately poor, simple minded or depressed. Many were middle class and, unless they were fugitives, held paying jobs. More than half of them were refugees from what is now Israel. Two were the sons of millionaires.

In a *New York Times* article, researcher Scott Atran (2003) reported:

Officials with the Army Defense Intelligence Agency who have interrogated Saudi-born members of Al Qaeda being detained at Guantánamo Bay, Cuba, have told me that these fundamentalists, especially those in leadership positions, are often educated above reasonable employment level; a surprising number have graduate degrees and come from high-status families.

In an account of the July 7, 2005, London Public Transport System, the first suicide terrorist bombing in Western Europe, Glen M. Segell (2005) reported:

This was especially the case in the July 7, 2005 attacks on the London Public Transport system where the bombers were young, middle-class, British citizens with good prospects.

In an article in the *Telegraph* on the August 2006 plot to use liquid bombs against airliner jets flying from the United Kingdom to America, Caroline Davies, John Steele and Catriona Davies (2006) reported:

Twenty-four terrorist suspects being held last night over an alleged plot to blow up as many as 10 transatlantic jets include middle-class, well-educated young men born in Britain. . . . among those arrested were the white son of a former Conservative Party worker, the son of an architect and an accountant and a heavily pregnant woman. Some had studied at university and came from families that owned several properties or ran their own businesses.

Similarly, in an account on *Fox News* of the July 2007 car bomb plot in which several medical doctors took part, David Stinger, a writer of the Associated Press, interviewed Paul Cornish, a former British army officer and director of defense studies at London's Chatham House think tank, who said:

This case could be the final proof that the idea that those involved in these types of attacks are all young, angry and poorly educated is a mistake.

Finally, in an article unraveling Chechen "Black Widows," Nabi Abdullaev (2007) concludes:

> The identified suicide bombers had not been living in abject poverty, nor were they known to have been raped or otherwise tortured and humiliated. . . .

Of course, sound empirical conclusions cannot be based on news reports or anecdotal evidence. Nonetheless, the evidence was building and more scientific evidence was being rediscovered or newly emerging as well.

The Empirical Evidence. The empirical evidence (by which I mean systematically developed empirical evidence) collected so far gives little reason to believe that improving individuals' material or educational circumstances would help reduce their desire to participate in terrorist activities.[4] If anything, the findings suggest that those with higher educational attainment and higher living standards are *more* likely to participate in terrorist activity. Some of the reports date back decades. For example, Russell and Miller (1983) attempted to draw a sociological profile of the modern urban terrorist, using a compilation of information on more than 350 terrorists from Argentinean, Brazilian, German, Iranian, Irish, Italian, Japanese, Palestinian, Spanish, Turkish, and Uruguayan terrorist groups active during 1966–1976. They found that

> . . . approximately two-thirds of those identified terrorists are persons with some university training, university graduates or postgraduate students. (p. 55)

Hudson and Majeska (1999) reinforced this in a report produced by the U.S. Library of Congress's Federal Research Division concerning the sociological characteristics of terrorists in the Cold War period. They concluded:

> Terrorists in general have more than average education, and very few Western terrorists are uneducated or illiterate. . . . Older members and leaders frequently were professionals such as doc-

tors, bankers, lawyers, engineers, journalists, university professors, and mid-level government executives.

Similarly, Singapore's Ministry of Home Affairs issued a white paper entitled *The Jemaah Islamiyah Arrests and the Threat of Terrorism*.[5] Among other things, the paper describes Jemaah Islamiyyah prisoners. Notably, it says:

> These men were not ignorant, destitute, or disenfranchised outcasts. . . . Like many of their counterparts in militant Islamic organizations in the region, they held normal, respectable jobs.

Another example is provided by Marc Sageman,[6] who concluded, on the basis of interviews with more than 400 al-Qaeda–affiliated terrorists from the Middle East, Southeast Asia, Northern Africa, and Europe:[7]

> The vast majority of terrorists in the sample came from solid middle class backgrounds, and its leadership came from the upper class. . . . Although al-Qaida justifies its operations by claiming to act on behalf of its poor brothers, its links to poverty are at best vicarious. . . . About two-thirds of the sample had attended college. . . . About 60 percent of al-Qaida terrorists in the sample have professional or semi-professional occupations. (2006)

This evidence was not yet conclusive for various reasons, so a series of further studies tightened the investigation. The earlier research was drawn from unrepresentative samples of terrorists, mainly famous leaders.[8] News reports could be similarly biased, since they emphasize the sensational and might neglect to report those instances in which economically desperate individuals participate in terrorist activity. This proved not to be a problem, however, judging by recent empirical analyses of the characteristics of terrorists. The groundwork was laid with the work of Krueger and Maleckova (2003), who investigated the links between poverty, low education, and participation in Hizballah militant activity. Using biographical data of 129 Hizballah members killed in paramilitary actions in the late 1980s and early 1990s, they found that having both a standard of living above the poverty line and a secondary-school education or higher were *positively* associated with participation in Hizballah. The U.S. State Department and the British Home Secretary have declared Hizballah to be a terrorist organization, but during the period studied by Krueger and Maleckova, Hizballah could, arguably, have been termed a resistance organization (Krueger, 2007).

I performed a similar analysis on members of Hamas and Palestinian Islamic Jihad (PIJ)—organizations that are, and were during the period studied, on the U.S. State Department list of terrorist organizations, and for good reason. In August 1988, Hamas published the Islamic Covenant, in which it declared jihad (holy war) against Israel, with the stated purpose of destroying Israel and creating a Palestinian state between the Mediterranean Sea and the Jordan River.[9] Since then, Hamas has taken responsibility for the deaths of more than 500 Israeli civilians and soldiers in addition to thousands of injuries and tens of thousands of mortar shell attacks against Israeli cities.[10] PIJ calls for an armed Islamic war against Israel to free Palestine and create an Islamic state in place of Israel. During its existence, PIJ has claimed responsibility for over 150 Israeli deaths and more than 1,000 injuries.[11]

I have been able to collect and translate information from the biographies of 335 Palestinian terrorists. Of these, 285 came from a representative sample of operational terrorists. To find these data, I tracked down Shahid ("martyrs") publications from Web sites and online journals of Hamas and PIJ between 1987 and early 2002. I turned these translations into a dataset and then combined it with data on more than 40,000 Palestinian males ages 15 to 56 obtained from the "Labor Force Surveys in Judea, Samaria and Gaza."[12] These data are described in greater detail in Berrebi (2007). The data, to be sure, have serious limitations. Most of the deceased terrorists died as part of a terrorist attack, but some died as a result of Israeli-targeted assassinations. Since targeted terrorists are presumably of higher rank, and thus of higher income or education, the results might suffer from a bias that would be introduced by the overrepresentation of relatively better-off terrorists. To evaluate this potential bias, the study repeated all tests using only the 157 observations in which it was clear from the biography that the deaths were as part of planned terrorist attacks. The results were identical in sign and statistically significant.

Since information at that time was available only for those hailed as "martyrs," it included only dead terrorists who had been able to successfully execute their missions. It did not include terrorists who had failed or who had been caught.[13] It is reasonable to suspect that successful terrorists will also be abler terrorists, potentially not representative of the entire population of terrorists, and therefore the results could not be generalized beyond successful terrorists.[14] Reporting bias was a legitimate concern, as was the fact that in most cases the poverty status of terrorists was inferred from descriptors indicating wealth, whereas the population data provided information about earnings rather than accumulated wealth. Despite the limitations, these data are informative. First, summary statistics revealed that 31 percent of the Palestinians, compared with only 16 percent of the terrorists, were considered impoverished.[15] Second, out of 208 observations in which information about the terrorist's education was available, 96 percent of the terrorists had at least a high school education and 65 percent had received some higher education, compared with 51 percent and 15 percent, respectively, in the Palestinian population of the same age, sex, and religion. I used these data to estimate a logistic equation to model participation in Hamas and PIJ, controlling for several factors simultaneously. The results from the simple summary statistics held up in the more-sophisticated analyses. Namely, both higher education and standard of living appear to be *positively* associated with membership in terror organizations, such as Hamas or PIJ, and with becoming a suicide bomber (Berrebi, 2007).

In a later study, Efraim Benmelech and I were able to obtain detailed information on all suicide attacks by Palestinian terrorists against Israeli targets in Israel, the West Bank, and the Gaza strip between September 2000 and August 2005. The information, collected from reports provided by the Israeli Security Agency, was culled into a dataset that covers 151 suicide bombing attacks carried out by 168 suicide bombers. These attacks killed 515 Israelis and injured 3,428. More important, the data also contained detailed information about failed attacks. As before, we reaffirmed that suicide bombers were on average more educated than the general Palestinian population (Benmelech and Berrebi, 2007). These data, however, suggested lower estimates than the ones estimated in Berrebi (2003, 2007). Since previous data did not include suicide bombers who were caught or failed in their mission, or suicide bombers who did not succeed in killing others—and who tend to be less educated than those who succeed in their killing missions—we suspect that selection bias may be the main reason for these differences in the estimates of education among suicide bombers. Table 5.1 reports the name, age, education, and terror organization affiliation of the top five suicide bombers, measured by the number of people they killed and injured in their attacks, and provides detailed information about the date, location, and number of casualties. Three of the top five suicide bombers had academic degrees, two were master's candidates, and one had a degree in law.

Another potential explanation for the difference in the magnitude of the estimates is that Berrebi (2003, 2007) uses data on suicide bombing attacks between 1993 and early 2002, and it is possible that during

Table 5.1
Top Five Palestinian Suicide Bombers, 2000–2005

Name	Age	Education	Organization	Attack Date and Location	Number Killed	Number Injured
'Abd al-Baasit 'Awdeh	25	High school	Hamas	3/27/2002 Netanya	29	144
Raa'id 'Abd al-Hamid 'Abd al-Razzaaq Misk	29	Master's candidate	Hamas	8/19/2003 Jerusalem	23	115
Sa'eed Hasan Husayn al-Hutari	22	High school	Hamas	6/1/2001 Tel-Aviv	21	83
Hanaadi Taysir 'Abd al-Malik Jaraadaat	29	Law school graduate	PIJ	10/4/2003 Haifa	21	48
Muhammad Hazzaa' 'Abd al-Rahmaan al-Ghoul	22	Master's candidate	Hamas	7/18/2002 Jerusalem	19	50

SOURCE: Benmelech and Berrebi (2007).

the al-Aqsa intifada terror organizations faced an excess demand for suicide bombers and became less selective in their recruiting during the 2001–2005 period. In either case, we were able to confirm earlier findings that Palestinian suicide bombers are more educated than the average in Palestinian society.

A new study about human capital and participation in domestic Islamic terrorist groups in the United States is based on a comparative analysis of the characteristics of 63 alleged domestic Islamic terrorists, who were indicted or convicted for involvement in terrorist activities, with those of the population of Muslims residing in the United States (Krueger, 2008). The study reveals that the alleged terrorists were somewhat better educated and younger, on average, than the general population of Muslim Americans. They were about as likely to be idle (neither working nor enrolled in schools) as were other American Muslims and overall did not appear especially deprived.

In summary, the preexisting literature, whether relying on biographical interview information, case studies, or more sophisticated econometric models analyses of the comparative population, typically agrees in its findings with respect to the socioeconomic status and education of individual terrorists. Namely, terrorist are rarely characterized by poverty or lack of education.

Evidence about individual terrorists does not necessarily indicate that poor economic conditions are not a source of terrorism. It could well be argued that poor macroeconomic conditions are the drivers behind the choice to engage in terrorism. Under this hypothesis, individuals can become terrorists because of poverty in their country, even if they are not themselves impoverished. Evidence to that effect would align closely with the literature on conflicts and civil wars.[16] However, the literature on terrorism typically suggests that macroeconomic conditions have little if anything to do with the amount of terrorism produced by countries.

In a study of terrorist incidents and casualties in 96 countries from 1986 to 2002, Piazza (2006) considers the significance of poverty, malnutrition, inequality, unemployment, inflation, and poor economic growth as predictors of terrorism, along with a variety of political and demographic control variables. This study's findings are that, contrary to popular opinion, no significant relationship between any of the measures of economic development and terrorism can be determined. Rather, such variables as population, ethno-religious diversity, increased state repression, and, most significantly, the structure of party politics are found to be significant predictors of terrorism.

Similarly, Abadie (2006) uses country-level data on terrorism risk from the World Market Research Center's Global Terrorism Index (WMRC-GTI); this covers 186 countries in 2003 and 2004 and studies the effect of poverty, measured by gross domestic product (GDP) per capita, on the intensity of international and domestic terrorism combined. Terrorist risk ratings have obvious limitations. They provide only a summary measure of an intrinsically complex phenomenon. However, they have the advantage of reflecting the total amount of terrorism intensity for every country in the world (Abadie, 2006). The empirical results of a regression analysis, using instrumental variables to correct for reverse causation, show that terrorism risk is not significantly higher for poorer countries once country-specific characteristics have been controlled for.

The unit of observation in these studies is the country in which the terrorist attack occurred (or was expected) rather than the country from which the terrorists originated. Arguably, economic conditions in the country of origin should be of greater importance to the terrorists and their organizations.

Krueger and Laitin (2008) examine the link between macroeconomic conditions and terrorism by looking not only at the target country but also at the attackers' country of origin. The analysis in this study relies on two datasets. The first dataset contains information on 781 worldwide significant events that, according to the U.S. Department of State's annual report, *Patterns of Global Terrorism,* occurred between 1997 and 2003. The second dataset contains information on 236 recorded suicide attacks in 11 countries since 1980. Variables describing the country, such as GDP per capita, GDP growth, and measures of terrain, religious affiliation, and literacy, were added to the data based on either the country of origin or target country. Using a myriad of econometric models and specifications, the study concludes:

The most salient patterns in the data on global terrorism that we presented suggest that, at the country level, the sources of international terrorism have more to do with repression than with poverty. The regression analysis showed that neither country GDP nor illiteracy is a good predictor of terrorist origins. . . . Thus terrorist perpetrators are not necessarily poor. But those who are repressed politically tend to terrorize the rich, giving international terrorist events the feel of economic warfare. Suicide attacks reveal much less on the interstate level. . . . in the suicide dataset, we see as with international terrorism, the origins are more likely to be in countries that deny civil liberties. . . .

In his book, *What Makes a Terrorist*, Alan Krueger (2007), after reviewing the macro evidence on terrorism, at the society or country level, concludes:

Education and poverty probably have little to do with terrorism. There are many reasons for improving education and reducing poverty around the world, but reducing terrorism is probably not one of them. (p. 90)

Later, when concluding the analysis of the national origins of foreign insurgents in Iraq, Krueger adds:

Economic circumstances in the countries of origin of foreign fighters do not seem to be particularly important predictor variables. . . . (p. 103)

. . . the occurrence of terrorism is mostly unrelated to GDP in the origin country and positively related to GDP in the target country. . . . (p. 104)

Similarly, Krieger and Meierrieks (2008), who reviewed the existing evidence based on 26 cross-country macroeconomic studies for which they assessed the influence of economic, political, demographic, international and geographic factors, concluded:

. . . no convincing evidence is found that economic factors— for example, economic growth, poverty, income disparity or the like—are closely connected to terrorism. Richer countries only seem to be more often targeted by transnational terrorism. . . . Additionally, higher levels of education or democratic or political system do not guard effectively against terrorism. . . . From our review, the most important determinants of terrorism are found to be political and demographic but not of economic nature.

The above-mentioned studies rely on cross-country analyses and examine the effect of macroeconomic conditions on the amount of terrorism at the state or country level. Arguably, cross-country studies have serious limitations. The underlying assumption in such analyses is that changes in countries' economic conditions share a common effect on the quantity of terrorism that those countries produce, once all the observed country characteristics have been accounted for.[17] However, this assumption is implausible because many features of individual countries cannot be feasibly controlled for in a multivariate regression analysis framework. These relate to, for example, underlying institutions; social, cultural, or psychological sensitivity to economic conditions or to violence in general and terrorism in particular; and to variations in how economic activity and terrorism activities are classified and reported.

Because of such concerns, I—together with my coauthors Efraim Benmelech and Esteban Klor—studied the effect of macroeconomic conditions on suicide terrorism at the regional level.[18] We used quarterly district economic data from the Palestinian Labor Force Surveys for 2000 to 2005 and merged these data with district data on suicide terrorism for the same period, employing additional control variables to account for grievances and local counterterrorism efforts. We were then able to asses whether economic conditions have an effect on the number of suicide terrorists originating from each district. Our findings, although still preliminary, are representative of the cross-country studies' findings mentioned above: notably, that any link between wages or unemployment and the number of suicide attacks is either weak or nonexistent.

Religion

According to the Merriam-Webster dictionary the word *religion* can be interpreted in several ways. The first is "the service and worship of God or the supernatural"; a second is "a personal set or institutionalized system of religious attitudes, beliefs, and practices;"[19] the third is "a cause, principle, or system of beliefs held to with ardor and faith." According to the third definition, communism and extreme nationalism could well be defined as religions, as could other secular systems of beliefs. For this analysis, and to be able to draw from the economics of religious sects as portrayed by Berman (2000, 2003), I have chosen to restrict discussion to the main organized religions, following the spirit of the first and second definitions presented above.

It would be out of scope for this section to deal with the complexity of the potential interlinkages between terrorism and religion; the intention here is rather to provide the reader with a brief overview of studies that attempted to examine the correlation between observed religions and terrorism. Other perspectives regarding the effects of religion are provided in Cragin (2009).

Examining the micro, individual-level, data Krueger and Maleckova (2003) find that none of the largest religious affiliations[20] seem to be distinctively prone to terrorism. Similarly, when studying the macro, country-level, terrorism data, Krueger and Laitin (2008) find that:

> We cannot reject that the same shares affiliated with the various religions jointly have no effect on terrorism at any of the levels of analyses. No religion appears to have a monopoly on terrorism; countries with very different religious faiths have all experienced terrorism, as target, origins, and hosts.

This evidence shows that no specific religion is more linked to terrorism than other religions. However, it does not indicate that religions play a less significant role than secularism.

In his 2005 book, *Dying to Win*, Robert Pape (2005) studied 315 suicide terrorist attacks from 1980 to 2003. He found little connection between suicide terrorism and Islamic fundamentalism or any one of the world's religions. Pape's claim relies mainly on the Tamil Tigers, which he describes as a group influenced by a Marxist/Leninist ideology,[21] which is largely atheistic and disavows any connection with the Hinduism practiced by many of the people in the region of Sri Lanka where this group operates. The Tamil Tigers were responsible for more suicide attacks over the studied period than any other group.

Similarly, after his 2004 study of al-Qaeda–affiliated individual terrorists, Sageman (2006) clarifies:

> In my sample, only 13 percent of terrorists went to madrassahs, and this practice was specific to Southeast Asia, where two school masters, Abdullah Sungkar and Abu Bakar Baasyir, recruited their best students to form the backbone of the Jamaah Islami- yah, the Indonesian al-Qaida affiliate. This means that 87 percent of terrorists in the sample had a secular education. . . . The vast majority of al-Qaida terrorists in the sample came from families with very moderate religious beliefs or a completely secular out- look. Indeed, 84 percent were radicalized in the West, rather than in their countries of origin. Most had come to the West to study, and at the time they had no intention of ever becoming terrorists. Another 8 percent consisted of Christian converts to Islam, who could not have been brainwashed into violence by their culture.

In conflicts where both secular and religious organizations engage and compete in the amount of terrorism they produce,[22] as is the case for Palestinian terrorism, attacks tend to originate equally from both. Figure 5.1 provides a breakdown of suicide attacks by terror orga- nizations from 2000 to 2006. Notably, the share of suicide attacks initiated by religious organizations (such as Hamas and PIJ) is only slightly greater than the share perpetrated by the remaining secular organizations.

Feldman and Ruffle (2008) analyzed 23,360 domestic ter- rorism attacks between 1998 and 2007. They find that religious terror groups actually carry out fewer attacks on average than do groups of other ideologies (for example, nationalist and communist). However, aside from the Tamil Tigers,[23] the remainder of the five deadliest terrorist organizations currently in operation are all radi- cal Islamists (that is, al-Qaeda, Hizballah, Taliban, and Hamas).[24] Moreover, religious groups claim at least as many victims as non-

Figure 5.1
Suicide Bombing Attacks by Terror Organizations, 2002–2006

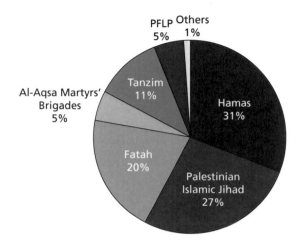

SOURCE: Benmelech, Berrebi, and Klor (2008).
RAND *MG849-5.1*

religiously motivated attacks for almost all tactics, not just suicide bombings as commonly perceived.[25]

Krieger and Meierrieks (2008) in their review of the terrorism cross-country analyses literature conclude:

> Although religion in popular discourse has been suggested as an important determinant of terrorist activity, empirical evidence tells a different story. . . . the nature of this linkage does not appear to be clear, as both a negative and positive connection between spiritual ideology and terrorism can be detected.

Dealing with religious terrorism can be confusing, since it is difficult to know whether terrorist organizations, which on the face of it are considered religious, are using religion to attract an audience while primarily motivated by secular goals. Further confusing to the outside observer is when political goals are claimed by terrorist organizations in the name of religion, despite the fact that religion was not at the source of these claims. Clearly, it is possible to find religiously motivated terrorists and terror organizations, and there is indication that religious terrorist organizations are potentially more effective in recruiting operatives, particularly committed volunteers for suicide missions. However, relying solely on religion-based explanations to study terrorism in general would leave a hole in our ability to understand the behavior of the many secular terrorists.[26] Despite the evidence of the increased lethality of religious terrorism, further research on the link between terrorism and religion is warranted.

Mental Health and Irrationality

Mental health is crucially important in evaluating whether rational-choice behavior is a good model. If terrorists were disproportionately mentally ill, there would be no point in searching for indications of rational behavior or in using rational-choice theory to analyze such behavior. In such a case, evidence about the characteristics of terrorists that seemingly contradicted potential rational-choice explanations would not be puzzling. On the other hand, if we were to find out that terrorists, including suicide terrorists, are not typically mentally ill, we would be compelled to continue our search for better explanations, keeping in mind that costly behavior does not equal crazy behavior.[27] Therefore, I have searched for evidence regarding the mental health of terrorist operatives.[28]

Martha Crenshaw (1981) has concluded from her studies that:

> No single motivation or personality can be valid for all circumstances. What limited data we have on individual terrorists . . . suggest that the outstanding common characteristic of terrorists is their normality.

Ariel Merari, a psychologist who has studied the psychological profiles of suicide terrorists since 1983 through media reports that contained biographical details, interviews with the suicides' families, and interviews with jailed would-be suicide attackers, concluded that they were unlikely to be psychologically abnormal (Merari, 2006). Hudson and Majeska (1999) also suggest that the-terrorists-as-mentally-ill approach appears to be contradicted (pp. 20–21).

Similarly, in a study of suicide terrorism, Scott Atran (2004) finds that:

> Overall, suicide terrorists exhibit no socially dysfunctional attributes (fatherless, friendless, jobless) or suicidal symptoms. Inconsistent with economic theories of criminal behavior, they do not kill themselves simply out of hopelessness or a sense of having nothing to lose.

Marc Sageman (2006) finds a near-total lack of mental disorders in his sample of al-Qaeda–affiliated individual terrorists. He explains that this makes sense, as individuals with mental disorders are usually weeded out early from any clandestine organization for security reasons.

Anat Berko, a criminologist and colonel in the IDF who studied the inner world of suicide bomber terrorists through a series of prison interviews she conducted with 'would-be' suicide bombers whose mission was foiled either directly by the IDF or by some technical failure in the mechanism of the explosives they were carrying, noted (Berko, 2007):

> . . . many of the suicide bombers do not have financial difficulties . . . not only do they generally not have economic problems, but most of the suicide bombers also do not have an emotional disturbance that prevents them from differentiating between reality and imagination. . . . (p. 9)

In their work on the psychology of terrorism, Kruglanski and Fishman (2006), reach similar conclusions:

> Terrorists do not seem to be characterized by a unique set of psychological traits or pathologies. . . . The vast heterogeneity of terrorism's users is consistent with the "tool" view, affording an analysis of terrorism in terms of means-ends psychology. The "tool" view implies conditions under which potential perpetrators may find terrorism more or less appealing. . . .

In an article that reviews the state of the art of available theories and data regarding the psychology of terrorism and relies on data and theoretical material gathered from the world's unclassified literature, Victoroff (2005) concludes that terrorists are psychologically extremely heterogeneous. He explains that whatever the stated goals and group identity, every terrorist, like every person, is motivated by his own complex of psychosocial experiences and traits. I interpret his conclusion to mean that we should not expect terrorists to be disproportionally insane.[29]

Summary

In summary, individual terrorists do not fit the profile of poor, ignorant, or religious individuals with low opportunity cost and no valued marketable skills; nor are they mentally unstable. The various "root causes" that have long been discussed may well be at work, but in complicated and sometimes nonintuitive ways, and apparently not in decisive ways. Other explanations appear to be needed. The next sections of the paper discuss whether the economist's rational-choice model, suitably adapted, may be more appropriate.

Rational-Choice Approach

Defining Terms: What Is a Rational-Choice Model?

"Rationality," as that term is used here, is based on rational-choice theory, which serves as a framework for understanding and often modeling social and economic behavior. It is the dominant theoretical paradigm in microeconomics and is also central to modern political science.[30] However, even within these narrow guidelines, one could distinguish between at least three levels of rationality. In the weakest sense, all actions are rational so long as the individual is using them to achieve predetermined ends. A stronger definition requires that individuals choose the *best* action according to stable utility functions[31] and the constraints facing them.[32] Finally, an even stronger definition of rationality requires that individuals respond to incentive[33] and behave according to rational expectations (that is, the individual's beliefs are correct on average).[34] In all of these cases, of course, choices and behaviors may prove ineffective because of erroneous information or perceptions, lack or information, or unpredictable complexity in the external world—the ingredients of what Nobelist Herbert Simon called "bounded rationality" (Simon, 1982) the result of which may be to settle for solutions that appear to be "good enough" whether or not truly the best.

Of interest to us is to what extent terrorists, including suicidal terrorists, satisfy the stronger definitions of rationality. Arguably, in the case of terrorism research in general and terrorists' behavior in particular, one needs to allow for a flexible form of utility function that could include satisfaction from perceived altruism and intangible psychological or social rewards, including expected rewards in the afterlife.[35]

The main argument favoring a rational-choice model is that, if terrorists and terror organizations behave rationally, knowledge of their beliefs and preferences should help us understand and predict their behavior. However, if they are irrational, their behavior cannot be explained through rational-choice models, and no systematic trends based on these models should be observed or sought.[36]

Are there any indications to suggest that terrorists and their organization behave rationally? To be sure, before searching for rational-choice explanations, it would be useful to observe behavior that suggests, or at least anecdotally supports, economic rational decisionmaking on the part of terrorists and their organizations. In looking for evidence for the rational-choice model, we should look at several levels of organization. Often, this is thought of as the level of individuals versus groups, but we can think also of tactical, operational, and strategic levels. Rational behavior might well exist at some, but not other, levels. For example, terrorists are often pragmatically risk-averse in conducting operations, even if the rationality of their overall strategy is questionable.[37]

Evidence

Let me consider evidence separately for tactical- and operational-level issues, and strategic issues.

Tactical- and Operational-Level Rationality. *Reasonably Chosen Targets.* At the group level, the evidence tends to support economic rational decisionmaking. For example, Darius Lakdawalla and I used comprehensive terrorism data from the Israeli-Palestinian conflict between 1949 and 2004 to study spatial and temporal determinants of terrorism risk (Berrebi and Lakdawalla, 2007). Specifically, we used detailed data about the exact location and timing of all fatal attacks against Israeli civilian targets[38] and merged those data with information about the targeted localities. We then explored how the spatial risk of terrorism differs with measures of target value and attack cost and analyzed the spacing, or the "waiting time," between terrorist attacks in a given locality. Doing so, we were able to assess whether or not terrorists behave rationally when they decide *which* targets to attack most often and whether there was an empirical pattern in terrorists' decisions about *when* to attack.

Four factors stand out as key determinants of spatial variation in risk: proximity of terrorist home bases, proximity of international borders, the presence of a Jewish population, and the presence of a center of government administration.[39] The first two probably improved access for terrorists and lowered the cost of attack; the latter two probably raised the expected benefit of attacks in the eyes of terrorist groups. Our analysis indicates that when distance to a terrorist home base doubles, the frequency of attacks falls by around 30 percent. International border localities are more than twice as likely to be hit. Areas with a Jewish population are three times as likely to be hit as other areas, as is Jerusalem and localities with a regional capital. It would seem, then, that attacks are hardly random: They are chosen for a combination of target attractiveness, feasibility, effectiveness, and cost.

Explainable Attack Timing. The analysis of attack timing also leads to several important conclusions. First, in the wake of a terrorist attack, the risk of a subsequent attack climbs in the hours following it and peaks the following day. After that point, risk decays for eight weeks. In fact, if a locality survives for eight weeks without an attack, it returns to its low, preattack risk level. That is, localities that have experienced an attack within the past eight weeks are at greater risk of an attack than other localities, but after eight weeks, their risk is no longer elevated. It is interesting to note that although this subsidence of risk occurs on average, patterns are very different for politically sensitive localities that are seats of government. For such localities, risk subsides within the first eight weeks but then begins a noticeable climb upward: Apparently, terrorists are not content to leave such high-profile areas untouched, even though they may choose to do so for less-attractive cities. The analysis of waiting time between attacks experienced by localities is consistent with a reasonable interpretation of the benefit value seen by the terrorists.

Reasonably Chosen Attack Tactics. Berman and Laitin (2005) explain that terrorists use suicide tactics primarily against "hard targets" against which the probability of apprehension is high using a conventional attack technology and targets are well protected—reducing significantly the expected success of a conventional attack, which altogether indicates a clear calculus in the terrorist's choice of attack tactics and targets.

Recognition of Human Capital Considerations in Suicide Bombing. Perhaps more indicative of tactical rationality are the findings from a recent study (Benmelech and Berrebi, 2007). In this study of the relationship between the human capital of suicide bombers and the outcomes of suicide attacks, we used, as noted above, detailed biographical data on 151 suicide bombing attacks carried out by 168 suicide bombers in Israel, the West Bank, and Gaza. The data contained detailed information about the characteristics of the attackers and also about the targets they were assigned. We were able to estimate which characteristics were likely to increase the productivity of terrorists and whether terrorist organizations seemed to be using these characteristics in assigning terrorists to targets, as projected based on rational expectations. In other words, we identified the characteristics that statistically increase the ability of individual terrorists to kill or injure and statistically decrease their probability of getting caught or failing in their attack mission. We then analyzed the characteristics of those sent to the most-valued and lucrative targets.[40] We found that the two key explanatory variables were the academic background and age of the suicide bomber. Both education and age indicate ability and experience. First, in terms of performance, we found that suicide bombers who had more than a high school education were 56.4 percent less likely to be caught, relative to the sample mean. An additional year of age is associated with a decrease of 17.6 percent, relative to the sample mean, in the probability of being caught. Similarly, older and better-educated suicide bombers, when assigned to more important targets, were more effective killers. For example, an educated suicide bomber killed roughly four to six more people when attacking a large city.

Given these results, rational-choice theory would suggest that terrorist organizations should assign their older and more-educated terrorists to attack larger, more-important, lucrative, civilian targets. Indeed, analyzing the connection from higher-ability suicide bombers to more important targets, we find that the effect of one year of age is large and represents an increase of 4 percentage points in the probability that a suicide bomber will be assigned to a target in a large city. In terms of economic magnitude, this coefficient implies that a 25-year-old suicide bomber has a 28 percentage point higher probability of being assigned to a target in a large city (representing an increase of 53.1 percent relative to the unconditional mean) than an 18-year-old suicide bomber. Similarly, educated suicide bombers are 62.8 percent less likely, compared with the unconditional mean, to be assigned to military targets. In short, assignment of terrorists to targets is statistically unlikely to be random. To the contrary, terrorist organization seems to behave rationally, since they do take into account their success, performance probabilities, and target values when considering assignments of terrorists to targets.

These cases strongly suggest short-term, tactical (and operational) rationality. I will next discuss the available evidence about organizations' behavior with respect to their long-term, officially stated goals.

Strategic-Level Rationality. *Economic Warfare.* According to a videotape of Osama bin Laden, released to the Arabic-language network Al-Jazeera on November 1, 2004, the head of al-Qaeda said that his group's goal is to force America into bankruptcy (CNN.com, 2004). As part of the "bleed-until-bankruptcy plan," he cited a British estimate that it cost al-Qaeda about $500,000 to carry out the attacks of September 11, 2001, an amount that he said paled in comparison with the costs incurred by the United States. In this example, it seems that al-Qaeda's leader behaved in a rational and calculated fashion and was extremely successful in the pursuit of his goal. After all, he claims to have forced the United States into implementing expensive counterterrorism measures that affected its entire economy and into pursuing a war in Afghanistan while spending significant amounts to help Pakistan capture terrorists on its ground. Some might even argue that the war in Iraq was a reaction to the aftermath of bin Laden's September 11, 2001, attack against the United States.

Accordingly, if we believe that an ultimate goal of terrorist organizations is to maximize economic hardship on its enemies, we should observe that attacks cause a serious, maybe even disabilitating, cost on the targeted economies, or at least a disproportionally higher cost than that incurred by the terrorist organization in organizing the terrorist attack or campaign. And, indeed, the evidence seems to mostly support this hypothesis. Abadie and Gardeazabal (2003) were probably the first to convincingly estimate the economic effects of terrorism, using Euskadi Ta Askatasuna (ETA) terrorism in the Basque region of Spain as a case study. They find that, after the outbreak of terrorism in the late 1960s, per capita GDP in the Basque region declined by about 10 percentage points relative to a "synthetic control region" without terrorism.[41] Eckstein and Tsiddon (2004) analyzed the effect of terrorism on consumption, investment, exports, and GDP per capita in Israel. They concluded that if Israel had not suffered from terrorism between 2000 and 2003, its GDP per capita would have been 10 percent higher than its actual level. In another study that empirically assessed the effect of terrorism on the stock-market valuation of Israeli companies that are traded in American markets, Esteban Klor and I find that, although the effect differed across industries, terrorism had a significant negative effect overall of 5 percent on nondefense-related companies.[42] We use data on Israeli and matching U.S. stocks that were traded on Amex, the New York Stock Exchange, and Nasdaq.[43] We collected daily end-of-the-day share prices for the sample period between January 1, 1998, and September 10, 2001,[44] and merged the data with daily terrorism data for this period.[45] We then employed an event study approach and estimated the divergence of the abnormal returns between Israeli and matching

U.S. stocks to quantify the effect of terrorism on stock returns. The magnitude of the losses caused by terrorism is on the order of $84.6 million in market capitalization for the average Israeli company not related to the defense sector, as measured in July 2007 (Berrebi and Klor, 2005, 2009). Similarly, Karolyi and Martell (2007) looked at the consequences of terrorism on targeted publicly traded firms, such as Royal Dutch Shell, British Petroleum-Aamoco Corp., Coca-Cola, McDonalds, and American Airlines.[46] Overall, they identified 75 attacks between 1995 and 2002 in which publicly traded firms were targeted and performed an event study to uncover a significant 83 basis point decline, which constitutes an average loss in market capitalization per firm per attack of $401 million. Is this enough to handicap an economy? Probably not, but it certainly could cause a significant economic hardship and the economic consequences would be several orders of magnitude greater than the cost of perpetrating the attacks. One could argue that terrorism has only a small effect on the economy, particularly when compared with the effect of external wars or natural disasters.[47] However, it is important to keep in mind that it is far less costly to perpetrate a terrorist campaign than to wage a war. To summarize, it seems that we can find evidence for a relatively significant effect of terrorism on the economy, although most can be attributed to psychological reactions in the aftermath of the attacks, rather than to the actual damages caused by the attacks.[48] From the point of view of terrorist organizations, the economic effect of terrorism reasonably supports rational-choice behavior, since it achieves serious "bang for the buck."

Pursuit of Territorial and Liberation Goals. Also shared by many terrorist organizations are territorial goals.[49] Territorial goals are often termed "liberation of territories from occupation." These goals typically reflect sincere beliefs or perceptions that territories that they believe to have rightful historical or religious claims on are subject to occupation. At times, however, an organization will make such claims in a deliberate manipulation intended to attract support from a targeted audience. In either case, it is relatively easy to find examples of terrorist organizations seeking territorial gains.

Following are some examples.[50]

- The charter of the Palestinian terrorist organization Hamas clearly has as a goal—what it perceives as the liberation of the land of Palestine.[51]
- The Al-Aqsa Martyrs' Brigade (AAMB), an armed Palestinian terrorist faction composed of Fatah-affiliated "Islamic Nationalists," has set as an objective the establishment of an independent and sovereign Palestinian state and an end to the occupation of what it sees as occupied Palestinian territories.
- Hizballah, as can be derived from its February 16, 1985, political manifesto,[52] includes among its goals the removal of all Western influences from Lebanon and from the Middle East, as well as the destruction of the state of Israel and the liberation of all Palestinian territories and Jerusalem from what it sees as Israeli occupation.
- Jaish-e-Mohammed (JeM), a Pakistan-based radical Islamist terrorist organization, advocates the liberation and subsequent integration of Jammu and Kashmir from Indian control into Pakistan.
- The Liberation Tigers of Tamil Eelam (LTTE), a Sri Lanka–based terrorist organization, advocates what it sees as the liberation of its homeland in the north and northeastern part of Sri Lanka, which it has called "Tamil Eelam."
- The ETA is a Basque terrorist group with the goal of liberating the Basque homeland region from what its members perceive as Spanish occupation.
- The Kurdistan Workers' Party (PKK) is a Kurdish terrorist organization that has a goal of liberating Kurdistan, an area that comprises parts of southeastern Turkey, northeastern Iraq, northeastern Syria and northwestern Iran, from what its members perceive as foreign occupation.

Similarly, as will be discussed in greater details below, terrorist activities regarding political goals and their influence on public opinion and electoral outcomes are yet another indication of rational-choice behavior.

Does Terrorism as a Strategy Work? With so many terrorist organizations sharing territorial claims, it should be possible, at least anecdotally, to document territorial concessions in response to terrorist campaigns.[53] Perhaps most convincing is a study of 188 suicide terrorist attacks worldwide from 1980 to 2001 (Pape, 2003),[54] which concluded:

> This study shows that suicide terrorism follows a strategic logic, one specifically designed to coerce modern liberal democracies to make significant territorial concessions. Moreover, over the past two decades, suicide terrorism has been rising largely because terrorists have learned that it pays. Suicide terrorists sought to compel American and French military forces to abandon Lebanon in 1983, Israeli forces to leave Lebanon in 1985, Israeli forces to quit the Gaza Strip and the West Bank in 1994 and 1995, the Sri Lankan government to create an independent Tamil state from 1990 on, and the Turkish government to grant autonomy to the Kurds in the late 1990s. In all but the case of Turkey, the terrorist political cause made more gains after the resort to suicide operations than it had before.

It is important to note that some terrorism researchers maintain that terrorists do not, on average, achieve their ultimate objectives (Abrahms, 2006) and accordingly challenge the rational terrorist thesis (Abrahms, 2004),[55] claiming that, "Terrorism has a habit of eliciting the opposite of the intended policy response."[56]

Nobel laureate Thomas Schelling, in his work on international terrorism, suggested that, although terrorists frequently accomplish intermediate means toward political objectives, they fail to achieve long-term objectives (Schelling, 1991).[57]

Allegedly, these last arguments cast doubt in the rational behavior of terrorist organizations.[58] However, when evaluating these claims, one should remember that terrorist organizations have different long-term goals[59] and each is likely to have several internally competing goals.[60] Therefore, it is difficult to estimate the extent to which they are successful in achieving their goals.[61] Moreover, as outside observers, the ability to assess whether terrorists achieve their goals can be established only if we know what those goals are. Nevertheless, we anecdotally observe terrorist organizations pursuing short-run and long-run objectives along cost and benefit considerations that directly influence their activities, which enables us to argue more comfortably in support of rational-choice behavior despite the inherent lack of predictive power. Counterterrorism expert Boaz Ganor suggests in his book, *The Counter-Terrorism Puzzle* (2005), that:

> In general, terrorist organizations usually conduct rational considerations of costs and benefits, but they often attribute different weight to the values taken into account in their cost-benefit calculations, and occasionally, may even consider values that are different from those of the ones coping with terrorism, thus making a decision that appears irrational to an outside observer. In most cases, though, the leadership of a terrorist organization will not make a decision whose cost is perceived to outweigh its benefits, that is, an irrational decision.

According to the evidence from the previous discussions, it is likely that terrorist organizations respond to incentives and more often than not use rational expectations in their calculations—thus conforming to the behavior expected in the rational-choice theory.

Ability to Explain Puzzles? Is there any rational-choice explanation to the seemingly contradictory evidence with respect to an individual's characteristics? Recall that individual terrorists tend to be wealthier and better educated than the population from which they are drawn, not particularly religious, and most likely in good mental health.

One possible explanation is that terrorists, although initially in good mental health, are unwittingly "programmed" or "brainwashed," potentially even through the educational system that is controlled or influenced by the leaders of terrorist organizations. If this is the case, we should observe rational behavior at the organizational level; however, it would suggest that terrorists are unlikely to follow consciously calculated behavioral choices. On the contrary, they would be unwittingly manipulated into joining the terrorist organization. This hypothesis, if true, would explain why educational attainment potentiates terrorist activity or is potentiated by it. Accordingly, the educational system is used to brainwash potential terrorists. Educational content that advocates particular political or religious messages would therefore increase an individual's propensity to join terrorist organizations and participate in terrorist activity, encouraging radical thought while only on the margin increasing productive opportunities in the labor market.

Although educational content is likely to be a factor influencing individual behavior, past experience suggests that we should be extremely careful of prematurely adopting an explanation that relies on the absence of free will.[62]

Another possible explanation is that terrorism is a high-skill occupation.[63] As such, individuals who want to volunteer must first have the necessary skills and show their ability to commit. If so, causality could be reversed. Accordingly, it is not those who are highly educated and hold lucrative jobs who disproportionally want to join terrorist organizations but rather that those who are initially interested in joining terrorist organizations must get more education and show their ability to hold a job, in an attempt to become an active terrorist. The limitation of this argument is that it requires that individuals decide relatively early in life that they want to become terrorists, as investments in education and job market skills are acquired from early age.[64]

What Might Rational-Choice Models Look Like?

It has been argued that we need more creative approaches to rational-choice behavior models in ways that move from the straightforward notion of self-interest into a notion of self-interest that is driven from pride, dignity, self-respect, or recognition (Varshney, 2003). These approaches can help us understand participation in terrorist activity. I am inclined to borrow from the concept of "value rationality" first proposed by Max Weber (1978). However, I consider an explanation to be within the framework of rational-choice theory only if a cost-benefit calculus can be applied, necessitating the abandonment or adjustment of goals if the costs of realizing them are too high, even if the goals consist of or are driven by ethical, aesthetic, religious, or other beliefs and the costs and benefits are measured in terms of satisfaction, suffering, hardship, or discomfort. Nevertheless, it seems to me that various alternative explanations based on rational-choice theory, as defined here, could support the literature's empirical individual-level and country-level findings.

Altruism. One such creative approach is provided in an intergenerational model in which the current generation is linked to the next one by some altruism, as in standard dynastic family models, and terrorist attacks in the current period increase the probability of the benefit of some public good accruing to the next generation. According to this model, although above-average education and wealth are expected to increase the opportunity cost of participation in terrorism, and in particular in a suicide attack, it is suggested that it probably also increases the sensitivity and feeling of responsibility to the future generation's welfare.[65] The latter effect might offset the deterrent effect of the former (Azam, 2005). The limitation of this explanation is that it relies on factors that are difficult to empirically observe or measure, such as altruism.[66]

Demand Versus Supply. So far, I have considered only the supply side of this equation (that is, the willingness of individuals to engage in terrorist activities). Suppose that differential participation of wealthier or better-educated individuals in terrorist activities was not a matter of differential motivation so much as choice on the part of terrorist organizations (that is, the terrorism market is mostly driven by demand-side forces). Such organizations may be faced with an excessive supply of potential participants and might therefore choose the select few they desire. Consequently, it may be that the terrorists selected by these groups are highly educated and in good socioeconomic status even though, on average, the education and wealth of those willing to join such organizations may be no greater than average. Ethan Bueno de Mesquita (2005) has developed a theoretical model of the interaction between a government, a terrorist organization, and potential terrorist volunteers in which, as a result of an endogenous choice, individuals with low ability or little education are most likely to volunteer to join the terrorist organization. However, the terrorist organization screens the volunteers for quality and, as a consequence, actual terrorist opera-tives are not poor or lacking in education.

Along similar lines, Iannaccone (2006) presents a compelling model that explains why, under certain conditions, a market for suicide terrorist attackers ("martyrs") can flourish. According to Iannaccone,

> It is on account of limited *demand* rather than limited supply that markets for "martyrs" so rarely flourish. Suicidal attacks almost never benefit the group best fit to recruit, train, and direct the potential martyrs. Once established, however, a market for mar-tyrs is hard to shut down. Supply-oriented deterrence has limited impact because: In every time, place, and culture, many people are willing to die for causes they value. . . . Demand-oriented deterrence has greater long-run impact because: The people who sacrifice their lives do not act spontaneously or in isolation. They must be recruited, and their sacrifices must be solicited, shaped, and rewarded in *group* settings. Only very special types of groups are able to produce the large social-symbolic rewards required to elicit suicide. Numerous social, political, and economic patholo-gies must combine in order to maintain the profitability of (and hence the underlying demand for) suicidal attacks.

Iannaccone's (2006) model provides a rational-choice explanation in which largely symbolic rewards provided to the individuals, and a profit obtained by the terrorist organization from the suicidal attacks, make this market possible. Similarly, the behavior of terrorists moti-vated by religion, including suicide attackers, could be persuasively explained through rational-choice models (Berman and Laitin, 2008). Berman and Laitin use "clubs," "club goods," and religion and empha-size the function of voluntary religious organizations as efficient pro-viders of local public goods to persuasively model participation in sui-cide terrorist attacks. According to this model, the sacrifices that these groups demand are economically efficient and make them well suited for solving extreme principal-agent problems in recruiting candidates for suicide attacks who will not defect. The predictions of this model are consistent with the evidence observed in the data on religious ter-rorist organizations and do not require appeal to irrationality or utter fanaticism. However, the main limitation of this model is that it does not apply to some important examples of terrorist campaigns (including suicide terrorist campaigns), such as the one launched by the Tamil Tigers. For one, LTTE is completely secular; second, there is no record of providing essential services to the poor to reduce members' reliance on the state; finally, the Tigers have not sent their most valuable cadres to perform suicide missions, as predicted by the model.[67] A modified model is consequently proposed by the authors. In the modified model, an alternative type of club is introduced; this club threatens the general population while protecting its members from that threat. That is, it is bad for nonmembers but it provides a local public good for members (relative to nonmembers). Rather than reducing the risk of defection by augmenting the inside options of members with benign local public goods, it reduces the risk of defection by destroying outside options of members with a pervasive public bad—the threat of attack from the club itself (Berman and Laitin, 2008). Although the combination of the two club models presented here provides a good explanation for most suicide terrorism campaigns, it leaves out examples of secular terrorist organizations, which fail to provide community services, and yet, to the best of our knowledge, do not rely on coercion for recruiting and threats to avoid defections, such as the Palestinian Al-Aqsa Martyrs' Brigades[68] or the Kurdistan Workers' Party.[69]

Another somewhat parallel explanation is that terrorists are rational people who use terrorism primarily to develop strong affective ties with fellow terrorists (Abrahms, 2008).[70] Accordingly, individuals join terrorist organizations to develop a sense of solidarity with other like-minded people.[71] The limitation of this explanation is that there seem to be many less-dangerous alternatives to develop these kinds of social bonds.

Political Activism. Perhaps most convincing is the idea that terrorism is an extreme and violent form of political activism. After all, terrorism is often defined as "the systematic use of violence to create a general climate of fear in a population and thereby to bring about a particular *political* objective" [emphasis added by author].[72] As such, we should observe that participation in terrorist activity is akin to political activism, attracting the more-educated and wealthier individuals, therefore settling the seemingly puzzling evidence described above. Accordingly, poorer individuals are more likely to be preoccupied with daily matters, such as providing for their families, and end up devoting less attention to militant struggles, and less-educated individuals are less likely to hold the convictions necessary to express opinions,[73] even more so to act on them. Political activism is as likely to be based on religion as it is to be rooted in secular ideologies, and politically active individuals are not expected to be mentally ill, both matching the micro-level evidence about terrorist operatives. An explanation based on the premise that terrorism is an extreme version of political activism could therefore be supported by rational-choice theory at both the individual and organizational levels.[74]

Besides providing one added alternative explanation, I find it useful to stipulate which observations would support this hypothesis and which would contradict it.

For example, according to this hypothesis, terrorists should be more likely to originate from areas where political freedom or civil liberties are limited,[75] to attack politically sensitive targets, or intensify their terrorist campaigns during politically sensitive periods. Perhaps most important, to support rational expectations, terrorism should have a clear political effect on its targeted population.

Political Freedom and Civil Liberties. Krueger and Laitin (2008) used the Freedom House Index, which rates various countries on the basis of civil liberties and political rights, to study the characteristics of the countries of origin of terrorists involved in 956 terrorist events that occurred from 1997 to 2003. They found that origin countries tend to have low levels of civil liberties. Similarly, Piazza (2006), who studied terrorist incidents and casualties in 96 countries from 1986 to 2002, found that increased state repression and, most significantly, the structure of party politics are significant predictors of terrorism. Alberto Abadie, in his analysis of the roots of terrorism, determines that political freedom has a significant, but nonlinear, effect on terrorism risk (Abadie, 2006). Finally, Krueger (2007) concludes his analysis of foreign insurgent in Iraq by saying:

The results for civil liberties were the same as what I found in the international terrorism results: countries with fewer civil liberties were more likely to be source countries of foreign insurgents in Iraq. If we measured political rights instead of civil liberties, we found that foreign insurgents were coming from more totalitarian regimes. On the other hand, civil liberties were more powerful as a predictor.

Political Sensitivity of Targets. It could be argued that the overwhelming number of political figures or targets, such as embassies, obviates this point. The RAND-Memorial Institute for the Prevention of Terrorism (RAND-MIPT) terrorism chronology database reveals that between January 1, 1968, and January 1, 2007, of the 30,611 recorded terrorist attacks worldwide, over a quarter (7,739) were against government or diplomatic targets. In fact, political targets were attacked more often than any other target category, more than religious figures and institutions, educational institutions, journalists and media, telecommunication, food or water supplies, utilities, transportation, tourists, airports, airlines and aviation, nongovernmental organizations, maritime or military, abortion-related, or even other terrorists and former terrorists targets *all combined*. Notably, half of the assassinations perpetrated by terrorists, and about 60 percent of the terrorist hostage and barricade attacks, were against government and diplomatic targets.

Careful use of logistic probability and count model regression analyses of the Palestinian case study shows, as mentioned above, that politically sensitive areas, such as localities with a regional capital, were three times more likely than other localities to be targeted. The same study also revealed that terrorists are unlikely to leave politically sensitive areas calm for long periods, whereas they might chose to do so for other comparable but politically insensitive areas (Berrebi and Lakdawalla, 2007).

Timing of Attacks and Electoral Outcomes. An interesting feature of the timing of terrorist attacks is that they tend to be concentrated within well-defined campaigns. Robert Pape, in his work about suicide terrorist attacks, finds that nearly all suicide attacks occur in organized, coherent campaigns, not as isolated or randomly limited incidents (Pape, 2005). Contrary to popular beliefs, terrorists are unlikely to attack in an uncalculated reaction to events or grievances. Jaeger and Paserman, in their study about the dynamics of violence in the Palestinian-Israeli conflict since the outbreak of the Second Intifada in September 2000, find that the conflict has followed an uneven pattern, with periods of high levels of violence and periods of relative calm. The estimated reaction functions for both Israelis and Palestinians reveal evidence of unidirectional Granger causality from Palestinian violence to Israeli violence, but not vice versa. Although Israelis react systematically to Palestinian attacks, Palestinian attacks are not caused by or in response to Israeli violence. The authors conclude that, despite the popular perception that Palestinians and Israelis are engaged in "tit-for-tat" violence, there is no evidence to support that notion (Jaeger and Paserman, 2008).

In a study of the interaction between terror attacks and electoral outcomes, Esteban Klor and I used the Israeli-Palestinian conflict to develop and analyze a game-theoretical dynamic model of reputation, whose predictions about the interaction between terrorism and electoral outcomes we tested empirically (Berrebi and Klor, 2006). The unique pure-strategies Markov-perfect equilibrium of our model (which takes place in an environment characterized by well-defined preferences and limited information, and which incorporates strategic behavior derived from beliefs that are in turn updated according to Bayes rule following the actual realization of terrorism and election outcomes) predicts two important empirical outcomes. First, we expect relative support for the right-wing party to increase after periods with high levels of terrorism and to decrease after periods of relative calm. Second, perhaps paradoxically, the model predicts that the expected *short-term* level of terrorism will be higher during the left-wing party's term in office than during that of the right-wing party. Notably, these predictions follow from the Palestinians' strategic considerations and not from different deterrence policies that the Israeli government might implement. The intuition behind the empirical predictions is that, when Israelis believe that there is a high probability that attacks cannot be prevented through concessions, they expect a high level of terrorism whether territorial concessions are granted or not. Therefore, Israelis, who, everything else equal, benefit from greater territorial control, vote for the right-wing party.

Within this range of beliefs, Palestinians realize that further attacks will not bring about territorial concessions and will only strengthen the Israeli voters' conviction that there is no point in making any. It is then that there is an effort by the Palestinians to scale down terrorist attacks to establish a reputation as a reliable partner for peace. Once such a reputation is established, if terrorism is kept at a low level, Israelis would not suffer a cost from maintaining the occupation and would thereby try to perpetuate it. Therefore, to impose a cost on the Israelis and force them to give up the perceived occupied territories, terrorism is ramped up again. Israelis, who now believe in the ability of the Palestinians to reliably control terrorism, then realize that the continued control over the claimed territories will lead to a stream of high-level terror attacks and therefore vote for the left-wing party.[76] Following the implications of the theoretical model, our empirical estimation concentrates on the striking variability in the level of terrorism for periods that precede Israeli elections. Accordingly, the Palestinians' optimal level of terrorism before an Israeli election varies depending on the identity of the incumbent political party in Israel. We therefore expect to observe a higher level of preelection terrorism when Labor (the left-wing party) holds office than when the Likud (the right-wing party) is in power. We test the hypotheses that our theoretical model elicits by combining data on terrorism in Israel, the West Bank, and Gaza between 1990 and 2003 with electoral outcomes data for the same period.

To determine whether preelection terrorism is relatively higher when Labor (the left-wing party) holds office, we use a combination of event-study methods and likelihood-ratio tests. The event-study method treats the ideology of the elected Israeli government as exogenous and studies its effect on the level of terrorism. To conduct an event-study analysis, we define the day on which the forthcoming Election Day is announced as the day of the event. Our sample contains four events: the elections of 1996, 1999, 2001, and 2003. For each event, we define an event window that spans from the day of the event until the end of the tenure of the corresponding government. The event-study method compares the level of terrorism during the event window with the level of terrorism during a previously specified estimation window. We define our estimation window as the event window of the preceding government. For each event, we calculate the weekly abnormal number of deaths from terrorism, defined as the difference between the observed weekly number of deaths (during the event window) minus the average number of weekly terror fatalities during the preceding government (the estimation window). We interpret the abnormal number of deaths from attacks during the event window as a measure of the effect of the ideology of a given government on terrorist activity. We aggregate the abnormal deaths into the number of cumulative abnormal deaths (CAD) to draw overall inferences. If the CAD graph oscillates around zero, then the studied event does not have an effect on the level of terrorism. On the contrary, if the theoretical predictions of our model are correct, then CAD should be positive and increasing for a left-wing government that succeeds a right-wing government and negative and decreasing for a right-wing government that succeeds a left-wing government. Figure 5.2 provide the results of this analysis.

Figure 5.2
Terrorist-Attack Intensity Versus Time, Relative to Announcement of Early Elections

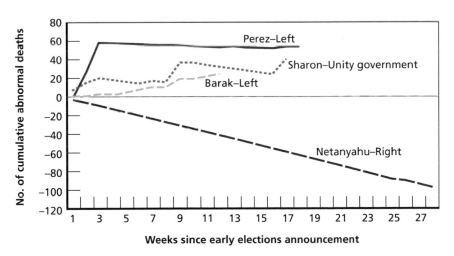

NOTE: The comparison period is from the preceding elections (under the previous government).
RAND *MG849-5.2*

The evidence obtained from the event-study analysis supports the hypothesis about the expected level of terrorism. The standard statistical test applied in event studies assumes that CAD is normally distributed. This is clearly not the case in our study, since terror fatalities are count data and are best described by a Poisson distribution. Therefore, we perform the more conventional likelihood-ratio test, assuming that deaths from terror attacks do follow a Poisson distribution. The findings of the likelihood-ratio tests also support the conclusions adduced from the event-study analysis.

Accordingly, there is a statistically significant increase in the level of terrorism during the left-wing party's term in office and a statistically significant decrease in terrorism during the tenure of the right-wing party. Therefore, we conclude that the timing of terrorist attacks is strategically set and oscillates around election periods.

When considering the political effect of terrorist attacks, a good non-Palestinian example is the March 2004 Madrid train bombings. Ten bombs exploded on three commuter trains full of passengers on their way to Madrid. The attack resulted in 191 deaths and 1,500 wounded. The terrorist group that carried out the attack sought to compel Spain to withdraw its troops from Afghanistan and especially Iraq. A study by Jose Garcia Montalvo (2008) compares absentee ballots cast before the bombing with votes cast after them and convincingly shows that the aftermath of the attack mobilized voters to elect a new government led by the Socialist Party because, in large part, this party campaigned on the promise to pull Spanish troops from Iraq.

For a rigorous analysis of voters' sensitivity to terrorism, Esteban Klor and I went on to identify the causal effects of terrorism on the preferences of the Israeli electorate (Berrebi and Klor, 2008). The assumption that voters' preferences are significantly affected by terrorism is of crucial importance and warrants careful examination. Our empirical strategy is based on a difference-in-differences approach that uses the variation of terror fatalities across time and space to control for possible time- or location-specific effects. Specifically, this methodology allows us to estimate the causal effects of terrorism by comparing changes in consecutive electoral results of localities that suffered terror attacks (treated group) with changes in electoral results of localities that did not suffer from terror attacks (control group). The key identifying assumption of this approach is that, in the absence of terrorism, the trends of the electoral preferences of treated and control localities would be the same. We use electoral data at the level of the polling station, provided by the Israeli Central Bureau of Statistics (ICBS), which include the total number of eligible voters, voter turnout, and the support for each political party in the parliamentary elections of 1988, 1992, 1996, 1999, and 2003. We then geographically divide the data into localities according to ICBS guidelines, divide the political parties with representatives in the parliament into right-left bloc vote, and combine the data with information on the number of noncombatant Israeli fatalities from terror attacks during the respective period. We incorporate into the analysis additional political, socioeconomic, and demographic variables. We find that one terror attack causes an increase of roughly 1.35 percentage points in the relative support for the right bloc. This effect is of a significant political magnitude, to the extent that the occurrence of a terror attack before an election (or the lack thereof) can clearly determine the electoral outcome. A calibration of the effect of terrorism on the distribution of seats in the Israeli parliament shows that terrorism not only affected the composition of every Israeli parliament during the time period at issue, but it may have very well determined which party obtained a plurality in two of the elections analyzed and could have shifted the majority of the parliament from the left to the right bloc of parties in one more election if another attack had occurred before that election. This study also reveals that terrorism can cause the ideological polarization of the electorate.

Many additional studies report a correlation between terrorism (or the threat thereof) and the electorate's political preferences.[77] Even if these correlations cannot be interpreted causally, they contribute to the evidence about the link between terrorism and electoral outcomes and provide numerous examples of cases in which terrorism was likely to have influenced political preferences and electoral outcomes.

At first glance, the effect of terrorism on voters' preferences may seem paradoxical from the terrorists' standpoint. Terrorism fatalities, with few exceptions, increase support for the bloc of parties associated with a more-intransigent position toward terrorism and territorial concessions. In other words, terrorism supposedly undermines the terrorist faction's goal. Some scholars may interpret this as further evidence that terrorist attacks against civilians do not help terrorist organizations achieve their stated goals (Abrahms, 2006). Other scholars place more emphasis on the complex structure of terrorist factions, who tend to have a number of objectives (Kydd and Walter, 2006) and are therefore likely to face trade-offs between their main objectives, with the risk that a chosen strategy in pursuit of some of them will undermine the likelihood of achieving others. An alternative explanation is that terrorist organizations perpetrate attacks with the goal of provoking reaction from the targeted government into a forceful response against the population whose interests they are supposedly representing, in the hope of radicalizing the population and increasing overall support for terrorist actions (Bueno de Mesquita and Dickson, 2006; Jaeger and Paserman, 2008; Siqueira and Sandler, 2006). These studies suggest that violence is used to radicalize the population. Jaeger et al. (2008) directly test this hypothesis, using public-opinion polls taken regularly in the West Bank and Gaza Strip since the beginning of the second intifada in 2000 and merging them with data on Israeli-inflicted Palestinian fatalities. They find that, although Palestinian fatalities immediately increase the radicalization of the Palestinian population, this effect is fleeting. In fact, the shift in opinion toward more radical views rarely persists more than a few weeks and disappears completely after 90 days. Moreover, there is no statistically significant radicalization effect in the aftermath of targeted killings.

According to Berrebi and Klor's (2006) model, discussed above, it is possible that, even if the electorate's support for the right bloc increases as a consequence of terror attacks, the political position of the right bloc (although still more hawkish than that of the left bloc) may be affected as well and may become less intransigent over time. Therefore, they rationalize not only the behavior of terrorist factions but also that of the targeted electorate (or government).

The discussion about the effect of terrorism on politics and electoral outcomes is also tied to the ability to induce territorial concessions, as discussed above. This can be viewed as political objectives in the same way that influencing voters' preferences are. Whether true or perceived, many withdrawals, from partial to complete pullouts, are attributed to the success of terrorism. Examples abound over the entire course of history, from the French pullout from Algeria in 1962 following the terrorist activities of National Liberation Front (FLN) and the National Algerian Movement (MNA) to the more recent Israeli pullouts from Lebanon in 2000 following Hizballah terrorist activity and the pullout from the Gaza Strip in 2005 following Palestinian, mainly Hamas, terrorism. According to Krueger (2007), in some writings in Britain at the end of the 18th century, George Washington was considered a terrorist for fighting the British military. Regardless of the obvious differences, terrorists do perceive these and numerous other examples as proof that terrorism can achieve political goals, influence electoral outcomes, and induce concessions, so they rationally chose their actions based on it.

To summarize, I find that a limited set of objectives to describe terrorist organizations or individuals is likely to produce a flawed representation of complex phenomena. As with other organizations, terrorist organizations have multiple, sometimes competing, objectives.[78] Having reviewed the literature and researched the issue myself, it seems that two objectives, shared by numerous terrorists and terrorist organizations, stand out as empirically grounded and concurrently provide hypotheses that conform to rational-choice behavior, therefore, the most compelling hypothesis. The first is based on political objectives along the lines presented by Krueger (2007) and Berrebi and Klor (2006, 2008), which has the "bonus" feature of actually matching the terrorists' own stated goals and our accepted definition of terrorism. The second is based on a combination of social objectives along the lines presented by Berman and Laitin (2008), Iannaccone (2006), and Abrahms (2008). In addition to the main factors, which stand at the core of these explanations of terrorists' behavior, numerous permissive elements are likely to play a significant role; for example, the availability of a breached educational system that allows for indoctrination and recruiting by the terrorist organizations or their supporters. Furthermore, these hypotheses should not be regarded as mutually exclusive, as they probably reflect many factors affecting terrorists and terrorist organizations' behavior.

Potential Implications for Counterterrorism

Given the evidence, it is not realistic to put much stock in "root-cause" explanations of terrorism, although the factors in question may indeed be contributors to the beginning, maintaining, and ending of terrorism (see also Noricks, 2009; Paul, 2009; and Gvineria, 2009). This conclusion is about as robust as they come in social science. The evidence gathered so far suggests that the pursuit of political power is a more likely motivating determinant and, to the extent possible, should guide us when devising counterterrorism policies.

Terrorists and terrorist groups should be assumed to be rational, at least in the sense of taking actions they believe are consistent with their goals, sometimes in the stronger sense of being "smart" (that is, locally optimal) and sometimes in the even stronger sense of being consistent with a credible assessment of prospects.[79]

While addressing the issue of rationality in counterterrorism strategy, Ganor (2005) suggests that:

> It is a common error to judge the enemy's rationality through the subjective mirror of those coping with terrorism. Cost-benefit considerations are the result of several variables—history, culture, sociological and psychological aspects, etc. An act that is perceived as beneficial to one, may not necessarily be perceived as such by someone else. The rational judgment must be based, therefore, on the cost-benefit considerations as perceived by the enemy alone.

Surely, our findings indicate that we should consider that terrorists are sensible to cost-benefit considerations and we ought to use this information to our advantage. Indeed, there are indications that counterterrorism methods devised to increase the cost considerations of terrorists can be effective. For example, although the results are still preliminary, a study that examines the effectiveness of home demolitions as a counterterrorism strategy against suicide terrorists suggests that by carefully targeting the attacker's homes for punitive demolitions, it is possible to deter future potential suicide attack volunteers (Benmelech and Berrebi, 2008).[80] Moreover, if we did not think that terrorists were rational and respond to incentives, we would be left with a "capture and kill" only strategy for counterterrorism. Establishing that incentives (at least in their weaker form) can potentially work in our favor is not only useful but a very hopeful message as well.

The model of rational choice needs to be applied with an extended concept of utility that allows for valuing causes greater than the individual and for valuing developments that may or may not occur in the individual's lifetime.[81]

We should be careful when considering potential concessions, since each concession is later incorporated into the terrorists' rational expectations, providing them with further support for the effectiveness of their tactics. Likewise, we should recognize the importance of psychological considerations in deterrence.[82]

Facing a rational opponent, it is only natural to expect terrorists to adapt to the counterterrorism measures we develop, as they reevaluate their cost-benefit calculus every period. Accordingly, no counterterrorism tool or method should be expected to last forever. Michael Intriligator, in his work on the economics of terrorism, cautions that terrorists will substitute other forms of terrorism if one form becomes more expensive or less valuable. They will substitute one target for another as it becomes harder to hit that target (Intriligator, 2008). Indeed, in a time-series analysis of various attack modes used by terrorists and after further examination of counterterrorism methods, Enders and Sandler (1993, 2002) find that terrorists both substitute attack modes and complement them. Accordingly, policies designed to reduce one type of attack can increase other attack modes.

Efforts to reduce the supply of terrorists may have low leverage because the phenomenon is actually driven by demand, with the terrorist organization requiring relatively few recruits but ones of relatively high quality.

Finally, although terrorism has often been considered a tactic, it is sometimes a conscious, rational strategy. This conclusion is especially well substantiated with respect to efforts to fight against perceived occupations or foreign influences.

Bibliography

Abadie, A., "Poverty, Political Freedom, and the Roots of Terrorism," *American Economic Review,* Vol. 96, No. 2, 2006, pp. 50–56.

Abadie, A., and J. Gardeazabal, "The Economic Costs of Conflict: A Case Study of the Basque Country," *The American Economic Review,* Vol. 93, No. 1, 2003, pp. 113–132.

Abdullaev, Nabi, "Unraveling Chechen 'Black Widows,'" *Homeland Defense Journal,* Vol. 5, No. 5, May 2007, pp. 18–21.

Abrahms, Max, "Are Terrorists Really Rational? The Palestinian Example," *Orbis,* Vol. 48, No. 3, Summer 2004, pp. 533–549.

———, "Why Terrorism Does Not Work," *International Security,* Vol. 31, No. 2, Fall 2006, pp. 42–78.

———, "What Terrorists Really Want: Terrorist Motives and Counterterrorism Strategy," *International Security,* Vol. 32, No. 4, Spring 2008, pp. 78–105.

Angrist, Joshua D., "The Economic Returns to Schooling in the West Bank and Gaza Strip," *The American Economic Review,* Vol. 85, No. 5, December 1995, pp. 1065–1087.

Atran, Scott, "Who Wants to Be a Martyr?" *New York Times,* May 5, 2003, p. A23.

———, "Mishandling Suicide Terrorism," *The Washington Quarterly,* Vol. 27, No. 3, 2004, pp. 67–90.

Azam, Jean Paul, "Suicide-Bombing as Inter-Generational Investment," *Public Choice,* Vol. 122, No. 1, 2005, pp. 177–198.

Becker, G. S., *The Economic Approach to Human Behavior,* Chicago,Ill.: University of Chicago Press, 1976.

Berko, Anat, *The Path to Paradise: The Inner World of Suicide Bombers and Their Dispatchers*, Westport, Conn.: Praeger Security International, 2007.

Benmelech, Efraim, and Claude Berrebi, "Human Capital and the Productivity of Suicide Bombers," *Journal of Economic Perspectives,* Vol. 21, No. 3, 2007, pp. 223–238.

———, "Counter-Suicide-Terrorism: Evidence from House Demolitions," Cambridge, Mass.: Harvard University, unpublished, 2008.

Benmelech, Efraim, Claude Berrebi, and Esteban F. Klor, "Economic Conditions and the Quality of Terrorism," Cambridge, Mass.: Harvard University, unpublished, 2009.

Berman, Eli, "Sect, Subsidy, and Sacrifice: An Economist's View of Ultra-Orthodox Jews," *Quarterly Journal of Economics,* Vol. 115, No. 3, 2000, pp. 905–953.

————, "Hamas, Taliban and the Jewish Underground: An Economist's View of Radical Religious Militias," *National Bureau of Economic Research Working Paper,* Cambridge, Mass., No. 10004, 2003.

Berman, Eli, and D. David Laitin, "Hard Targets: Theory and Evidence on Suicide Attacks," *National Bureau of Economic Research Working Paper,* Cambridge, Mass., No. 11740, 2005.

————, "Rational Martyrs vs. Hard Targets: Evidence on the Tactical Use of Suicide Attacks," in Eva Meyersson Milgrom, ed., *Suicide Bombing from an Interdisciplinary Perspective,* Princeton, N.J.: Princeton University Press, forthcoming, 2009.

Berrebi, Claude, "Evidence About the Link Between Education, Poverty and Terrorism Among Palestinians," Princeton, N.J.: Princeton University, Industrial Relations Section Working Paper, No. 477, 2003.

————, "Evidence About the Link Between Education, Poverty and Terrorism Among Palestinians," *Peace Economics, Peace Science and Public Policy,* Vol. 13, No. 1, 2007.

Berrebi, Claude, and Esteban F. Klor, "The Impact of Terrorism Across Industries: An Empirical Study," CEPR Discussion Paper 5360, London, U.K., 2005.

————, "On Terrorism and Electoral Outcomes: Theory and Evidence from the Israeli-Palestinian Conflict," *The Journal of Conflict Resolution,* Vol. 50, No. 6, 2006, pp. 899–925.

————, "Are Voters Sensitive to Terrorism? Direct Evidence from the Israeli Electorate," *American Political Science Review,* Vol. 102, No. 3, August 2008, pp. 279–301.

————, "The Impact of Terrorism on the Defense Industry," *Economica,* forthcoming, 2009.

Berrebi, Claude, and Darius Lakdawalla, "How Does Terrorism Risk Vary Across Space and Time? An Analysis Based on the Israeli Experience," *Defence and Peace Economics,* Vol. 18, No. 2, 2007, pp. 113–131.

Bloom, Mia, "Palestinian Suicide Bombing: Public Support, Market Share, and Outbidding," *Political Science Quarterly,* Vol. 119, Spring 2004, pp. 61–88.

————, *Dying to Kill: The Allure of Suicide Terrorism,* New York: Columbia University Press, 2005.

Bueno de Mesquita, Ethan, "The Quality of Terror," *American Journal of Political Science,* Vol. 49, No. 3, July 2005, pp. 515–530.

Bueno de Mesquita, Ethan, and Eric Dickson, "The Propaganda of the Deed: Terrorism, Counterterrorism and Mobilization," *American Journal of Political Science,* Vol. 51, April 2006, pp. 364–381.

Caplan, B., "Rational Irrationality: A Framework for the Neoclassical-Behavioral Debate," *Eastern Economic Journal,* Vol. 26, No. 2, 2000, pp. 191–211.

———, "Terrorism: The Relevance of the Rational Choice Model," *Public Choice,* Vol. 128, No. 1, 2006, pp. 91–107.

CNN.com, "Bin Laden: Goal Is to Bankrupt U.S.: Al-Jazeera Releases Full Transcript of al Qaeda Leader's Tape," November, 1, 2004. As of July 9, 2008: http://www.cnn.com/2004/WORLD/meast/11/01/binladen.tape/

Collier, P., and A. Hoeffler, "On Economic Causes of Civil War," *Oxford Economic Papers,* Vol. 50, No. 4, 1998, pp. 563–573.

———, "Greed and Grievance in Civil War," *Oxford Economic Papers,* Vol. 56, No. 4, 2004, pp. 563–595.

Cragin, Kim, "Cross-Cutting Observations and Some Implications for Policymakers," in Paul K. Davis and Kim Cragin, eds., *Social Science for Counterterrorism: Putting the Pieces Together,* Santa Monica, Calif.: RAND Corporation, 2009. As of January 20, 2009: http://www.rand.org/pubs/monographs/MG849/

Crenshaw, M., "The Causes of Terrorism," *Comparative Politics,* Vol. 13, No. 4, 1981, pp. 379–399.

Davies, Caroline, John Steele, and Catriona Davies, "Middle-Class and British: The Muslims in Plots to Bomb Jets," *Telegraph,* December 8, 2006.

Davis, D., and B. Silver, "The Threat of Terrorism, Presidential Approval, and the 2004 Election," paper presented at the Annual Meeting of the American Political Science Association, Chicago, Ill., 2004, pp. 2–5.

Davis, Paul K., and Brian Michael Jenkins, *Deterrence & Influence in Counterterrorism: A Component in the War on Al Qaeda,* Santa Monica, Calif.: RAND Corporation, 2002. As of January 6, 2009: http://www.rand.org/pubs/monograph_reports/MR1619/

Davis, Paul K., Jonathan Kulick, and Michael Egner, *Implications of Modern Decision Science for Military Decision-Support Systems,* Santa Monica, Calif.: RAND Corporation, 2005. As of January 6, 2009: http://www.rand.org/pubs/monographs/MG360/

Eckstein, Z., and D. Tsiddon, "Macroeconomic Consequences of Terror: Theory and the Case of Israel," *Journal of Monetary Economics,* Vol. 51, No. 5, 2004, pp. 971–1002.

Enders, Walter, and Todd Sandler "The Effectiveness of Antiterrorism Policies: A Vector-Autoregression Intervention Analysis," *American Political Science Review,* Vol. 87, No. 4, 1993, pp. 829–844.

————, "What Do We Know About the Substitution Effect in Transnational Terrorism?" Tuscaloosa, Ala.: University of Alabama, unpublished, 2002.

Fearon, J. D., and David D. Laitin, "Ethnicity, Insurgency, and Civil War," *American Political Science Review,* Vol. 97, No. 1, 2003, pp. 75–90.

Feldman, Naomi E., and Bradley J. Ruffle, "Religious Terrorism: A Cross-Country Analysis," Beer-Sheva, Israel: Ben-Gurion University, unpublished, 2008.

Ganor, Boaz, *The Counter-Terrorism Puzzle,* New Brunswick, N.J.: Transaction Publishers, 2005.

Guilmartin, E., "Terrorist Attacks and Presidential Approval from 1949–2002," U.S. Military Academy, West Point, N.Y., unpublished, 2004.

Gupta, Dipak, *Understanding Terrorism and Political Violence,* New York: Routledge, 2008.

Gvineria, Gaga, "How Does Terrorism End?" in Paul K. Davis and Kim Cragin, eds., *Social Science for Counterterrorism: Putting the Pieces Together,* Santa Monica, Calif.: RAND Corporation, 2009. As of January 20, 2009: http://www.rand.org/pubs/monographs/MG849/

Hamermesh, D. S., and N. M. Soss, "An Economic Theory of Suicide," *Journal of Political Economy,* Vol. 82, No. 1, 1974, p. 83.

Harel, A., and A. Isacharoff "The Seventh War: How We Won and Why We Lost the War with the Palestinians," Tel-Aviv, Israel: Yedi'ot Aharonot & Hemed Books, 2006 [Hebrew].

Hassan, Nasra, "An Arsenal of Believers," *The New Yorker,* November 19, 2001, pp. 36–41.

Hoffman, Bruce, *Inside Terrorism,* New York: Columbia University Press, 2006.

Hudson, Rex. A., and Marilyn Lundell Majeska, "The Sociology and Psychology of Terrorism: Who Becomes a Terrorist and Why?" 1999. As of July 1, 2008: http://purl.access.gpo.gov/GPO/LPS17114

Iannaccone, Laurence R., "The Market for Martyrs," Fairfax, Va.: George Mason University, unpublished, 2006.

Iannaccone, Laurence R., and Eli Berman, "Religious Extremism: The Good, the Bad, and the Deadly," *Public Choice,* Vol. 128, No. 1, 2006, pp. 109–129.

Intriligator, Michael D., "The Economics of Terrorism," *Western Economic Association International,* Honolulu, Hawaii, 2008.

Intriligator, Michael D., and Dagobert L. Brito, "The Potential Contribution of Psychology to Nuclear War Issues," *American Psychologist,* Vol. 43, 1988, pp. 318–321.

Jackson, Brian A., Lloyd S. Dixon, and Victoria A. Greenfield, *Economically Targeted Terrorism: A Review of the Literature and Framework for Considering Defensive Approaches*, Santa Monica, Calif.: RAND Corporation, TR-476-CTRMP, 2007. As of January 6, 2009: http://www.rand.org/pubs/technical_reports/TR476/

Jaeger, David A., and M. Daniele Paserman, "The Cycle of Violence? An Empirical Analysis of Fatalities in the Palestinian-Israeli Conflict," *American Economic Review*, Vol. 98, No. 4, September 2008, pp. 1591–1604.

Jaeger, David A., Esteban F. Klor, Sami H. Miaari, and Daniele M. Paserman, "The Struggle for Palestinian Hearts and Minds: Violence and Public Opinion in the Second Intifada," Williamsburg, Va.: College of William and Mary Working Paper, No. 72, 2008.

Jenkins, Brian Michael, *Unconquerable Nation: Knowing Our Enemy, Strengthening Ourselves*, Santa Monica, Calif.: RAND Corporation, 2006. As of January 6, 2009: http://www.rand.org/pubs/monographs/MG454/

Karolyi, Andrew G., and Rodolfo Martell, "Terrorism and the Stock Market," Columbus, Ohio: Ohio State University, unpublished, 2007.

Krieger, Tim, and Daniel Meierrieks, "What Causes Terrorism?" Paderborn, Germany: University of Paderborn, unpublished, 2008.

Krueger, Alan B., *What Makes a Terrorist: Economics and the Roots of Terrorism*, Princeton, N.J.: Princeton University Press, 2007.

———, "What Makes a Homegrown Terrorist? Human Capital and Participation in Domestic Islamic Terrorist Groups in the U.S.A.," Princeton, N.J.: Princeton University, unpublished, 2008.

Krueger, Alan B., and Jitka Maleckova, "Education, Poverty and Terrorism: Is There a Causal Connection?" *The Journal of Economic Perspectives,* Vol. 17, No. 4, 2003, pp. 119–144.

Krueger, Alan B., and David D. Laitin. "Kto Kogo? A Cross-Country Study of the Origins and Targets of Terrorism," in Philip Keefer and Norman Loayza, eds., *Terrorism, Economic Development, and Political Openness*, New York: Cambridge University Press, 2008, pp. 148–173.

Kruglanski, A. W., and S. Fishman, "The Psychology of Terrorism: 'Syndrome' Versus 'Tool' Perspectives," *Terrorism and Political Violence,* Vol. 18, No. 2, 2006, pp. 193–215.

Kydd, Andrew H., and Barbara F. Walter, "The Strategies of Terrorism," *International Security,* Vol. 31, Summer 2006, pp. 49–80.

Llussa, Fernanda, and Jose Tavares, "The Economics of Terrorism: A Synopsis," *The Economics of Peace and Security Journal,* Vol. 2, No. 1, 2007.

————, "Economics and Terrorism: What We Know, What We Should Know, and the Data We Need," in Philip Keefer and Norman Loayza, eds., *Terrorism, Economic Development, and Political Openness,* New York: Cambridge University Press, 2008, pp. 233–253

Merari, A., "Psychological Aspects of Suicide Terrorism," in Bruce Bongar et al., eds., *Psychology of Terrorism,* New York: Oxford University Press, 2006.

Miguel, E., S. Satyanath, and E. Sergenti, "Economic Shocks and Civil Conflict: An Instrumental Variables Approach," *Journal of Political Economy,* Vol. 112, No. 4, 2004, pp. 725–753.

Mill, John Stuart, *Utilitarianism,* London, U.K.: Longmans, Green and Co., 1879.

Montalvo, Jose G., "Voting After Bombing: The Electoral Effect of Madrid's 11M Terrorist Attack," Barcelona, Spain: Universitat Pompeu Fabra, unpublished, 2008.

Noricks, Darcy M.E., "The Root Causes of Terrorism," in Paul K. Davis and Kim Cragin, eds., *Social Science for Counterterrorism: Putting the Pieces Together,* Santa Monica, Calif.: RAND Corporation, 2009. As of January 20, 2009: http://www.rand.org/pubs/monographs/MG849/

Pape, Robert A., "The Strategic Logic of Suicide Terrorism," *American Political Science Review,* Vol. 97, August 2003, pp. 343–361.

————, *Dying to Win: The Logic of Suicide Terrorism,* New York: Random House, 2005.

————, "Methods and Findings in the Study of Suicide Terrorism," *American Political Science Review,* Vol. 102, No. 2, May 2008, pp. 275–277.

Paul, Christopher, "How Do Terrorists Generate and Maintain Support?" in Paul K. Davis and Kim Cragin, eds., *Social Science for Counterterrorism: Putting the Pieces Together,* Santa Monica, Calif.: RAND Corporation, 2009. As of January 6, 2009: http://www.rand.org/pubs/monographs/MG849/

Piazza, J. A., "Rooted in Poverty? Terrorism, Poor Economic Development, and Social Cleavages," *Terrorism and Political Violence,* Vol. 18, No. 1, 2006, pp. 159–177.

Public Safety Canada, Web site. As of July 10, 2008: http://www.publicsafety.gc.ca/prg/ns/le/cle-eng.aspx

Republic of Singapore, Ministry of Home Affairs, "The Jemaah Islamiyah Arrests and the Threat of Terrorism," white paper, 2003. As of July 5, 2008: http://www.mha.gov.sg/get_blob.aspx?file_id=252_complete.pdf

Richardson, Harry W., Peter Gordon, and James E. Moore II, *The Economic Costs and Consequences of Terrorism,* Cheltenham, U.K.: Edward Elgar Publishing, 2007.

Russell, Charles, and Bowman Miller, "Profile of a Terrorist," in Lawrence Zelic Freedman and Alexander Yonah, eds., *Perspectives on Terrorism*, Wilmington, Del.: Scholarly Resources Inc., 1983, pp. 45–60.

Sageman, Marc *Understanding Terror Networks*: Philadelphia, Pa.: University of Pennsylvania Press, 2004.

———, "Common Myths About al-Qaida Terrorism," *eJournal USA*, August 2006.

Schelling, Thomas C., *Arms and Influence*, Henry L. Stimson Lectures Series, New Haven, Conn.: Yale University Press, 1966.

———, "What Purposes Can 'International Terrorism' Serve?" in Raymond Gillespie Frey and Christopher W. Morris, eds., *Violence, Terrorism, and Justice*, New York: Cambridge University Press, 1991.

Segell, Glen M., "Terrorism: London Public Transport—July 7, 2005," *Strategic Insights,* Vol. 4, No. 8, August 2005.

Shalev, A., *The Intifada: Causes and Effects*, Jerusalem, Israel: *Jerusalem Post*; Boulder, Colo.: Westview Press, 1991. As of July 21, 2008: http://www.btselem.org/English/Punitive_Demolitions/Index.asp

Shambaugh, G., and W. Josiger, "Public Prudence, the Policy Salience of Terrorism and Presidential Approval Following Terrorist Incidents," Washington, D.C.: Georgetown University, unpublished, 2004.

Silberman, Jonathan, and Garey Durden, "The Rational Behavior Theory of Voter Participation: The Evidence from Congressional Elections," *Public Choice,* Vol. 23, Fall 1975, pp. 101–108.

Simon, Herbert, *Models of Bounded Rationality*, Cambridge, Mass.: MIT Press, 1982.

Siqueira, Kevin, and Todd Sandler, "Terrorists Versus the Government: Strategic Interaction, Support and Sponsorship," *The Journal of Conflict Resolution,* Vol. 50, December 2006, pp. 878–898.

Stinger, David, "Doctors Among the Arrested in U.K. Terror Sweep," *Fox News,* July 2, 2007.

Stout, Mark E., Jessica M. Huckabey, and John R. Schindler, *The Terrorist Perspectives Project: Strategic and Operational Views of al Qaida and Associated Movements*, Annapolis, Md.: Naval Institute Press, 2008.

"Terrorism," Encyclopædia Britannica, Encyclopedia Britannica Online, 2008. As of July 13, 2008:
http://www.britannica.com/EBchecked/topic/588371/terrorism

Treverton, Gregory F., Justin L. Adams, James Dertouzos, Arindam Dutta, Susan S. Everingham, and Eric V. Larson, "The Costs of Responding to the Terrorist Threats: The U.S. Case," in Philip Keefer and Norman Loayza, eds., *Terrorism, Economic Development, and Political Openness*, New York: Cambridge University Press, 2008.

U.S. Department of State, "Patterns of Global Terrorism," annual. As of April 6, 2009:
http://www.state.gov/s/ct/rls/pgtrpt/

————, Bureau of International Information Programs, "Common Myths about al-Qaida Terrorism," *eJournal USA*, August 2006. As of January 12, 2009:
http://www.ciaonet.org/olj/fpa/fpa_aug06/fpa_aug06_myths.pdf

Varshney, Ashutosh, "Nationalism, Ethnic Conflict, and Rationality," *Perspectives on Politics*, Vol. 1, No. 1, 2003, pp. 85–99.

Victoroff, J., "The Mind of the Terrorist: A Review and Critique of Psychological Approaches," *Journal of Conflict Resolution*, Vol. 49, No. 1, 2005, p. 3.

Weber, M., *Economy and Society: An Outline of Interpretive Sociology*, Los Angeles, Calif.: University of California Press, 1978.

Wikipedia, "Al Aqsa Martyrs' Brigades," undated. As of January 20, 2009.
http://en.wikipedia.org/wiki/Al-Aqsa_Martyrs'_Brigades
http://en.wikipedia.org/wiki/List_of_Al-Aqsa_Martyrs%27_Brigades_suicide_attacks

————, "Hizballah and Its Manifesto," undated. As of July 11, 2008:
http://en.wikipedia.org/wiki/Hezbollah

———— "Kurdistan Workers' Party," undated. As of January 20, 2009:
http://en.wikipedia.org/wiki/Kurdistan_Workers_Party

Wilgoren, Jodi, "After the Attacks: The Hijackers; A Terrorist Profile Emerges That Confounds the Experts," *New York Times*, September 15, 2001, p. A2.

Wintrobe, R., "Can Suicide Bombers Be Rational?" *Comparative Politics Workshop*, Chicago, Ill.: University of Chicago, unpublished, 2003.

Yale Law School, "The Avalon Project: Documents in Law, History and Diplomacy," 2009. As of January 12, 2009:
http://avalon.law.yale.edu/20th_century/hamas.asp

Endnotes

[1] The conventional wisdom follows even though, in most variants of the Beckerian model, one cannot determine the relationship between criminal participation and the variables of interest without making assumptions about the individual's risk aversion.

[2] This applies if we believe that the dynamic that brings suicide bombers to volunteer is comparable with what brings individuals to commit suicide.

[3] Further background details on the likely sources for widespread embrace of the theory that poverty is behind terrorism can be found in Berrebi (2003, 2007).

[4] An underlying assumption in this paper is that what causes terrorism in one context is relevant to other contexts. Clearly, as the similarities in circumstances weaken, so does this assumption's validity.

[5] Republic of Singapore, Ministry of Home Affairs (2003).

[6] Marc Sageman is a forensic psychiatrist, former Central Intelligence Agency (CIA) officer, senior fellow at the Foreign Policy Research Institute in Philadelphia, and a senior associate at the Center for Strategic and International Studies in Washington.

[7] Quotes are from U.S. Department of State (2006).

[8] We should expect the educational attainment and wealth of leaders in any complex organization to be higher than that of its members.

[9] *Jihad* has multiple interpretations, including that of internal and personal struggle rather than holy war. Here I use *jihad* to mean waging a war that could result in fatalities and physical injuries.

[10] The source here is a database that I have constructed containing daily information on all fatal terrorist attacks against noncombatants that occurred on Israeli soil from 1949 to January 31, 2003. Every attack is described by date, method of operation, location, terrorist organizations claiming responsibility, and additional data about the victims, such as age, gender, and place of residence. The information was gathered from the Israeli Foreign Ministry, the National Insurance Institute of Israel, and Israeli newspapers *Ha'aretz* and *Ma'ariv*. The information was checked for accuracy against data from the Israeli Defense Forces (IDF). For further details about the terrorism chronology database, see Berrebi (2003).

[11] Further information about Hamas and PIJ can be found at the Public Safety Canada Web site.

[12] These data were collected and used under the supervision of Joshua Angrist. See Angrist (1995) for details.

[13] Terrorists who failed in their attack would most likely not be considered Shahid or be hailed; consequently, no biography was published on their behalf.

[14] A more recent study repeated this analysis on a more comprehensive dataset, which included failed attacks. It reached similar qualitative, though quantitatively slightly weaker, results (Benmelech and Berrebi, 2007).

[15] See Berrebi (2007) for how an individual's economic status was inferred in each of the populations and a discussion of potential problems introduced by this method.

[16] See Fearon and Laitin (2003), Collier and Hoeffler (1998, 2004), and Miguel, Satyanath, and Sergenti (2004) for examples of a positive link between civil wars, conflicts, and economic conditions.

[17] To alleviate this problem, economists often construct specifications that include countries and years' fixed effects.

[18] See Benmelech, Berrebi, and Klor (2009) for further details.

[19] To interpret the word *religious* included in the second definition provided for *religion*, the reader is directed to the definition of *religious*, namely: "relating to or manifesting faithful devotion to an acknowledged ultimate reality or deity."

[20] The four major religious faiths considered in this study were Islam, Christianity, Buddhism, and Hinduism.

[21] Hoffman (2006) argues that the Tamil Tigers are best described as nationalist-separatist and that, despite its apparent secularism, the group operates more like a cult than a secular terrorist group.

[22] See Bloom (2004, 2005) for a model of outbidding between terrorist factions.

[23] See Bloom (2004, 2005).

[24] This puzzle will be addressed later in this paper as a part of the discussion of rational choice.

[25] These groups, arguably, constitute a greater threat to government than does the threat emanating from secular terrorist groups characterized by higher incidence of less-deadly attacks.

[26] Including a significant share of suicide attackers.

[27] Even when the cost is extremely high, such as giving up one's life.

[28] In the past, arguments about ill-mindedness and irrationality were easily accepted by the media, professionals, and academics with respect to individuals who joined cults such as the International Society of Krishna Consciousness or the Bhagwan movement of Shree Rajneesh, only to be proven wrong later. The profile of the typical cultist ended up including normal background and circumstances, normal per-

sonalities and relationships, and a normal subsequent life (Iannaccone, 2006). We should make every effort to avoid falling into this "convenient" trap in the research of those who engage in terrorist activities.

[29] Although the empirical evidence suggests that mental health is unlikely to predict involvement in terrorism, there is a possibility that sustained involvement with terrorism would cause detrimental effects to one's mental health. To that extent, long periods of involvement with terrorism could arguably lead to less than rational decisionmaking.

[30] The "rationality" described by rational-choice theory is different from the colloquial and most philosophical uses of rationality. For more on rational-choice theory, see Becker (1976).

[31] A utility function is a conceptual device for summarizing the factors that influence a person's overall well-being. Rational people are assumed to maximize their utility subject to the constraints they face.

[32] Notably, this definition requires at minimum a consistency and transitivity of choice (that is, if I prefer A over B, and B over C, I must prefer A over C). Models of rational choice are diverse and exist in various forms, adding complexity and assumptions to this simplified definition.

[33] In economics jargon, this would be referred to as facing a negatively sloped demand curve.

[34] Caplan (2006) discusses different types of rationality and analysis with respect to terrorists.

[35] This form of utility function relaxes the need for strict individual self-interest, narrowly construed. An alternative way to partially reconcile selfishness, rational expectations, and rational-choice theory is through the theory of "rational irrationality" (Caplan, 2000). However, this approach suggests that volunteer suicide attackers are irrational (Caplan, 2006). Caplan's "rational irrationality" should be distinguished from what Schelling (1966) calls "the rationality of irrationality," in which an act that is seemingly irrational for individual attackers is meant to demonstrate credibility to a democratic audience that still more and greater attacks are sure to come.

[36] Alternative approaches have been suggested for dealing with irrational terrorists (for example, Caplan, 2006 and Wintrobe, 2003).

[37] In many instances, tactical rationality could suffice to help us develop deterrence tools based on rational-choice behavior. For a discussion of operational rationality and resulting deterrability, see Davis and Jenkins (2002).

[38] Detailed descriptions of each terrorist event in our data enabled us to associate a latitude-longitude coordinate with each attack, which we lincorporated into Geo-

graphic Information System (GIS) software for mapping and distance computations. Noncombatant military personnel who at the time of the incident are unarmed or not on duty were counted as civilians.

[39] A locality is defined as having a regional capital or a center of government administration if one or more of its cities or villages hosted an official bureau of the Israeli Ministry of Interior in 2004. See Berrebi and Lakdawalla (2007) for additional details about the data.

[40] For further details about the data and empirical estimations used in this study, see Benmelech and Berrebi (2007).

[41] By using the different regions in Spain that did not suffer from terrorism and constructing a weighted average comparison to the Basque region, Abadie and Gardeazabal (2003) effectively created a "synthetic control region" that mimics the Basque region in the absence of terrorism.

[42] It is interesting to note that companies related to the defense, security, or anti-terrorism industries economically benefit from terrorism (Berrebi and Klor, 2005, 2009).

[43] See Berrebi and Klor (2005; forthcoming, 2009) for further details about the matching methodology.

[44] Our analysis stops on September 10, 2001, since after that date, U.S.-traded companies are no longer valid controls for Israeli companies affected by terrorist attacks. To allow a comparison of the before and after Intifada effect, only companies traded both before and after September 28, 2000, were included.

[45] See Berrebi (2003) for details about the Israeli daily terrorist attack data.

[46] These are the five publicly traded firms that have suffered the biggest cumulative losses, according to Karolyi and Martell (2007).

[47] See Llussa and Tavares (2007) for a synopsis of studies that discuss the effect of terrorism on aggregate output. For an example of the economy adjusting its allocation of resources to the circumstances imposed by the event of terrorism, see Berrebi and Klor (2005, 2009).

[48] For a good discussion of small versus big effect views of the consequences of terrorism, see Krueger (2007). For a review of the cost of responding to terrorist threats, see Treverton et al. (2008), Richardson, Gordon, and Moore (2007), and Jackson, Dixon, and Greenfield (2007).

[49] I distinguish between territorial goals and the more-inclusive category of political goals, which often comprise territorial ones.

[50] The examples are only that: partial and arbitrary. There was no intent to systematically review the universe of terrorist groups with territorial claims. That would be beyond the scope of this paper.

[51] For an English translation of Hamas's charter, see Yale Law School (2009).

[52] For further details about Hizballah and its manifesto, see Wikipedia.

[53] One problem that arises when evaluating the success in achieving ultimate territorial goals is that, as long as a terrorist campaign is ongoing, it is impossible to confidently determine a failure. Rational expectation would be consistent if, on average, out of those terrorist organizations that ceased their terrorist activities, more achieved concessions than not.

[54] Pape conducted a follow-up study (2005) that expanded and updated this analysis. The follow-up study adds more data on the global patterns of suicide terrorism through the end of 2003 and also tests the main hypotheses against all of the other causal factors that are prominent in the literature across several domains, relying on methods that include variation between cases of suicide terrorism and cases without variation (Pape, 2008). The main findings remain unchanged.

[55] Abrahms (2004) recognizes that terrorists behave according to what I have termed *tactical rationality* on several dimensions, including purposiveness, logic, timing, target selection, and learning. However, he based his irrationality argument on the inability of terrorists to achieve their ultimate stated political goals, such as furthering territorial concessions.

[56] Note that the article was written on the basis of the Palestinian terrorist's example before Israel's pull-out (that is, complete evacuation) of the Gaza Strip in 2005, and Hamas's de facto control of the strip shortly thereafter. These events (considered major achievements to the terrorist organizations in pursuit of their objectives) put Abrahms's irrationality thesis into question.

[57] "With a few exceptions it is hard to see that the attention and publicity have been of much value except as ends in themselves" (Schelling, 1991), p. 20.

[58] Brian M. Jenkins argues that al-Qaeda sees its cause as a process rather than as the efficient pursuit of concrete objectives: "Allah will decide what outcomes will be, but the process of jihad is worthy" (Jenkins, 2006). This might explain the doggedness with which al-Qaeda and predecessor organizations have pursued their cause despite incredible setbacks. Some contrary evidence can be seen in recent recounting of discussions among al-Qaeda members and associates (Stout, Huckabey, and Schindler, 2008).

[59] For example, within the category of political goals, nationalist groups might seek autonomy or secession, whereas religious groups seek the replacement of secular with religious law, and social revolutionary groups seek to overthrow capitalism.

[60] A multitude of competing goals does not contradict stable preferences in the framework of a well-behaved utility function. For example, rational individuals can have preferences over money and leisure at the same time, despite the obvious trade-off between the two, and allocate their time to best achieve both.

[61] As we seek a probabilistic rather than a deterministic measure, the relevant measure in this context would rely on the appropriate weighted average of their goals given their preferences over differing and potentially competing goals and the a priori probability of achieving each. Notably, it is possible to observe seemingly inconsistent behavior over time because of changes in success probabilities while preferences remain stable.

[62] See Iannaccone (2006) for refutation of claims that cultists lack freedom of choice.

[63] See Benmelech and Berrebi (2007) for supporting evidence to that effect.

[64] Further limiting are cases where the terrorist organization did not even exist during the time the individual had to invest in his education and therefore could not have influenced investment decisions with respect to aspired educational attainment.

[65] See also Gupta (2008).

[66] Arguably, it is impossible to empirically single out altruism from other potential factors. However, we should not dismiss a potentially valid explanation just because we lack the tools to empirically validate it.

[67] As acknowledged by the authors, this is an anomaly unsettled by their basic model.

[68] For more details about the al-Aqsa Martyrs' Brigades, see Wikipedia.

[69] For more details about the Kurdistan Workers' Party, see Wikipedia.

[70] Economists, social scientists, and sociologists routinely treat social objectives as rational. As noted above, the definition of rationality in this chapter accords with this interpretation of rationality.

[71] This study also challenges the notion that terrorists are rational actors who attack civilians for political ends. However, the political reality mostly contradicts the authors' interpretation of the terrorists' achievements, which leads me to believe that both political and social goals could be at work simultaneously.

[72] "Terrorism" (2008).

[73] See Krueger (2007) for empirical evidence to that effect.

[74] For empirical support for the rational-behavior theory of voter participation, see, for example, Silberman and Durden (1975). Note that there is an ongoing debate

with respect to the adequacy of rational-choice theory to explain voters' behavior and political parties' operations.

[75] This assumes that terrorism is not only an extreme version of, but also a substitute for, peaceful political activism.

[76] For a formal mathematical presentation of the game theoretical equilibrium results, along with further intuition, see Berrebi and Klor (2006).

[77] For examples relating to the United States, see Davis and Silver (2004), Guilmartin (2004), and Shambaugh and Josiger (2004).

[78] I have omitted objectives that I believed would not be feasibly quantified, such as revenge in response to immediate or long accumulated grievances. Such objectives are likely to be important in and of themselves.

[79] For a contrast between rational-analytic methods for decisions and intuitive decisionmaking based on heuristics and suggested steps toward a synthesis, see Davis, Kulick, and Egner (2005).

[80] These findings contradict those of IDF's Brigadier General Ariyeh Shalev who, in his book on the first intifada (Shalev, 1991), examined the effect of house demolitions on the scope of violence and found that "House demolitions during the [first] Intifada revealed no correlation between the number of houses demolished and the number of violent incidents reported." Many have interpreted this to mean that ". . . the number of violent events did not diminish following house demolitions, and at times even rose." Shalev (1991) used seven data points and no formal analysis to reach his conclusion. Similar findings were reached in an internal IDF report on house demolitions during the al-Aqsa intifada (according to Harel and Isacharoff, 2006). I am unaware of any *empirical* analysis that supported this decision.

[81] "Utility" was initially understood to include far more than just selfish materialistic self-interest (Mill, 1879).

[82] For a good example, see Intriligator and Brito (1988).

CHAPTER SIX

Organizational Decisionmaking by Terrorist Groups

Brian A. Jackson

Introduction

Why do groups choose violence rather than other possible modes of protest or political action? How and when do groups alter their strategic choices to move away from violence or negotiate with the governments they have previously targeted? Having decided to stage terrorist attacks, why do groups choose one type of campaign, operation, or target over another? At the tactical level, why to they pursue one weapon or tactic over others that are available—perhaps more available—to them?[1] How organizations make these decisions and the factors that affect their behavior are major shapers of the threat of terrorism, since the terrorist group is the "unit of production" for terrorist violence: Without a group deciding to stage a terrorist attack and deciding what kind of attack it should be, there will be—almost by definition—no terrorism.

Understanding how and why groups make the choices that they do requires understanding

- how they see their goals and interests and link actions they can take to advance them
- how they get and interpret information about the environment in which they operate
- their understanding of external audiences or constituencies that influence them or that they hope to influence

- their internal group dynamics and how the relationships between members, internal conflicts, or overall group cohesion can shape choices
- other group preferences, including such factors as their tolerance for risk, desire for information and deliberation before acting, or other elements that affect the relative attractiveness of different courses of action.

At a given point in time, groups have a vision of their options—from the strategic down to the tactical level—and, shaped by these factors, of the decision processes that drive their choice among those options.

Groups also face constraints that may get in the way of successfully achieving what they want. For example, the information they use as the basis for decisions may be wrong, communications problems may hamstring their efforts to coordinate action, divergence of interests among individual group members or leaders may lead them to act in ways that undermine the interests of the group, or simple "bad luck"—their environment changing in ways they could not have foreseen—can cause their efforts to fail.

Individual choices are made by terrorist groups at specific points of time, using whatever decision process is in place at that point. However, over its operational career, a terrorist group will make many choices and the decision processes that it uses can also evolve. Each action a group takes provides the opportunity for the group to assess its assumptions about how its choices are—or are not—advancing its goals. Looking at the perceived success or failure of particular actions may change group preferences or demonstrate the need for other activities—for example, improved training or weapons in response to operations the group views as ineffective. A group's ability to learn from its experiences and alter its behavior defines the extent to which the analyst must understand the group as a dynamic rather than as a static adversary (Jackson, 2005; Jackson et al., 2005).

How Different Disciplines Approach This Question

To understand how terrorist organizations make decisions, different fields of social science bring different approaches and perspectives, use different sources of data, and inform different elements of the problem. Economic approaches look at how choices seek to increase a group's "utility" based on their assumed costs and benefits (Sandler and Enders, 2004). Approaches that frame groups as rational economic decisionmakers have their limits, of course, and more-sophisticated behavioral-economics approaches examine how such factors as imperfect information, limits on decisionmaking, and other factors can shape group choices. Approaches based on individual and group psychology examine the factors that shape how group members or leaders make choices and the interpersonal and organizational dynamics that can shift decisions away from either any individual's likely choice or what might be viewed as the "average" choice of all the people involved (see, for example, Moscovici and Zavalloni, 1969). Organization theory looks at how differences in organizational design and functioning can shape the choices a group makes (Chai, 1993; Combating Terrorism Center, 2006; Oots, 1989). Historical, political science, and sociological approaches frequently take a broader view, examining the contexts in which groups operate and other organizations or factors that can contribute to shaping a group's preferences or constraining its options (McCormick, 2003). Game theory explicitly examines the effects of competition and other dynamics on group choices. Competitors of a terrorist group can include both the security forces that seek to limit their effectiveness or take them on directly and other terrorist groups (Sandler and Arce, 2003).*

Relationship to Companion Papers

Consideration of group decisionmaking links directly into several companion papers:

- Root-cause factors define in large part how particular groups assess the costs and benefits of turning to terrorist violence (Noricks, 2009).

---

* Although social-science factors shape group decisionmaking, nonsocial-science factors also are important, such as availability of information or equipment and the physical vulnerability of potential targets.

- Individual-motivation factors drive individuals to believe that terrorism is appropriate or to join an existing terrorist organization (Helmus, 2009). As a result, they shape the "labor pool" available to the group, defining the individuals who will be in the group to make decisions, provide technical insight into the potential consequences of making one choice or another, and bring their own views and assumptions about which strategies, operations, or tactics would best serve the group's interest.
- The dynamics of sympathy for terrorism in broader populations similarly links directly into a group's decision calculus (Paul, 2009). To take a metaphor many analysts have used to describe terrorist organizations—as "sharks in the water" that must keep moving to survive—and combine it with Mao's classic observations on insurgency, the population is the school of fish through which the terrorist shark swims and its tolerance or assistance can be a major contributor to the group's viability. Although different groups have different needs for it, to the extent it is needed, maintaining that support becomes an important shaper of group preferences and choices (see, for example, Mascini, 2006).
- Similarly, questions about how terrorism ends (Gvineria, 2009) have a major nexus with thinking about how and why groups make choices: For many groups, maintaining their own survival is a major driver and, conversely, as a particular organization begins to experience changes in structure and functioning that may lead to its demise, the way it approaches decisionmaking can change as well.

Understanding Terrorist Group Decisionmaking

Social-science approaches have been applied to a variety of terrorist decisionmaking problems. At the broadest level, strategic questions, such as why organizations choose to cross the threshold of staging terrorist violence at all, have been examined (see, for example, Post, Ruby, and Shaw, 2002, and the references therein). Other similar "threshold questions" have been asked on more specific subjects, such as the factors that cause groups to choose to use unconventional rather than conventional weapons or carry out mass casualty attacks.[2] Rather than examining "all or nothing" questions that focus on specific thresholds, other studies ask questions of degree—for example, what affects the portion of a terrorist group's activities that focus on one target set as opposed to other types of sites it could attack.

The approaches applied in studies of organizational decisionmaking differ. Broader theoretical work draws on such fields as organizational theory and psychology that attempt to build an overall picture of the processes of and influences on group decisionmaking. More focused work "zooms in" on individual variables, looking for correlation between, for example, such variables as group structure or ideology and the choices of terrorist targets. Case studies look at individual groups' decisionmaking processes, choices, and behavior and often examine how they have changed over time.[3] The empirical work on organizational choices is somewhat limited by the available data. Since terrorist groups are violent, clandestine organizations, collecting data on their internal functioning is challenging under the best of circumstances (Silke, 2004).

As a result, many studies are based on the most easily observable data on terrorist activities—counts and characteristics of their attacks—and focus on the subset of choices that shapes them. The nature and number of attack operations is one important consequence of group decisionmaking. However, studies that rely only on attack data have limits: There are well-documented problems with the related datasets and, more seriously, looking at number of attacks represents only one window into one category of group choices (Drakos, 2007; Drakos and Gofas, 2006; Schulze, 2004; Silke, 2001). As a result, there is a significant body of work that can best be described as theory supported by evidence, rather than theory that has been *proven*. Much of the work is done by making analogies to other types of organizations that are easier to study and for which a deeper body of theory and research therefore exists.[4] Sometimes, such argument by analogy is sensible; at other times, it may be more problematic.

The Implications of Group Structure and Functioning on Decisionmaking

An understanding of terrorist decisionmaking must begin with the observation that terrorist groups differ in ways that have important implications for the way they make choices. Looking across terrorist groups, small cells of individuals, such as the Red Brigades or Red Army Faction, isolated as clandestine groups and relying only on their own members, made choices driven by the internal dynamics and pressures of their existence. There have been much broader violent movements, such as the Earth Liberation Front or various radical right-wing organizations, which exist as loosely coupled entities whose "decisions" are made through sympathizers publishing their views and ideas for consideration by the movement as a whole. What "decisionmaking" means in each of these contexts is very different, with the factors shaping it in the former potentially being irrelevant in the latter case.

Even focusing only on the current Salafi jihadi–inspired terrorist threat facing the United States, there is still wide heterogeneity, ranging from concern about the dynamics of opinion and activity in a global network of people and groups linked via the Internet to the choices made by small groups of individuals who may radicalize largely in isolation. As a result, before digging into specific factors that might influence decisionmaking in one group or another, the first step must be to examine the ways terrorist organizations can differ that affect what decisionmaking means and where it occurs. I break these differences into two classes—those relating to the structure of terrorist groups and those relating to their functioning.

The Structure of Terrorist Groups. In considering how organizations do anything—including make decisions—a common starting point is to ask how they are structured. Starting from organizational theory, this approach to terrorism analysis asks questions about how different functions, capabilities, or authorities are broken up within a group. For example, for a group operating in multiple geographic areas, are its activities managed centrally or are they broken up into area commands? Are all members of the group expected to have similar skills or are there different parts of the organization with functional specialties?[5] Are there different levels of authority within the organization where decisions are made or specific actions taken?* Are there formal positions within the organization (for example, a position in charge of fundraising or logistics) or are roles more fluid? The answers to all of these questions can have implications for group decisionmaking and the factors that will shape it.[6]

The fact that terrorist organizations have structured themselves differently has led to analytical efforts to build groups' "organization charts" to map their functioning. These studies have identified some organizations, such as the Provisional Irish Republican Army, which developed complex structures with considerable differentiation in both functions and areas of responsibility,[7] whereas other smaller groups have been much less complex or differentiated in the way they have designed themselves.[8] Studies have also documented groups' making significant changes in the way they have structured themselves over time in response to external pressures and demands.[9]

In the literature that approaches terrorist groups as formally structured organizations, the focus has frequently been on the behavior of "the terrorist cell" either as an isolated entity or as the smallest unit within larger group structures.[10] Defined as a small number of collocated individuals who were all involved in terrorist activities, much of the focus of analysis has been on how such small groups function and how their circumstances drive behavior (see, for example, Post, 1987). Cells might be linked to one another through their leadership, with this interaction drawing on thinking built from examination of such comparatively hierarchical organizations as commercial firms. Different cells might have different functions; for example, some might focus on attack operations whereas others play logistical roles.

Since the late 1990s, there has been a countervailing movement in the analytical community that, rather than focusing on groups as formally structured entities, instead looks at terrorism as created by less structured, more-fluid networks of individuals and organizations (notably, Arquilla and Ronfeldt, 1999, 2001). In such networks, coor-

---

* This might differ from decision to decision; for example, "high-stakes" decisions might be made centrally, whereas authority to make routine decisions is delegated to lower-level leaders or even to individual members.

dination of activities happens through loose links among individuals or entities with resources coming together (or "swarming") to carry out tasks rather than through formalized organizational relationships. With al-Qaeda as the most prominent terrorist threat on the world stage and its evolution to a loosely linked transnational organization after the loss of its Afghan safe haven after the September 11, 2001, attacks, this "network-centric" approach to considering terrorist organization and behavior has been a dominant model in thinking.[11] The new focus on networks and terrorism was viewed by some as a change in the nature of terrorism—a "new terrorism" contrasted against an older, more traditional political violence (Lesser et al., 1999).[12] In subsequent years, examination of al-Qaeda has expanded the network concept even more to consider the organization as a broad social movement that plays more of an inspirational role (particularly via modern media tools, such as the Internet[13]) than being directly involved in all the activities of sympathizers or actions taken in its name.[14]

Reflecting this shift in focus away from groups' formal organizational structures, a significant body of terrorism analysis in recent years has applied such techniques as social network analysis, which focuses on individual terrorists and their activities. These analytical methods attempt to determine not "who the organizational chart says are connected to whom," but "who *really* is connected to whom" and how those informal interpersonal connections shape behavior.[15] These analyses assess structures within organizations not based on the centrality or importance of a particular position or unit but of particular individuals based on their links to others within and outside the organization.[16]

The differences between structural and "network-focused" approaches have created some disagreement in the terrorism literature. Furthermore, analyses approached from one perspective or the other can produce different conclusions about what factors drive group behavior or choices. These differences in views through the late 1990s and into the years after the September 11, 2001, attacks on the United States centered on the importance of more tightly linked hierarchical terrorist groups versus more loosely coupled networked versions. Subsequently, over the course of 2008 during the time of this research, these differences have persisted and were reflected in scholarly disagreement about the contemporary threat—whether it largely comes from loose, mainly self-organized groups of individuals (a position identified predominantly with Marc Sageman (Sageman, 2008) or whether groups with more centralized authority involved in recruitment of operatives and operational planning were still important (an argument identified centrally with Bruce Hoffman (Hoffman, 2008)). I have sought to build a framework in this paper for considering decisionmaking that would be applicable to organizations at either end of this spectrum, because I believe that implying that there is an "either/or" choice between formal organization and looser network- or movement-driven influences on group behavior is, in fact, largely artificial.[17] Recent studies, informed both by the behavior of the broad global Salifi jihadi movement and by the activities of individual groups within it, have recognized that the activities of small cells of individuals, formal structures within defined organizations, and the influences of broader global networks and ideological movements are all important shapers of behavior (for example, Atran and Sageman, 2006; Cragin and Daly, 2004; Krebs, 2002).[18] Strategists within the movement have even made this point for themselves:

> In 1989, strategist Abu Musab al-Suri . . . after discussing the relative merits and flaws of "pyramid hierarchy," "thread connection," and hybrid organizational structures, confessed: "Structuring an organization requires a lot of thought and foresight, it should take into account the nature and strengths of the enemy, the type and strengths of its security system, the geographical nature of the country, what has worked and what has failed in similar situations. . . . The particular conditions on the ground should determine the best structure for the organization" (Stout et al., 2008, p. 34).

This is similar to the study of other organizations, where it has been recognized that both formal relationships (that is, those appearing in an organization chart) and informal relationships among group members are important for understanding how an organization actually functions. Given the heterogeneity of terrorist group behavior, it is likely that both self-organized, loosely connected groups of individuals and more-structured organizations with stronger command and control will contribute to the threat, so treatments of group decisionmaking need to address both organizational types.

The Functioning of Terrorist Groups. Although a more inclusive view of terrorist organizational structure better captures such groups' complexity, it makes analyzing "terrorist group decisionmaking" more difficult. What *decisionmaking* could mean in a small group of individuals is very different from what it might signify in a delocalized movement that exists globally and is connected only via the Internet. Similarly, decision dynamics could be very different in a small group where there is a clear leader from one whose members are linked by common ideals but where no one of them has any formal authority over the rest. In spite of these obvious differences, "decisions" made by the full range of terrorist entities are important and can have implications for threat analysis and modeling of terrorist behavior. What is needed is an approach to think about decisionmaking that accommodates the structural diversity but still makes it possible to think through terrorist decisionmaking in a systematic way; we need a way to define reasonable "decisionmaking units" within larger, potentially much more complex terrorist organizations. Doing so makes it possible to think through the factors that can shape decisions in a variety of groups, without the real differences among them creating confusion or leading the analyst to false conclusions.

To do so, I drew on past work that examined authority and influence relationships within terrorist organizations involved in making and implementing decisions. This work was driven by the basic observation that, within a terrorist group, network, or even a broad movement, authority or influence is exerted by some members over others and that the scope of that influence varies (Jackson, 2006). With respect to decisionmaking, the relevant observation is that decisions are shaped by the actors involved in making them and the authority or influence* rela-

---

* I use the terms *authority* and *influence* together to emphasize that the ability of one member of a group to tell another member what to do, or to affect a decision they are making, could come from formal authority (for example, the command and control of a leader over a fol-

tionships that exist among them. This approach represents a hybrid in some respects between the social network type approach to terrorism (which looks at relations between individuals) and more-organizational approaches (which look at formal organizing and authority structures). By looking at the authority and influence relationships among individuals, it asks which actors should be viewed as part of a "decision-making unit."[19] The basic premise of this approach is that the stronger the linkages and influence relationships that exist among actors, the more appropriate it is to view them as part of a single decisionmaking unit, since (a) the stronger the links, the more effect the interactions among individuals are likely to have on choices and (b) the stronger the links, the greater chance that the decisions will actually lead to action by those individuals. To look at the different strengths of interaction among members of a terrorist group, I used influence or control at the strategic, operational, and tactical levels to organize my thinking.

According to this logic, elements of an organization that interact with each other only at the strategic level—over general goals and aims of action—should not be considered part of a single decisionmaking unit. Relevant examples of these sorts of interactions are the exchanges among individuals that occur on jihadi Internet message boards or the more broad-based "interaction" that occurs between Osama bin Laden and sympathetic members of the global Salafi jihad through the release of speeches or videos. Although such interactions and exchanges are important and can inform thinking about the global terrorist threat, the individuals involved are sufficiently distant from one another that it is not productive to think about them as common participants in a decisionmaking process.[20]

In contrast, as the relationships between individuals get closer and tighter, for example, elements of groups where command or influence relationships among actors provide control at the operational level as well (such as relationships between leaders and individual units within a dispersed terrorist organization) or all the way down to the tactical level (such as relationships among members of a single terrorist cell

---

lower) or through less well-defined influence (for example, of a prominent or more-experienced member of a group over newer members).

making choices about its own activities), it becomes much more useful to view the individuals as a single decisionmaking unit. In these cases, the relationships among individuals are close enough that their nature is likely to affect how choices are made and will certainly dictate how the group acts on those decisions.[21]

The structures of influence relationships can result in significantly different decisionmaking units within groups. Where there are strong and directional authority and influence relationships, decisionmaking may involve largely group leadership. At one extreme are authoritarian groups in which a single leader makes the majority (or all) of the decisions. Cult-like terrorist groups (for example, Aum Shinrikyo in Japan led by Shoko Asahara) or other groups led by single charismatic leaders could be at or near this extreme. Other groups may have decision processes involving multiple leaders (for example, the Provisional Irish Republican Army's Army Council and associated command structure),* although there may be differences in the amount of authority and decisionmaking influence.[22] In groups in which influence or authority is more diffuse and the distinction between leadership and group members is less clear, decisionmaking may be more participatory. In such groups, choices and the "authority" to put them into action may come from the group as a whole through either democratic, majority-rule, or even consensus-based processes.[23] Tying these functional concerns back to the previous structural discussion, the different archetypes of group organizational decision structures are also related to the balance between the centralization of decisionmaking authority versus its delegation to elements of the group.[24]

As a result, in examining terrorist decisionmaking, *this paper focuses on groups of individuals in which specific operational choices and tactical decisions are made, and in which enough direct authority or influence exists to implement them, as the "decisionmaking unit" in terrorist organizations.* The subset of a given organization that should be appro-

---

* Whether groups are hierarchical (which is frequently cited as the opposite of networked) or not is largely a question about the directionality of authority and influence relationships in the organization. Organizations in which influence goes largely in one direction (top down) are traditional hierarchies. Networked groups may still have some directional authority relationships within the group overall or subcomponents.

priately viewed as a decisionmaking unit would therefore be determined by the relationships between individual members, and more complex organizations might have multiple distinct loci where decisions of different types are made.* Small terrorist cells will generally be appropriate to consider as single decisionmaking units, since they make tactical-level choices about the activities they undertake and the attacks they stage and they act on those choices. In some cases, actors with only operational-level influence over others might qualify for inclusion—for example, including group leaders with only partial authority over dispersed cells—when actual choices are being made at the operational level. Individuals who are involved in only general strategic and operational-level discussions—such as what occurs in most Internet venues or through the broad dissemination of strategic documents[25]—would not be considered part of a decisionmaking unit within a group.[26] It is not that I believe that the influence of such broader networks or movements is not important, just that it is more analytically productive to view it as an external factor that may constrain or inform a group's decision process rather than as an integral part of the process itself.

Factors Influencing Group Decisionmaking

Whether the relevant decisionmaking unit within a terrorist group is a single leader, a group, or the organization as a whole, a common set of factors can be identified that shape the likelihood of the group's choosing one course of action over another. At one time, there was debate in the analytical literature about whether it was appropriate to consider terrorist groups (and individual terrorists) as rational actors or whether individual pathology or other factors made their decisionmaking and choices entirely different from "normal" individuals and organizations (Ahmed, 1998; Anderton and Carter, 2005; Crenshaw, 1972, 2000; Dugan, Lafree, and Piquero, 2005; Lake, 2002; Miller, 2006; Muller and Opp, 1986; Ruby, 2002; Shapiro, 2005a; Sprinzak, 2000). Accord-

---

* This might differ from decision to decision—many individuals may be empowered to make certain types of choices about day-to-day operations but fewer for "higher-stakes" choices.

ing to studies of both individual terrorists and group behavior, it is clear that most members of terrorist groups resemble members of the general population in most ways and—concurrent with the requirements for being considered rational actors—that groups make choices based on their beliefs about their interests and goals.[27]

As a result, although the outcome of their decision processes may be abhorrent and deviate considerably from what most members of society would choose to do, terrorist group decision processes themselves are not inherently irrational. Understanding what shapes those decisions, therefore, requires cataloguing the factors that lead the group to conclude that a particular course of action is more in its interest than other possible actions (or no action at all). However, accepting terrorists as rational does not mean that their choices will always be good ones, even from their own point of view.[28,29]

To provide a framework to discuss the range of factors, I have broken down the elements that affect the likelihood of a group's choosing to act into five classes that capture both the reasons why a group might act and broader processes and preferences that might shape its decision to do so:

- group members' understanding of external audiences or constituencies that influence them or that they hope to influence, and assumptions about how they will respond to specific strategic, operational, or tactical choices
- the group's goals, interests, and values and how the group understands the actions it could take with respect to them[30]
- the group's internal group dynamics
- the group's view of the risks involved in a given action, and whether those risks are acceptable given what group members believe the action might achieve
- the resources the group is willing to commit
- whether group members believe they have enough information to make the decision to act.

For a given choice, a combination of these different factors will drive the likelihood that a group will decide to act in a particular way.[31] All elements will also be influenced considerably by a group's preferences—for example, some groups will be willing to take more risk than others—again emphasizing how analysis at this level must recognize and appropriately address differences among groups. The following sections drill into each of these categories in more detail and examine "subfactors" that shape their contribution to group choices.

Belief That Action Will Positively Influence Relevant Audiences. For terrorist groups whose capabilities are dwarfed by the states or organizations they are seeking to target, use of violence for influence purposes has long been a core element of terrorism (Bueno de Mesquita and Dickson, 2007; Decker and Rainey, 1980; Fleming, 1980; Martin, 1985). This dynamic, with a central purpose of terrorist attacks viewed as "violent propaganda"[32] to affect audiences, was the basis for Brian Jenkins's frequently quoted statement that terrorists "want a lot of people watching, not a lot of people dead" (Jenkins, 1987). Even as terrorism has changed such that some terrorists—whether because of their religious motivations or their other goals—do indeed "want a lot of people dead," the effect of terrorist violence on audiences is still an important part of groups' decisionmaking calculus. Use of the Internet and video by modern terrorist groups—and the level of effort these groups devote both to capturing their acts and delivering them globally with comparatively high-production values—demonstrates the importance of audience reactions in their activities (Kimmage and Ridolfo, 2007).

A significant amount of the focus in this area has been the effect of attacks on varied audiences, but other activities have the potential to affect the opinions of the group.[33] For example, groups' choices to broadly pursue criminal activity to support themselves, or to engage in such specific criminal acts as drug dealing or kidnapping, have been influenced by the groups' views of how that decision would be seen by outside audiences (see, for example, Makarenko, 2004; Roth and Sever, 2007).[34]

The effect of an action by a terrorist group on that group's targeted populations—political leadership, government organizations, commercial firms, or members of the public of the nations the group views as its enemy or are seeking to provoke—is one set of public perceptions that shape its choices; there are other audiences as well.[35] Support (or potential support) populations that the terrorist group identifies with[36] or relies on are critical (della Porta, 1995; Drake, 1998).[37] Arguments about how audiences react have been used to explain differences in groups' choices of targets in communities where their support base will be exposed to the violence rather than attacks staged far away from their "home territory."[38] Broader views about the perceived legitimacy of the actions groups are planning to take (for example, whether a particular means of attack will be viewed as abhorrent by the group's supporters or provoke too strong a reaction in its adversaries)[39] can constrain the group's choices.[40] A group's judgments about the likely reaction of the audiences that are important to it for any given action are based on the information that the group has available to it, so limits in that information's quality or completeness could result in unforeseen reactions—and therefore consequences—as a result.[41] Examples from recent al-Qaeda activities include backlashes that occurred after large attacks targeting hotels in Jordan, large sectarian and civilian targeted attacks by al-Qaeda in Iraq,[42] and other operations that caused significant Muslim casualties, such as the November 2003 attack in Riyadh, Saudi Arabia.

The views of external audiences can also change a terrorist group's incentives to act and the values it places on different choices in combination with other external events. The actions of competing terrorist groups[43] that threaten the group's stature or position as well as external factors (such as current events or political changes)[44] that the group feels it must respond to can provide a catalyst for action (Drake, 1998).

Belief That Action Will Advance Group Goals or Interests. The direct linkage of group actions—whether the staging of violent attacks or more mundane choices, such as logistical decisions—to group goals is the core of the "rationalist" approach to assessing terrorist behavior (see McCormick, 2003, and the references therein).[45] Although the other motivations for terrorist action can complicate the direct linkage of all activity to the group's goals (Crenshaw, 2001),[46] this factor shaping the chance of a group's making a particular choice centers on its *beliefs* about how acting will advance its interests.

A number of taxonomies of factors have been developed specifically focused on the *transition to violence* by groups—or the choice of terrorism versus other modes of protest or action. They include 11 characteristics put forward by Sprinzak (Sprinzak, 2000):

1. the intensity with which the group delegitimizes its opponents
2. the absence of moral inhibitions or antiviolence taboos in the group's culture
3. members' previous experience with violence
4. whether the group has rationally assessed the risks and opportunities of violent action
5. the level of its organizational, financial, and political resources
6. the group's sense of imminent threat
7. competition with other groups
8. the age of the activists involved, where younger groups are more likely to turn to violence
9. any external influence or manipulation of the group toward violence or support that makes the transition to violence easier
10. a sense on the part of the group of humiliation and the need to take revenge
11. the leader's past experience with violence.[47]

The last item is discussed in Post, Ruby, and Shaw (2002), which also puts forward a detailed framework of factors thought to contribute to the transition to violence. Paraphrasing, factors in that framework relevant to organizational decisionmaking are

1. group ideology and goals
2. group experience with violence
3. leadership characteristics
4. leadership style and decisionmaking processes
5. organizational processes
6. presence of groupthink and polarization in the group
7. group views of environment, adversary, and level of threat outside support
8. planning for violent action and other organizational changes.[48]

These factors, framed more generally to apply to decisions beyond simply the choice of terrorism are reflected in the factors discussed here.

In evaluating a potential action for whether it will advance the group's strategy or other interests, a central element will be the consistency of the action with the group's ideology. Looking across a variety of different groups, the effect of ideology is clear, for example, in how groups select targets for attack and which of many possible targets are considered most advantageous (Calle and Sánchez-Cuenca, 2007; della Porta, 1995; Drake, 1993, 1998) and what other activities are consistent with the group's approach to operations. A similar calculus applies to whether a particular action is consistent with advancing a group's other goals or interests (Drake, 1998).

Just as was the case for a group's ability to anticipate the reaction of an outside audience to its actions, a group's ability to predict how particular actions will advance its interests is neither perfect nor immune to error. For some terrorist groups that are simply seeking to produce destabilization or chaos, nearly any outcome of a terrorist attack may suffice. However, for groups with more subtle agendas, it may be difficult to anticipate whether a given attack or other action will be beneficial and whether or not it is may depend on the reactions of others or on events that are outside the group's direct control. As a result, the history of terrorism is replete with choices made by groups that believed at the time that the choice would be advantageous but, with more complete information and the benefit of hindsight, that proved to be ill-advised. In many cases, actions undermining rather than advancing a group's interests are driven by the response to the action, either alienating sympathetic populations (discussed in the previous section) or catalyzing action by the group's direct opponents that hurt it over the longer term.[49]

Beliefs will be influenced by group preferences that shape how consistency of different actions with goals, interests, values, ideology, and so on are judged, as well as on the environmental conditions that shape what the group sees as its available choices and their relative merits (McCormick, 2003, pp. 481–482). It is also the case that there may be disagreement within a group over particular actions,[50] since judging the ideological consistency of a particular attack or whether an action that is under consideration is a good idea is hardly an exact science.

Alignment of action with the preferences of any external influences can also affect the relative attractiveness of different choices available to a group, including state sponsors,[51] cooperating groups, or strategic and operational influences of networks or movements the group is associated with.[52]

Although much of this discussion has been framed in terms of organizational interests related to the political and other goals of the group, it is important not to lose sight of the fact that groups can have other interests as well. Some terrorist groups or members have used violence and criminal means as a way to amass resources to support their own standard of living.[53] Furthermore, as individuals become more invested in the organization, the survival of the group may become a goal in and of itself—separate from its pursuit of its stated goals—which could push decisionmaking in other ways.[54]

Belief That Action Will Produce Positive Reactions Inside the Group. Although groups take action to advance their goals, it is also true that they take actions for reasons that might be viewed as "entirely their own": to address internal organizational needs that may or may not directly relate to the ideological or other agenda the group supposedly exists to pursue (Crenshaw, 2001). This reality means that simply thinking through how particular actions do or do not mesh with a group's stated ideology or goals may not be enough to understand the choices it makes. An example of a wholly internal motivation for terrorist violence is early anarchic terrorist groups, which McCormick has labeled as "expressionist" because their violence was not aimed at achieving particular ends but instead as a "means of individual expression [that] served the individual and collective psychological needs of the terrorists themselves" (McCormick, 2003, p. 477). Such expressive actions might involve a much less deliberative decision process than acts intended to achieve particular goals. This description of an earlier generation of terrorists echoes the characterization of some that modern Salafi jihadi terrorism is less about the goals it is seeking to achieve and more about the process of struggle that those goals enable (see, for example, Jenkins, 2005).

The characteristics of the individuals who lead or are members of a terrorist group can therefore produce actions that are viewed as desirable inside the group but may not be wholly intelligible to the outside observer. These types of idiosyncratic preferences could push toward particular actions (for example, for Aum Shinrikyo leader Shoko Asahara's love of poisons pushing toward the group's use of chemical weapons or individual group members' tendencies toward sectarian brutality in terrorist groups in Northern Ireland undermining the organizations' pursuit of their political objectives). A variety of authors have noted a "bias to action"[55] among terrorist groups that can create a push to act rather than wait or reflect on individual decisions (Crenshaw, 1985; Drake, 1998).[56] This bias to action can be stronger in some members of the organization than others, producing conflict.[57] This push toward action over inaction has led others to liken terrorist groups to "sharks in the water," which must remain in motion to maintain group cohesion and survive (Sper, 1995). Individual members' past experience with violence has also been cited as a contributor to these tendencies in groups and a factor that pushes them toward violent rather than nonviolent activities (e.g., Post et al., 2002). The need to maintain positive "internal opinions" within a group may also constrain choices, foreclosing options that might produce dissent or undermine morale and cohesion (Strinkowski, 1985, p. 44).

Within small groups of individuals, it has also been observed that the dynamics that exist among members and the deliberative processes involved in decisionmaking can skew their results. Processes such as groupthink, where the pressures within a group and limits on the information available to it push decisions away from what might appear to an impartial observer as being the best choice, have been cited as a problem in terrorist organizations (Drake, 1998, pp. 168–169).[58] Other biases can be generated by internal loyalties or dynamics that limit dissent or questioning,[59] and groups' immediate past experience—e.g., recent successes or failures—can lead to biases either toward or away from similar future courses of action.[60]

Acceptability of the Risks Associated with Acting, Given What the Action Might Achieve. Terrorists do risky things, but they want their operations to succeed. Indeed, for groups whose relevance and influence depend on a reputation for effectiveness, repeated failures can have a cost well beyond not achieving the goals of individual operations. Although the advent of suicide terrorism has led some to question how consideration of risk influences terrorist choices (since the safety of operatives was a key part of many groups' risk calculus), groups that stage suicide operations still want their attacks to succeed, so risk is still a factor that shapes planning and decisionmaking.[61] One practical reason is that a substantial "logistics tail" exists behind a typical suicide attack. If the attack fails (for example, the attacker is captured), many members of the organization may be at risk. And, if the operation is implemented incorrectly, it may have the wrong effect.

Terrorist groups can differ considerably in their tolerance for risk, with some willing to "gamble" more than others (see Phillips, 2005, for a theoretical discussion). According to the statements of some group leaders and members, terrorist organizations have been characterized as relatively risk-averse (Hoffman, 1997). The risk tolerance of a group can be shaped by a variety of factors, including the characteristics and preferences if its members as well as their recent experiences, successes, and failures. Group dynamics can skew groups' perceptions of risk as well, resulting in what has been termed the "risky shift"—that is, groups often take riskier decisions than either individuals deciding alone or an "average" of those individual decisions.[62] Depending on the nature of the group, interpersonal dynamics might shift risk decisions in other ways as well.

What the group considers a success (that is, how permissive their "success criteria" are) will also have an effect. Studies have defined success for terrorist operations in different ways,[63] but the reality is that success has meant different things for different groups. For example, statements made by al-Qaeda suggest that at least parts of the group are designing operations in an attempt to maximize the casualties and costs that they produce.[64] This is in significant contrast to other terrorist organizations whose definitions of a successful operation were more modest and had different requirements.[65] How critical success is to the group at a given time will also likely play a role; for example, a group that stages five attacks a day in the context of a high-intensity terrorist campaign (which could, for example, be part of a ongoing insurgency or a wider civil war) may be far less concerned with the success of any individual operation than a group that "gears up" for a single attack over months or years. A group's judgment of the likelihood of success of an action will also be shaped by its own beliefs about "how good it is"—whether it has the capabilities and skills to be successful. Groups' conception of success will also change over time as its circumstances and aims shift.

Environmental characteristics, such as defenses at the types of targets the group wants to attack, the general counterterrorism effectiveness of the state in which the group operates* are also drivers of the perceived risk of different actions. Counterterrorism pressure can make action seem risky (since operatives might be caught in the process) but feelings of pressure and "threat" have also been cited as a factor that may push groups to act and to do so in ways that they otherwise might not (Post, Ruby, and Shaw, 2002). The group's perceptions of its ability to address any counterterrorism pressure or defenses at targeted sites, as well as its operational security skills to hide from security organizations, will also shape risk judgments.[66]

It is also the case that, for any given group, the acceptability of the risk involved can be shaped by how great the perceived gains are from acting. Although this realization implies a linkage between different factors that I have discussed separately, it is impossible to escape that risk judgments will almost certainly be shaped by what the "upside" is of action over inaction. Situations where the group seeks the potential for great gains—or its situation is sufficiently dire that the costs of inaction seem catastrophic—may lead to risks being viewed as tolerable that might not be otherwise.[67] Whether a group's decision that higher risks were justified is correct or not can be assessed only in hindsight, with that judgment frequently biased by whether the action ended up being successful.

---

*   State counterterrorism effectiveness also has effects on later elements that affect such group decisions as the amount of resources the group is willing to devote to operations.

Level of Resources the Group Is Willing to Commit to Action. Whether the group is willing to devote what it sees as "enough" resources to allow an operation to succeed will also contribute to the choices that it makes. Groups will have some view of how many people, how much money, and how many other resources (such as weapons for attack operations) will be required for a particular action. These judgments will be influenced in part by the group's risk tolerance (discussed above), since that will shape both whether it is willing to "cut it close" in planning operations (for example, using at or near the minimum amount of resources it thinks is needed) and also how much of its available resources it is willing to gamble on a particular operation or activity. Other internal preferences will shape this component of decisionmaking as well.

This decision will also likely be affected by the group's available stocks of resources—in money, technology (e.g., weapons), people, and even time for planning and training—where richer groups will be more likely than poorer ones to allocate resources to action. Elements of the group's environment (for example, the counterterrorism effectiveness of relevant states or how "underground" the group is operating [della Porta, 1995; Drake, 1998, p. 167–168; McCormick, 2003, and the references therein]) will affect the stocks of resources that the group has available to it. Other factors, such as whether the group has a safe haven from which it can operate, can increase its "stocks" of such things as planning and training time.[68] If the group has sources of resources from external entities (for example, money, capabilities, or technology provided by sponsors such as sympathetic states or other terrorist groups), then it may be willing to devote more resources to an action, since it may not be expending its own stocks.

Belief That the Group Has "Enough" Information to Decide to Act. Whether or not a group believes that it has "enough" information to make a decision to act will also influence decisionmaking. Terrorist groups have different thresholds for the amount of information the need before they decide to act. At one extreme, in the context of terrorist campaigns some groups have sent operational teams "out" with the authority to stage attacks if they see a good opportunity, setting a very low bar for how much information is demanded before an attack is staged. At the other are groups that stage extensive surveillance and preattack research before they are willing to even begin an operation. Preferences for required levels of information will differ from group to group and even from individual to individual within groups.*

A group's information requirements fall into two broad classes— situational awareness (understanding of its environment)[69] and the technical knowledge needed to both assess its choices and implement the courses of action it chooses. How much of both of these a group has will depend in part on its internal stocks of resources—what the people it has know, its ability to gather new information through experimentation (or, in the case of a specific attack operation, through intelligence, surveillance, and reconnaissance)—and any external sources of information it has available to it.[70] Relevant sources of information can include other groups, state sponsors, sympathetic individuals outside the group, and so on. If the group is deep underground, its need to remain clandestine can be a barrier to gathering new information of both types.[71]

Even if particular information is theoretically available to an organization, this does not necessarily mean that it will be available to the specific decisionmakers considering a choice. Linking this discussion back to the structural issues discussed at the opening of this section, if information and decisionmaking are located in different parts of a group, then a group's choices can be affected. For example, if decisions are made centrally, local information that is available to individuals in local operational units may not be available. Similar arguments could be made for specialized knowledge available in some parts of a group (for example, with its bomb-makers). For such knowledge to reach decisionmakers, groups must have sufficient—and sufficiently effective—communications capabilities to move either the information or the people who have it to where the decisions are actually being made. This can be a security challenge for terrorist groups. Just as there are differences among groups in how internal structures and processes affected how decisions were made, the effect of what group mem-

---

* See, for example, the rationale for relocation to Afghanistan by Ayman al-Zawahiri described in Gerges (2005).

bers know and how that knowledge will affect decisions will differ. For example, in groups that must accept recruits based on "enthusiasm rather than skill" to gain the necessary manpower, individuals might be accepted—and potentially involved in decisionmaking— whose influence would not improve the quality of those choices. In contrast, in other groups, particularly those where suicide operations are a central component of the group's activities, such individuals can be given assignments where their effect on the group's choices will be both circumscribed and temporary.

Nature of the Evidence on Terrorist Group Decisionmaking: Agreements and Disagreements

Within the literature on factors shaping terrorist decisionmaking, there is broad consensus on a set of factors that influence group decisions, even though there might not be extensive datasets truly substantiating the link. Some of this includes "reasonable theory" on the influence of group members' or leaders' views on decisionmaking or the influence of organizational structure, some illustrated by analogy to other organizations that are more amenable to study. This theory is also frequently illustrated by examples drawn from individual terrorist group members' or leaders' statements on how decisions were made inside their respective groups or on reporting on individual terrorist organizations. As a result, there is not a great deal of disagreement in the literature on individual factors and their influence—especially in view of the wide range of approaches taken (from studying coupled groups through very loosely coupled networks) and issues of terminology. Some of the other factors that I have described as most exploratory throughout the discussion are based solely on analogies to organizations other than terrorist groups.

Relationships and Hierarchies

In thinking about terrorist group decisionmaking, previous studies have sought to bring together all of the factors potentially shaping decisions into a single "mechanistic" model. Reflecting the interest in terrorist group targeting behavior, one focus has been creating models of how groups select targets and plan attacks. Two notable examples can be found in Drake (1998). In the first, the various factors that could potentially determine which targets are desirable to terrorist groups are viewed as the gradually narrowing of a "funnel" from all possible targets to those meeting all the criteria. In the second, the influence of each factor is even more strict, with such criteria as the ideological justification for a particular terrorist attack or protection at the proposed target framed as a series of "yes/no" choices (requiring sequential "yes" decisions before an attack is staged) (Drake, 1998, pp. 176, 180).

Such constructions provide a systematic guide for thinking through how decision processes might occur in an idealized terrorist group. However, viewing the process as a more fluid combination of factors that determine a final choice provides a way to capture the broader set of influences that can shape decisions. As a result, rather than viewing the various factors identified in the literature as sequential elements of a decision process, I suggest a model (Figure 6.1) that organizes them as a set of influences that come together in different combinations and can change the way a group is likely to act.

In considering the factors shaping how a group makes choices, the central elements affecting the likelihood of a particular decision are that the risk of acting is viewed as acceptable, adequate resources can be allocated to it such that the group thinks it can succeed, the group has enough information to act, and one or more of the rationales for acting is in place: Acting will either advance group interests or goals, produce a positive reaction in a relevant external audience, or address internal group needs. Below these top-level elements are a series of factors that can increase or decrease the likelihood that they will be satisfied.

How a decisionmaking unit within a group weighs these factors differs from organization to organization. As discussed in the introductory sections of the paper, the structure of a group and the authority or influence relationships that define the decisionmaking units within it will also shape the relative importance of different factors. Other group characteristics matter as well. For groups that are early in their life cycle and are highly dependent on the support of a local population, how actions are expected to shape the views of the population may be of greatest importance. The choices of groups with complex internal dynamics or competing internal factions may be dominated by how they will "play" for internal constituencies. Some groups will be optimizers, seeking to make the best choices possible given their circumstances; others will satisfice, seeking what is "good enough" rather than optimal. Groups will also have different preferences and tolerances for such factors as operational risk or incomplete information. Although a given level of information or risk may reach an acceptable threshold for one organization and it will decide to act, another organization might instead decide to gather more intelligence or defer a decision until the risks of acting can be reduced.

Implications for Strategy and Policy

In considering strategies for counterterrorism aimed at the organizational level, the decisionmaking processes of terrorist groups are one potential target for action. In general, if options are unavailable to take on terrorist groups directly, action to complicate or shape their decisionmaking in ways that make it more difficult for them to plan and stage violent actions can be an alternative. To the extent that the factors shaping a particular group's decision processes can be identified and understood, defensive measures can also be better crafted to frustrate their operations or guide group choices in ways that are favorable for defense and security organizations. Clear models, based on available social-science understandings of terrorist organizational behaviors, can help to guide such policy design.

In considering terrorist group decisionmaking, and opportunities for counterterrorism action, there is an existing basis for thinking: Efforts at deterrence and influence are, at their most basic, efforts to shape the choices that groups make about the things they do and the ways they attempt to do them. From this perspective, security measures seek to shape group risk tolerance, as do "classic" attempts at deterrence through threats of punishment or retaliation for specific acts; information operations telegraphing how use of unconventional weapons would be viewed negatively in a group's

Figure 6.1
Factors Shaping Terrorist Group Decisionmaking

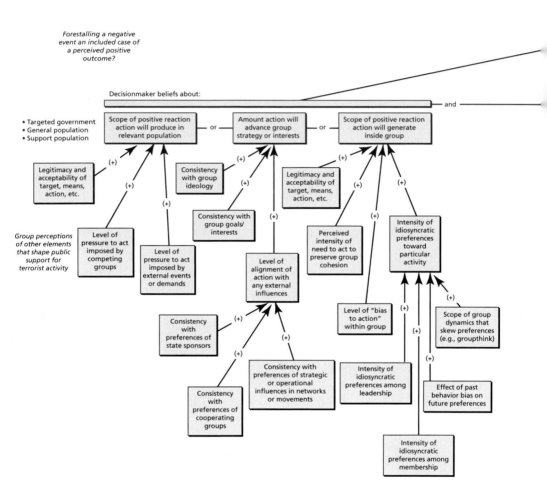

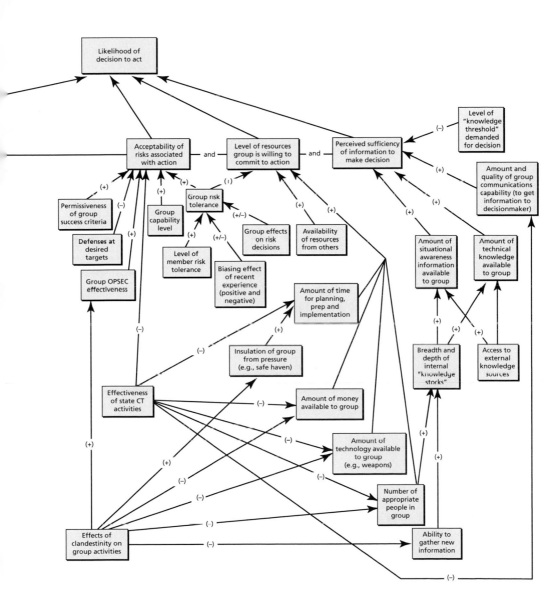

sympathizer community or by the international community seek to shape a variety of different elements of decisionmaking and also the apparent value of pursing them; diplomatic efforts to end state sponsorship seek to constrain group resources and reduce the chance that the group will be willing to devote enough people and technology to operations of concern; and so on.[72]

Drawing on current social-science literature and other available data on group decisionmaking, the model represented in Figure 6.1 captures the range of elements that shape group choices, from what those choices are intended to accomplish to the risk and information thresholds a group may impose on its on decisions. In framing these factors, some elements go back through decades of terrorism research and are supported by broad bodies of knowledge in related fields. For example, the influence of organizational dynamics on decisionmaking has been recognized for many years, has been examined in a variety of terrorist organizations, and—given the much broader interest in how all groups make choices—has been examined in a variety of other types of organizations as well. Others factors are more provisional, being based on smaller amounts of data or less direct study of the behavior in terrorist groups. An example of this latter class is questions about the effect of organizational risk tolerance on group decisionmaking. Although there are enough data to make a case that it is an important factor that needs to be considered and included, it has not itself been systematically studied. As a result, in some cases, concepts are framed just as they are in literature sources and my model summarizes those concepts in a uniform structure; in others, I have crafted categories to bring together separate (and sometimes still thin) strands of thought that have not yet been fully explored or placed into a useful framework.

When considering application of this model to guide thinking about strategy and policy, some caveats are appropriate. First, for simplicity the model is framed as a group considering a single choice, with the factors shaping the likelihood of making a decision. In reality, decisionmakers generally consider multiple options and make comparisons among them—even if only to compare the consequences of an action against the consequences of taking no action at all. As a result, the additional complication of making those comparisons among options, and the effect of which options groups choose to weigh against each other, is a factor that I have not considered explicitly.

Second, although understanding the factors that shape group choices can make it possible to make more informed projections about what groups may or may not do, real limits constrain the ability to do so. There are limits on the intelligence that is available about internal group deliberations. As a result, just as the terrorist group decisionmaker will invariably have limits on available data, so too will the policymaker or operator in counterterrorism.

Third, even a systematic and deliberative terrorist group that makes the best decision it can at a given time can simply be wrong, meaning that even with an excellent model that captures and integrates all available information on how choices are made, it may still not be possible to predict the group's future behavior. To be rational does not mean to be perfect. A variety of elements can affect the quality of a group's choices, including divergence between what the group thinks and what "truly is" (for example, a group's assumptions about how its actions will be interpreted by their target audiences). A group can similarly have misperceptions about its own levels of skills or knowledge that can lead to miscalculation and error (Drake, 1998). Incorrect knowledge, bad situational awareness, too little information, or too much data to assess in the time available to the decisionmakers can similarly degrade the quality of an organization's choices.[73] The situation the group faces may also simply change, making choices that might have been beneficial in one environment detrimental under new circumstances. As with all groups, terrorist organizations have bounds on their understanding and knowledge that make it difficult to project their own actions into the future and understand their consequences with certainty.

Fourth, although many analyses approach terrorist groups as "single decisionmakers," the reality is that they (and all organizations) are made up of individuals whose interests may or may not be entirely congruent with those of the group as a whole. Individual actors "looking out for themselves" can therefore inject different sets of preferences into a decision process. Some authors have framed this as a principal-agent problem given that central terrorist actors have difficulty controlling what other members do, thus opening the opportunity for self interested behavior (Chai, 1993; Shapiro, 2005a). Such individual behaviors also arise as a result of rivalry within groups and contests for power or resources.[74]

Finally, although a group may make "the right choice" to advance its interests in a specific situation, it is not certain that it will actually be able to implement its decisions. Some of the same factors that were discussed above—such as command and control issues,[75] organizational structure constraints, security concerns limiting communication, and the bringing together of resources—can also get in the way of a group's ability to translate thought into action.

As a result, even armed with the best that social-science research currently has to offer, the task of understanding terrorist group decisionmaking and anticipating group behavior must be approached with humility. In doing so, the analyst (and a decisionmaker relying on the analysis) must know something of the nature of the group itself. Examination must focus on the appropriate decisionmaking units within an organization, where the results of social-science research on group behavior are applicable, and should not assume that identical processes apply in groups as vastly different as individual terrorist cells and loosely coupled collections of individuals linked only by a common ideology. Although both the interplay of ideas and the deliberative processes that occur at all points on the spectrum of terrorist organizational behavior are important, variations in the processes in different types of organizations must be understood and taken into account.

The fact that decisionmaking and the factors that shape it are inherently idiosyncratic in individual terrorist groups means that general models that can predict "terrorist behavior" will be elusive at best. However, at the same time, even if models may not be predictive, laying out the range of factors that shape terrorist group decisions can still aid in understanding how groups of interest might behave and can guide counterterrorist thinking. In examining a specific group and the factors that are most important given the available information on its leadership, membership, and environment, such models can help to assess the possible effects of different strategies for deterrence or influence and to prioritize those likely to be more or less likely to be effective against it.

Bibliography

Ahmed, Fazal, *Terrorism as a Rational Tactic: An International Study*, Ann Arbor, Mich.: University of Michigan, 1998.

Al-Zawahiri, Ayman, letter to Abu Mus'ab al-Zarqawi, July, 9, 2005. As of January 5, 2008:
http://www.globalsecurity.org/security/library/report/2005/zawahiri-zarqawi-letter_9jul2005.htm

Anderton, Charles, and John Carter, "On Rational Choice Theory and the Study of Terrorism," *Defence and Peace Economics,* Vol. 16, No. 4, 2005, pp. 275–282.

Arquilla, John, and David F. Ronfeldt, "The Advent of Netwar: Analytic Background," *Studies in Conflict & Terrorism,* Vol. 22, No. 3, 1999, pp. 193–206.

————, eds., *Networks and Netwars: The Future of Terror, Crime, and Militancy,* Santa Monica, Calif.: RAND Corporation, 2001. As of January 5, 2009:
http://www.rand.org/pubs/monograph_reports/MR1382/

Atran, Scott, and Marc Sageman, "Connecting the Dots," *Bulletin of the Atomic Scientists,* Vol. 62, No. 4, 2006, p. 68.

Atwan, Abdel Bari, *The Secret History of al Qaeda*, Berkeley, Calif.: University of California Press, 2006.

Berrebi, Claude, and Darius Lakdawalla, "How Does Terrorism Risk Vary Across Space and Time? An Analysis Based on the Israeli Experience," *Defence and Peace Economics,* Vol. 18, No. 2, 2007, pp. 113–131.

Bloom, Mia, *Dying to Kill: the Allure of Suicide Terror,* New York: Columbia University Press, 2007.

Bueno de Mesquita, Ethan, and Eric S. Dickson, "The Propaganda of the Deed: Terrorism, Counterterrorism, and Mobilization," *American Journal of Political Science,* Vol. 51, No. 2, 2007, pp. 364–381.

Calle, Luis de la, and Ignacio Sánchez-Cuenca, "The Production of Terrorist Violence: Analyzing Target Selection in the IRA and ETA," unpublished, 2007.

Carley, Kathleen M., "Organizational Learning and Personnel Turnover," *Organization Science,* Vol. 3, No. 1, 1992, pp. 20–46.

Carley, Kathleen M., and David M. Svoboda, "Modeling Organizational Adaptation as a Simulated Annealing Process," *Sociological Methods & Research,* Vol. 25, No. 1, 1996, pp. 138–168.

Carter, Josh, "Transcending the Nuclear Framework: Deterrence and Compellence as Counter-Terrorism Strategies," *Low Intensity Conflict and Law Enforcement,* Vol. 10, No. 2, Summer 2001, pp. 84–102.

Casebeer, William, and Troy Thomas, "Deterring Violent Non-State Actors in the New Millennium," *Strategic Insights,* Vol. 1, No. 10, December 2002. As of January 5, 2009:
http://www.ccc.nps.navy.mil/si/dec02/terrorism2.asp

Chai, Sun-Ki, "An Organizational Economics Theory of Antigovernment Violence," *Comparative Politics,* Vol. 26, No. 1, 1993, pp. 99–110.

Combating Terrorism Center, *Harmony and Disharmony: Exploiting al-Qa'ida's Organizational Vulnerabilities,* West Point, N.Y.: U.S. Military Academy, 2006.

Connor, Robert, *Defeating the Modern Asymmetric Threat,* Monterey, Calif.: Naval Postgraduate School, 2002.

Cragin, Kim, and Sara A. Daly, *The Dynamic Terrorist Threat: An Assessment of Group Motivations and Capabilities in a Changing World,* Santa Monica, Calif.: RAND Corporation, 2004. As of January 5, 2009:
http://www.rand.org/pubs/monograph_reports/MR1782/

Crelinsten, Ronald D., "The Internal Dynamics of the FLQ During the October Crisis of 1970," in David C. Rapoport, ed., *Inside Terrorist Organizations,* New York: Columbia University Press, 1988, pp. 59–89.

Crenshaw [Hutchinson], Martha, "The Concept of Revolutionary Terrorism," *The Journal of Conflict Resolution,* Vol. 16, No. 3, September 1972, pp. 383–396.

Crenshaw, Martha "The Causes of Terrorism," *Comparative Politics,* Vol. 13, No. 4, July 1981, pp. 379–399.

———, "An Organizational Approach to the Analysis of Political Terrorism," *Orbis,* Vol. 29, No. 3, Fall 1985, pp. 465–489.

———, "The Psychology of Political Terrorism," in Margaret G. Hermann, ed., *Political Psychology,* San Francisco, Calif.: Jossey-Bass Publishers, 1986, pp. 379–413.

———, "The Logic of Terrorism: Terrorist Behavior as a Product of Strategic Choice," in Walter Reich, ed., *Origins of Terrorism: Psychologies, Ideologies, Theologies, States of Mind,* Washington, D.C.: Woodrow Wilson International Center for Scholars, 1990, pp. 7–24.

———, ed., *Terrorism in Context,* University Park, Pa.: Pennsylvania State University Press, 1995.

———, "The Psychology of Terrorism: An Agenda for the 21st Century," *Political Psychology,* Vol. 21, No. 2, 2000, pp. 405–420.

————, "Theories of Terrorism: Instrumental and Organizational Approaches," in David C. Rapoport, ed., *Inside Terrorist Organizations,* London: Frank Cass, 2001, pp. 13–31.

Davis, Paul K., and John Arquilla, *Thinking About Opponent Behavior in Crisis and Conflict: A Generic Model for Analysis and Group Discussion,* Santa Monica, Calif.: RAND Corporation, 1991. As of July 14, 2008: http://www.rand.org/pubs/notes/N3322/

Davis, Paul K., and Brian Michael Jenkins, *Deterrence & Influence in Counterterrorism: A Component in the War on Al Qaeda,* Santa Monica, Calif.: RAND Corporation, 2002. As of July 14, 2008: http://www.rand.org/pubs/monograph_reports/MR1619/

Decker, Warren, and Daniel Rainey, "Terrorism as Communication," paper presented at the annual meeting of the Speech Communication Association, New York, November 13–16, 1980. As of February 25, 2009: http://eric.ed.gov/ERICDocs/data/ericdocs2sql/content_storage_01/0000019b/80/35/f1/0b.pdf

Della Porta, Donatella, "Left-Wing Terrorism in Italy," in Martha Crenshaw, ed., *Terrorism in Context,* University Park, Pa.: Pennsylvania State University Press, 1995, pp. 160–210.

Dingley, James, "The Bombing of Omagh, 15 August 1998: The Bombers, Their Tactics, Strategy, and Purpose Behind the Incident," *Studies in Conflict & Terrorism,* Vol. 24, 2001, pp. 451–465.

Dolnik, Adam, and Anjali Bhattacharjee, "Hamas: Suicide Bombings, Rockets, or WMD?" *Terrorism and Political Violence,* Vol. 14, No. 3, Autumn 2002, pp. 109–128.

Don, Bruce W., David R. Frelinger, Scott Gerwehr, Eric Landree, and Brian A. Jackson, *Network Technologies for Networked Terrorists: Assessing the Value of Information and Communications Technologies to Modern Terrorist Organizations,* Santa Monica, Calif.: RAND Corporation, 2007. As of July 14, 2008: http://www.rand.org/pubs/technical_reports/TR454/

Drake, C. J. M., "The Role of Ideology in Terrorists' Target Selection," *Terrorism and Political Violence,* Vol. 5, No. 4, 1993, pp. 253–265.

————, *Terrorists' Target Selection,* New York: St. Martin's Press, 1998.

Drakos, Konstantinos, "The Size of Under-Reporting Bias in Recorded Transnational Terrorist Activity," *J.R. Statist. Soc A,* Vol. 170, No. 4, October 1, 2007, pp. 909–921.

Drakos, Konstantinos, and Andreas Gofas, "The Devil You Know But Are Afraid to Face: Underreporting Bias and Its Distorting Effects on the Study of Terrorism," *Journal of Conflict Resolution,* Vol. 50, No. 5, October 1, 2006, pp. 714–735.

Dugan, Laura, Gary Lafree, and Alex R. Piquero, "Testing a Rational Choice Model of Airline Hijackings," *Criminology,* Vol. 43, No. 4, 2005, pp. 1031–1065.

Fleming, Marie, "Propaganda by the Deed: Terrorism and Anarchist Theory in Late Nineteenth-Century Europe," *Studies in Conflict & Terrorism,* Vol. 4, No. 1–4, 1980, pp. 1–23.

Gerges, Fawaz A., *The Far Enemy: Why Jihad Went Global,* New York: Cambridge University Press, 2005.

Gurr, Ted Robert, "Terrorism in Democracies: Its Social and Political Bases," in Walter Reich, ed., *Origins of Terrorism: Psychologies, Ideologies, Theologies, States of Mind,* Washington, D.C.: Woodrow Wilson International Center for Scholars, 1990, pp. 86–102.

Gvineria, Gaga, "How Does Terrorism End?" in Paul K. Davis and Kim Cragin, eds., *Social Science for Counterterrorism: Putting the Pieces Together,* Santa Monica, Calif.: RAND Corporation, 2009. As of January 17, 2009: http://www.rand.org/pubs/monographs/MG849/

Hayden, Nancy Kay, *Terrifying Landscapes: A Study of Scientific Research into Understanding Motivations of Non-State Actors to Acquire and/or Use Weapons of Mass Destruction,* Fort Belvoir, Va.: Defense Threat Reduction Agency, 2007.

Helmus, Todd C., "Why and How Some People Become Terrorists," in Paul K. Davis and Kim Cragin, eds., *Social Science for Counterterrorism: Putting the Pieces Together,* Santa Monica, Calif.: RAND Corporation, 2009. As of January 17, 2009: http://www.rand.org/pubs/monographs/MG849/

Hoffman, Aaron, "Taking Credit for Their Work: When Groups Announce Responsibility for Acts of Terror, 1968–1977," paper presented at the annual meeting of the International Studies Association, Montreal, Quebec, March 17, 2004.

Hoffman, Bruce, "The Modern Terrorist Mindset: Tactics, Targets and Technologies," Center for the Study of Terrorism and Political Violence, St. Andrews. University, Scotland, October 1997. As of February 25, 2009: http://www.ciaonet.org/wps/hob03/

———, "The Myth of Grass-Roots Terrorism: Why Osama bin Laden Still Matters," *Foreign Affairs,* May/June, 2008.

Hoffman, Bruce, and Gordon McCormick, "Terrorism, Signaling, and Suicide Attack," *Studies in Conflict & Terrorism,* Vol. 27, No. 4, 2004, pp. 243–281.

Homeland Security Institute, *Underlying Reasons for Success and Failure of Terrorist Attacks: Selected Case Studies,* Washington, D.C., 2007.

Horgan, John and Max Taylor, "The Provisional Irish Republican Army: Command and Functional Structure," *Terrorism and Political Violence,* Vol. 9, No. 3, 1997.

Jackson, Brian, "Technology Acquisition by Terrorist Groups: Threat Assessment Informed by Lessons from Private Sector Technology Adoption," *Studies in Conflict & Terrorism,* Vol. 24, No. 3, 2001, pp. 183–213.

———, "Groups, Networks, or Movements: A Command-and-Control-Driven Approach to Classifying Terrorist Organizations and Its Application to Al Qaeda," *Studies in Conflict & Terrorism,* Vol. 29, No. 3, 2006, pp. 241–262.

Jackson, Brian A., John C. Baker, Kim Cragin, John Parachini, Horacio R. Trujillo, and Peter Chalk, *Aptitude for Destruction, Volume 1: Organizational Learning in Terrorist Groups and Its Implications for Combating Terrorism,* Santa Monica, Calif.: RAND Corporation, 2005. As of July 14, 2008:
http://www.rand.org/pubs/monographs/MG331/

———, *Aptitude for Destruction, Volume 2: Case Studies of Organizational Learning in Five Terrorist Groups,* Santa Monica, Calif.: RAND Corporation, 2005. As of July 14, 2008:
http://www.rand.org/pubs/monographs/MG332/

Jackson, Brian A., David R. Frelinger, Michael J. Lostumbo, and Robert W. Button, *Evaluating Novel Threats to the Homeland: Unmanned Aerial Vehicles and Cruise Missiles,* Santa Monica, Calif.: RAND Corporation, 2008. As of July 14, 2008:
http://www.rand.org/pubs/monographs/MG626/

Jenkins, Brian M., "The Future Course of International Terrorism," *The Futurist,* July–August 1987.

———, "The Jihadists' Operational Code," in David Aaron, ed., *Three Years After: Next Steps in the War on Terror,* Santa Monica, Calif.: RAND Corporation, 2005, pp. 3–8. As of July 14, 2008:
http://www.rand.org/pubs/conf_proceedings/CF212/

Jordan, Javier, Fernando M. Mañas, and Nicola Horsburgh, "Strengths and Weaknesses of Grassroot Jihadist Networks: The Madrid Bombings," *Studies in Conflict & Terrorism,* Vol. 31, No. 1, January 1, 2008, pp. 17–39.

Kimmage, Daniel, and Kathleen Ridolfo, *Iraqi Insurgent Media: The War of Ideas and Images,* Washington, D.C.: Radio Free Europe/Radio Liberty, 2007.

Kirby, Aidan, "The London Bombers as 'Self-Starters': A Case Study in Indigenous Radicalization and the Emergence of Autonomous Cliques," *Studies in Conflict & Terrorism,* Vol. 30, No. 5, 2007, pp. 415–428.

Koschade, Stuart, "A Social Network Analysis of Jemaah Islamiyah: The Applications to Counterterrorism and Intelligence," *Studies in Conflict & Terrorism,* Vol. 29, No. 6, 2006, pp. 559–575.

Krebs, Valdis, "Mapping Networks of Terrorist Cells," *Connections,* Vol. 23, No. 3, 2002, pp. 43–52.

Kydd, Andrew H., and Barbara F. Walter, "The Strategies of Terrorism," *International Security,* Vol. 31, No. 1, 2006, pp. 49–80.

Lake, David A., "Rational Extremism: Understanding Terrorism in the Twenty-First Century," *Dialog-IO,* Vol. 56, No. 2, Spring 2002, pp. 15–29.

Lesser, Ian O., Bruce Hoffman, John Arquilla, David F. Ronfeldt, Michele Zanini, and Brian Michael Jenkins, eds., *Countering the New Terrorism,* Santa Monica, Calif.: RAND Corporation, 1999. As of July 14, 2008: http://www.rand.org/pubs/monograph_reports/MR989/

Lustick, Ian S., "Terrorism in the Arab-Israeli Conflict: Targets and Audiences," in Martha Crenshaw, ed., *Terrorism in Context,* University Park, Pa.: Pennsylvania State University Press, 1995, pp. 514–552.

Magouirk, Justin, Scott Atran, and Marc Sageman, "Connecting Terrorist Networks," *Studies in Conflict & Terrorism,* Vol. 31, No. 1, January 1, 2008, pp. 1–16.

Makarenko, Tamara, "The Crime-Terror Continuum: Tracing the Interplay Between Transnational Organised Crime and Terrorism," *Global Crime,* Vol. 6, No. 1, 2004, pp. 129–145.

Martin, L. John, "The Media's Role in International Terrorism," *Studies in Conflict & Terrorism,* Vol. 8, No. 2, 1985, pp. 127–146.

Mascini, Peter, "Can the Violent Jihad Do Without Sympathizers?" *Studies in Conflict & Terrorism,* Vol. 29, No. 4, 2006, pp. 343–357.

McCartan, Lisa M., Andrea Masselli, Michael Rey, and Danielle Rusnak, "The Logic of Terrorist Target Choice: An Examination of Chechen Rebel Bombings from 1997–2003," *Studies in Conflict & Terrorism,* Vol. 31, No. 1, January 1, 2008, pp. 60–79.

McCormick, Gordon, "The Shining Path and Peruvian Terrorism," in David C. Rapoport, ed., *Inside Terrorist Organizations,* New York: Columbia University Press, 1988, pp. 109–128.

———, "Terrorist Decision Making," *Annual Review of Political Science,* Vol. 6, No. 1, 2003, pp. 473–507.

Melese, Francois, and Diana Angelis, "Deterring Terrorists from Using WMD: A Brinksmanship Strategy for the United Nations," *Defense & Security Analysis,* Vol. 20, No. 4, December 2004, pp. 337–341.

Merkl, Peter H., "West German Left-Wing Terrorism," in Martha Crenshaw, ed., *Terrorism in Context,* University Park, Pa.: Pennsylvania State University Press, 1995, pp. 160–210.

Miller, Laurence, "The Terrorist Mind: I. A Psychological and Political Analysis," *International Journal of Offender Therapy and Comparative Criminology,* Vol. 50, No. 2, April 1, 2006, pp. 121–138.

Moscovici, Serge, and Marisa Zavalloni, "The Group as a Polarizer of Attitudes," *Journal of Personality and Social Psychology,* Vol. 12, No. 2, 1969, pp. 125–135.

Muller, Edward N., and Karl-Dieter Opp, "Rational Choice and Rebellious Collective Action," *The American Political Science Review,* Vol. 80, No. 2, June 1986, pp. 471–488.

National Research Council, *Discouraging Terrorism: Some Implications of 9/11,* Washington, D.C.: National Academies Press, 2002.

Neumann, Peter R., "The Bullet and the Ballot Box: The Case of the IRA," *Journal of Strategic Studies,* Vol. 28, No. 6, 2005, pp. 941–975.

Noricks, Darcy M.E., "The Root Causes of Terrorism," in Paul K. Davis and Kim Cragin, eds., *Social Science for Counterterrorism: Putting the Pieces Together,* Santa Monica, Calif.: RAND Corporation, 2009. As of January 17, 2009: http://www.rand.org/pubs/monographs/MG849/

Oots, Kent Layne, "Organizational Perspectives on the Formation and Disintegration of Terrorist Groups," *Terrorism,* Vol. 12, 1989, pp. 139–152.

Parachini, John, "Comparing Motives and Outcomes of Mass Casualty Terrorism Involving Conventional and Unconventional Weapons," *Studies in Conflict & Terrorism,* Vol. 24, No. 5, 2001, pp. 389–406.

———, "Putting WMD Terrorism into Perspective," *Washington Quarterly,* Vol. 26, No. 4, Autumn 2003, pp. 37–50.

Paul, Christopher, "How Do Terrorists Generate and Maintain Support?" in Paul K. Davis and Kim Cragin, eds., *Social Science for Counterterrorism: Putting the Pieces Together,* Santa Monica, Calif.: RAND Corporation, 2009. As of January 17, 2009: http://www.rand.org/pubs/monographs/MG849/

Perrow, Charles, *Complex Organizations,* Glenview, Ill.: Scott, Foresman & Co., 1979.

Phillips, Peter J., "The 'Price' of Terrorism," *Defence and Peace Economics,* Vol. 16, No. 6, 2005, pp. 403–414.

Post, Jerrold M., "Rewarding Fire with Fire: Effects of Retaliation on Terrorist Group Dynamics," *Studies in Conflict & Terrorism,* Vol. 10, No. 1, 1987, pp. 23–35.

Post, Jerrold M., Keven G. Ruby, and Eric D. Shaw, "The Radical Group in Context: 1. An Integrated Framework for the Analysis of Group Risk for Terrorism," *Studies in Conflict & Terrorism,* Vol. 25, No. 2, 2002, pp. 73–100.

Roth, Mitchel P., and Murat Sever, "The Kurdish Workers Party (PKK) as Criminal Syndicate: Funding Terrorism Through Organized Crime, A Case Study," *Studies in Conflict & Terrorism,* Vol. 30, No. 10, 2007, pp. 901–920.

Ruby, Charles L., "Are Terrorists Mentally Deranged?" *Analyses of Social Issues and Public Policy,* Vol. 2, No. 1, 2002, pp. 15–26.

Sageman, Marc, *Understanding Terror Networks,* Philadelphia, Pa.: University of Pennsylvania Press, 2004.

————, *Leaderless Jihad: Terror Networks in the Twenty-First Century,* University of Philadelphia, Pa.: Pennsylvania Press, 2008.

Sandler, Todd, and Daniel G. Arce M., "Terrorism & Game Theory," *Simulation & Gaming,* Vol. 34, No. 3, 2003, pp. 319–337.

Sandler, Todd, and Walter Enders, "An Economic Perspective on Transnational Terrorism," *European Journal of Political Economy,* Vol. 20, No. 2, 2004, pp. 301–316.

Schulze, Frederick, "Breaking the Cycle: Empirical Research and Postgraduate Studies on Terrorism," in Andrew Silke, ed., *Research on Terrorism: Trends, Achievements, and Failures,* London: Frank Cass, 2004, pp. 161–185.

Shapiro, Jacob, "The Greedy Terrorist: A Rational-Choice Perspective on Terrorist Organizations' Inefficiencies and Vulnerabilities," *Strategic Insights,* Vol. 4, No. 1, January 2005a.

————, "Organizing Terror: Hierarchy and Networks in Covert Organizations," paper presented at the annual meeting of the American Political Science Association, Washington, D.C., September 1, 2005b.

————, "The Terrorist's Challenge: Security, Efficiency, Control," Center for International Security and Cooperation, Stanford University, April 26, 2007. As of February 25, 2009:
http://igcc.ucsd.edu/pdf/Shapiro.pdf

Sharif, Idris, *The Success of Political Terrorist Events: An Analysis of Terrorist Tactics and Victim Characteristics, 1968–1977,* Lanham, Md.: University Press of America, 1996.

Silke, Andrew, "The Devil You Know: Continuing Problems with Research on Terrorism," *Terrorism and Political Violence,* Vol. 13, No. 4, 2001, pp. 1–14.

————, ed., *Research on Terrorism: Trends, Achievements, and Failures,* London: Frank Cass, 2004.

Sper, Mary K., *Towards Understanding Terrorism: A Theoretical Examination of Internal Cohesion in Terrorist Groups and the Negative Dynamic of Violence,* Monterey, Calif.: Naval Postgraduate School, 1995.

Sprinzak, Ehud, "Rational Fanatics," *Foreign Policy,* No. 120, September 2000, pp. 66–73.

Stout, Mark E., Jessica M. Huckabey, John R. Schindler, and Jim Lacey, *The Terrorist Perspectives Project: Strategic and Operational Views of al Qaida and Associated Movements,* Annapolis, Md.: Naval Institute Press, 2008.

Strinkowski, Nicholas Charles, *The Organizational Behavior of Revolutionary Groups*, Evanston, Ill.: Northwestern University, 1985.

Tucker, David, "What's New About the New Terrorism and How Dangerous Is It?" *Terrorism and Political Violence,* Vol. 13, Autumn 2001, pp. 1–14.

Endnotes

[1] For example, why would groups choose to pursue unconventional weapons even though they have conventional alternatives, or why would organizations specifically seek out high-technology weapons such as surface-to-air missiles when they have other ways to attack aircraft?

[2] See, for example, Hayden (2007), Parachini (2001, pp. 389–406); 2003.

[3] For example, individual chapters in Crenshaw (1995).

[4] This has been the approach in some previous RAND work, e.g., Jackson (2001); Jackson (2005); Jackson, Baker, Cragin, Parachini, Trujillo, and Chalk, 2005).

[5] A variety of factors are related to whether organizations differentiate into specialized units; some terrorist groups have but others have not. See Strinkowski (1985).

[6] Structural issues have consequences for the functioning of groups that go beyond group decisionmaking. For example, organizations with more formalized and defined roles and structures may have an easier time replacing individuals as members (or leaders) come and go, but the formalization may also increase the security risk to the group by making it easier for security organizations to determine how the organization works and find ways to undermine its operation. See, for example, Gerges's discussions of al-Qaeda in this respect (Gerges, 2005, p. 41).

[7] See, for example, Horgan and Taylor (1997, pp. 1–32).

[8] See, for example, the discussion of the Weather Underground in Strinkowski (1985).

[9] See, for example, discussion of PIRA in Jackson et al. (2005).

[10] See, for example, case studies in Crenshaw (1995).

[11] For recent examples focused on al-Qaeda–inspired terrorism, see Jordan, Mañas, and Horsburgh (2008); Kirby (2007).

[12] However, others have questioned whether there has really been substantial change in the nature of terrorism, or whether the new focus on networks is more a function of new attention to behaviors that have always existed but were simply not broadly appreciated (for example, Tucker, 2001).

[13] Although the use of the Internet in particular has created significant changes in the way some terrorist groups operate, judgments about the breadth of its effects on all parts of the terrorist threat differ (see, for example, Don, Frelinger, Gerwehr, Landree, and Jackson, 2007).

[14] Stout, Huckabey, and Lacey (2008) discuss al-Qaeda's own preference to refer to itself as a movement rather than a network or other more structurally focused term.

[15] For example, Koschade (2006); Magouirk, Atran, and Sageman (2008); Sageman (2004); and many others.

[16] For examining the complementary and interacting effects of terrorist groups' structure, functioning, and the choices they make—and the effect of choices on performance—computational models simulating organizational behavior are particularly useful. See, for example, Carley (1992); Carley, and Svoboda (1996).

[17] The dynamic nature of terrorist groups' structure and the looseness of some ties among members has been recognized for some time (for example, della Porta, 1995). See also Strinkowski (1985, pp. 94–99) for a discussion that applies general hierarchical language to the analysis of terrorist groups but recognizes the limits of doing so.

[18] Indeed, elements of each of these organizational dynamics can exist within the same group, with looser network- or movement-like behavioral influences coexisting with more traditional, structured, and hierarchical group processes. See, for example, Jackson (2006).

[19] In my previous work (Jackson, 2006), I used these concepts to define the difference between a terrorist group, network, and movement and draw a "boundary" around each type of terrorist organization, which is a similar, although not identical, application to what I am doing here to define a reasonable "decisionmaking unit" for analysis.

[20] See, for example, the discussion of the nature of strategic and other discussions in the broad Salafi jihadi movement in Stout et al. (2008).

[21] It is important to note that at this level also a great deal of social-science research examining small group decisionmaking becomes relevant for helping to illuminate group behavior.

[22] See, for example, the discussion of differences of opinion on the al-Qaeda Shura Council regarding the September 11, 2001, attacks discussed in Gerges (2005, pp. 19, 37) and the dominance of charismatic personalities within individual elements of the organization (pp. 34–35).

[23] See, for example, Perrow (1979).

[24] See the discussion in Shapiro (2005a, 2007); and Strinkowski (1985).

[25] See, for example, the discussion in Atwan (2006, p. 142).

26 This is somewhere between considering an entire terrorist group as a single, unitary actor, as is done in some analyses, and much more complex examinations that would approach each group as a collection of potentially conflicting views and opinions (discussed in McCormick, 2003, p. 482).

27 Terrorist organizations may pursue goals and interests that differ significantly from what more mainstream groups and individuals might, and their decision process may be shaped by preferences and factors that make the resulting choices very aberrant compared with the general expectations of society.

28 Drake captured this aptly in an analysis of terrorists' choice of attack targets: "When examining the actions of terrorists, one must bear in mind that one is dealing with people, with all their imperfections and unpredictability, rather than some hypothetical, hyper-rational beings" (1998, p. 163). See also, Crenshaw (1990).

29 A variety of "constrained" versions of rationality have been defined by various scholars to capture the effects of other factors on an otherwise rational decisionmaking process. An earlier model of "limited rationality" framed by Davis and Arquilla was focused on opponent state leaders; it captures many of the elements discussed in this framework of terrorist group decisionmaking: "A leader with limited rationality will make decisions that bear a reasonable relationship to his objectives and values, but they may be flawed by misperceptions, miscalculations, cognitive biases, and the tendency to either accept more or fewer risks than complete 'rationality' would call for" (Davis and Arquilla, 1991, pp. v–vi).

30 In the classical view of terrorism as essentially "armed propaganda," the central mechanism through which a group tried to advance its goals by violent action was based on how a given attack would influence external audiences or constituencies. However, terrorist groups do choose to take actions that are not done solely for propaganda purposes. For example, logistical operations to acquire resources may be carried out in spite of the potential negative effects on public opinion of groups. As a result, I have separated actions to advance group interests or goals from purely influential effects of terrorist attacks to capture the full range of choices these groups make.

31 A particularly good description of the confluence of factors that can lead a terrorist to choose to act can be found in a description of decisionmaking in the Canadian terrorist group the FLQ (the Front de Libération du Quebec) before it staged a kidnapping operation. Different elements of the group felt pressured to act because of a range of factors (discussed below), whereas others felt that the group would be better served by remaining clandestine and building its capacity and capabilities (Crelinsten, 1988).

32 In contrast, for example, to a military strike where the destruction of a particular target was the central purpose.

33 See Neumann (2005) for a discussion of the Irish Republican Army (IRA) and its behavior after it chose to compete in the political process linking its perceived success directly to public opinion.

[34] The choice to pursue criminal activity could also be related to a group's values and other concerns discussed in the next section.

[35] See Crenshaw (1981) for a review of the varied audiences relevant to terrorist group activities and the different reactions groups seek to produce in them.

[36] See, for example, Lustick (1995, pp. 514–552).

[37] For example, one element of al-Zawahiri's thinking in considering al-Qaeda's targeting of the United States was reportedly based on the positive reaction among the umma. He was quoted as saying al-Qaeda "wins over the umma when we choose a target that it favours, one that means it can sympathize with those who hit it . . . it has responded favourably to the call for jihad against the Americans" (Atwan, 2006, p. 85).

[38] For example, PIRA targeting in Northern Ireland rather than on the British Mainland (see discussion of PIRA in Jackson,. Baker, Cragin, Parachini, Trujillo, and Chalk, 2005) and Chechen groups choices of targets in Russia rather than Chechnya (McCartan et al., 2008).

[39] See, for example, a description of the debate in the al-Qaeda Shura Council regarding pursuit of unconventional weapons in Gerges (2005, p. 196).

[40] "In revolutionary organizations the strategies and tactics of the group are often constrained by cultural norms from the relevant environment or the internal norms of the collectivity" (Strinkowski, 1985, p. 44).

[41] See Gurr (1990, pp. 86–102) and Hoffman and McCormick (2004, pp. 243–281).

[42] See a letter from Ayman al-Zawahiri to Abu Mus'ab al-Zarqawi (al-Zawahiri, 2005) or comments by Abu Muhammed al-Maqdisi quoted in Stout et al. (2008, p. 22).

[43] For example, suggestions of competition among Italian groups leading to changes in levels of violence (della Porta, 1995) or among Palestinian groups leading to different claims of responsibility (Hoffman, 2004); discussion in Bloom (2007). See also the discussion of other examples in Combating Terrorism Center (2006).

[44] Such reaction by groups can be observed in propaganda or other statements released on the Internet in response to current events.

[45] Kydd and Walter (2006) provide a review of the various "strategies of terrorism" that link violent action in particular to group goals. See Dolnik and Bhattacharjee (2002) for an example of a notional rational choice analysis of a group's weapons choices incorporating a number of cost-benefit and preference elements.

[46] A nuanced discussion of the multiple layers of motivations that drove decision-making in the Real IRA's Omaugh bombing operation can be found in Dingley

(2001). That the operation did not go as planned and the reactions and outcomes actually significantly undermined the group's interests is instructive.

[47] Paraphrased from the discussion in Post, Ruby, and Shaw (2002, pp. 73–100).

[48] Paraphrased from the discussion in Post, Ruby, and Shaw (2002, pp. 73–100).

[49] See the discussion of the Omaugh bombing in Dingley (2001).

[50] For example, Gerges discusses significantly different views within al-Qaeda of whether or not the asymmetry in forces between the group and its avowed enemies was important (2005, p. 199).

[51] For example, state sponsors have been seen for some time to impose certain limits on what is acceptable for any associated terrorist groups, since exceeding limits in violence or crossing thresholds in the types of weapons used (for example, transition to unconventional arms) was viewed as risking retaliation against the state.

[52] This last category includes the effect of broader influences in the global jihadist movement (from the direct statements of Osama bin Laden and Ayman al-Zawahiri down to discussions on relevant Internet posting boards) on the decisions by any individual group sympathetic to the goals and interests of the movement.

[53] For example, leaders of Loyalist terrorist groups in Northern Ireland were characterized to the author by interviewees in law enforcement organizations as using group resources for their own ends, in contrast to the Provisional Irish Republican Army where funds were viewed as group resources first. A more complete discussion of "misappropriation of funds" in terrorist groups is available in Shapiro (2005a).

[54] Reviewed in McCormick (2003, pp. 473–507).

[55] This drive can create very different decision dynamics in groups where it is dominant:

> Some (otherwise distinctive) groups, such as the contemporary Irish Republican Army (IRA), al-Qaeda, Hamas, and the Tamil Tigers, have largely managed to subordinate their actions to their political objectives. Like their predecessors in the People's Will, the Socialist Revolutionary Party, or the early Fenian movement, they employ violence largely as a means to an end. Others, such as the late November 17, the Popular Forces of 25 April (FP-25), the Justice Commandos of the Armenian Genocide, or any number of today's "amateur" terrorists, have effectively subordinated their political objectives to their need to act. As for their predecessors in the early anarchist movement or the more recent Japanese Red Army, Baader-Meinhof Gang, or Weather Underground, violence is an instrument of expression rather than an instrument of political change. The decision dynamics associated with each of these two sets of cases are distinct (McCormick, 2003, p. 480).

[56] For example, in his history of al-Qaeda and the international Salafi jihad, Gerges cites this sort of pressure on Ayman al-Zawahiri when he was still running Islamic

jihad: "But by the end of the Afghan war in the late 1980s, al-Zawahiri, his former associates said, came under tremendous pressure from Islamic jihad's rank and file to launch attacks in Egypt, and by the early 1990s he obliged" (2005, p. 121). Others have flagged differences in the preference for violence between "street level" operatives in PIRA and the commanders, as well as changes in how groups' new recruits' preferences for violence may shift over time as initial "ideologically driven" recruits are supplanted by later recruits attracted by the organizations' violent acts rather than the rationale for them (for example, see discussion of the Red Army Faction (RAF) in Merkl [1995, pp. 191–195], and a number of other groups, including al-Qaeda, in Combating Terrorism Center [2006]). McCormick (2003) provides a broader review of literature relevant to this factor.

[57] Stout et al. (2008) discuss the problem of a bias to action among rank-and-file members of al-Qaeda and the complications it created for group leaders executing their strategy (pp. 50–55). Similar issues have been observed in other groups in vastly different circumstances, such as the Provisional Irish Republican Army.

[58] For a discussion of groupthink in the Liberation Tigers of Tamil Eelam (LTTE), see Connor (2002, p. 102).

[59] Gerges cites this as a problem in al-Qaeda owing to individuals' loyalty to bin Laden (2005, p. 38). See also Sper (1995).

[60] For example, in an examination of Chechen attacks, successful bombings (particularly those with high victim counts) increased the probability of future bombings (McCartan, Masselli, Rey, and Rusnak, 2008). Although a number of possible mechanisms could explain such a correlation, one might be a bias in decisionmaking created by past "success."

[61] McCormick has framed this risk tradeoff as groups seeing to balance their need for security (which requires invisibility) and influence (which requires overt action) (2003, pp. 498–499).

[62] See the discussion in Drake (1998, p. 170). McCormick (2003) provides a broader review of literature relevant to this factor.

[63] For example, Homeland Security Institute (2007); Sharif (1996).

[64] For example, in his discussion of Ayman al-Zawahiri's argument for the use of suicide operations, Atwan quotes him as arguing that they are the way to "inflict maximum casualties on the enemy at the least cost in terms of casualties for the mujahedin" (2006, p. 84). Comments by bin Laden have reflected similar arguments about the level of costs al-Qaeda has been able to inflict on the West compared with the costs to the group.

[65] See, for example, the discussion of PIRA in Jackson, Baker, Cragin, Parachini, Trujillo, and Chalk (2005) and a broader discussion of various groups' success criteria in Jackson, Frelinger, Lostumbo, and Button (2008, pp. 64–65).

[66] Remaining underground or clandestine can negatively affect factors shaping group decisions, such as the availability of information or resources, even though operational security is helped.

[67] See, for example, the discussion in Berrebi and Lakdawalla (2007), which indicates that Palestinian groups are willing to travel farther (taking greater risk of apprehension) for higher-value terrorist attacks.

[68] See, for example, the rationale for relocation to Afghanistan by Ayman al-Zawahiri described in Gerges (2005).

[69] Extracted discussions of strategic intelligence from jihadi documents (quoted and described in Stout et al. (2008, pp. 74–79) are instructive in this area.

[70] These activities can be broadly captured as the groups' ability to learn new information over time. The author and RAND colleagues have reviewed these processes and the factors that shape groups' ability to carry them out in Jackson (2005) and Jackson, Baker, Cragin, Parachini, Trujillo, and Chalk (2005).

[71] See, for example, the discussion of al-Qaeda in Combating Terrorism Center (2006); this has been observed in a variety of other terrorist groups as well (for example, a similar description in a 1980s case study of the Shining Path is available in McCormick, 1988).

[72] For a review of deterrence issues and terrorism, see Carter (2001); Casebeer and Thomas (2002); Davis and Jenkins (2002); Melese and Angelis (2004); and National Research Council (2002).

[73] Discussed in McCormick (2003, p. 482).

[74] See, for example, Gerges's discussion of internal rivalries and factions within al-Qaeda that hurt the functioning of the organization (2005, pp. 104–105).

[75] See, for example, the discussion in Strinkowski (1985) of the sources of authority of group leadership that provide the ability to influence others' behavior. Shapiro (2005b, 2007) discusses the principal-agent issues in terrorist organizations that make it difficult for even leaders with authority to control the behavior of supposed subordinates. Al-Qaeda has had such problems when the actions of its affiliates have produced negative responses in public opinion. See also Combating Terrorism Center (2006). Stout et al. discuss views of al-Qaeda strategists of the problem of "uncoordinated and ill-considered action [by independent actors in the movement that] are at best unhelpful and at worst counterproductive" (2008, p. 118).

CHAPTER SEVEN

How Does Terrorism End?

Gaga Gvineria

Introduction

One of the most important terrorism-related questions for the world today is how terrorism *ends*, or at least diminishes. For those seeking to counter terrorism, the question is perhaps more salient than questions about causes. The answer to how terrorism ends is also interesting because the mechanisms of decline do not simply mirror the mechanisms of how terrorism arises. Recognizing this causes us to think differently about the factors and processes with which counterterrorism must be concerned.

How terrorism ends has been discussed in only a relatively small portion of the social-science literature on terrorism. Some articles approach the issue empirically (Alterman, 1999, p. 16; Crenshaw, 1987, 1991; Cronin, 2006; Gurr, 1990; Ross, 1995; Ross and Gurr, 1989; Jones and Libicki, 2008), but most articles subsume the end-of-terrorism issue under five broad investigative topics: (1) origins, root causes, and motivations of terrorism; (2) histories of individual terrorist groups; (3) effectiveness of counterterrorist policies; (4) evolution of terrorist groups as part of wider social movements; and (5) organizations' structure, needs and internal dynamics.

This paper mainly relies on the literature of the first type (origins, root causes, and motivations) but draws selectively from the vast general literature on terrorism and from the literatures on counterterrorism and the end of civil wars. The discussion in this paper is organized as follows. The next section presents a framework showing eight modes of terrorism decline, that is, eight major classes of cases. The next section describes processes, developments, and factors that seem to contribute to each mode of decline. The following section inverts the discussion: It draws on the preceding sections to summarize potential factors that help *sustain* terrorism (the undercutting of which factors, arguably, leads to decline) and indicates plausible relationships among them. A final section then summarizes government policies that work to undermine terrorism, relating them to the critical factors discussed earlier. An appendix critiques the state of the literature on the end of terrorism and suggests some directions for future work.

Modes of Decline

The literature on the end of terrorism suggests a number of ways terrorism has ended, characterizing them by what appeared to the authors to have been the dominant reasons in historical cases. The various papers do not typically relate well to each other and no integrated discussion seems to exist, so an organizing structure was constructed for this review. It drew heavily on suggestions by Crenshaw (1995), a review that covered Crenshaw's work (Alterman, 1999), and the insightful work of several other authors (Cronin, 2006; Ross and Gurr, 1989; Weinberg, 1991). The result was a set of eight modes of decline (that is, classes of cases), named by what was identified as the key element of the decline.[1]

Some explanation of terminology is important here. We refer to "elements" rather than using such words as "factor" because the monograph of which this paper is part uses "factor" in the context of causal models, where a factor is an independent variable. In contrast, the "elements" used to name the modes of decline are really labels for attributes of outcome or intermediate outcomes. Although they could be regarded as "causes" of the subsequent end of terrorism, they are not the independent variables of policy but rather intermediate outcomes to be sought.

The modes are not mutually exclusive in that in a given case of the end of terrorism, several of the elements may exist and it may even be unclear as to which, if any, is dominant.

The eight modes are itemized below and in Table 7.1, which also gives historical examples.

1. **Substantial Success.** Terrorist tactics succeeded in at least helping to achieve the terrorists' main goals, usually in conjunction with less-violent political actions.
2. **Partial Success.** Terrorists achieved public recognition for its organization and its causes and chose to abandon terrorism to avoid the costs of continuing violence and thereby alienating supporters, sponsors, and key third-party actors.
3. **Direct State Action, Including Repression.** The state's direct measures (including repression) succeeded in undermining the terrorists' coercive capabilities.
4. **Disintegration Through Burnout.** The terrorist organization disintegrated as members lost commitment to the organization and its mission because of, for example, collective perceptions of failure, physical exhaustion, availability of attractive paths out of violence, general disillusionment with violence, and growing disagreements with ideology and strategy.
5. **Loss of Terrorist Leaders.** Death or imprisonment of influential leaders occurred and the rest of the organization was unwilling or unable to continue functioning as a terrorist organization.
6. **Unsuccessful Generational Transition.** Terrorists were unable to pass their cause to the next generation when the first generation of leaders departed or were eliminated.
7. **Loss of Popular Support.** Terrorists lost support of their constituencies: populations, governments, and other organizations.
8. **Emergence of New Alternatives.** Better options for political change or organizational survival and enhancement emerged, including more traditional forms of warfare and revolution, mass-based protest movement, opportunities for legal political action, and organized criminal activity.

Table 7.1
Classes of Cases and Historical Examples

Mode of Terrorism Decline	Notable Historical Examples
Substantial success (primary objectives met, by whatever means)	Original Irish Republican Army (IRA) (circa 1921) EOKA (Cyprus) Croatian Ustasha African National Congress (ANC) Nepalese Maoists Irgun/Stern Gang (Israel)
Partial success	Palestinian Liberation Organization (PLO) Provisional Irish Republican Army (PIRA)
Direct state action (sometimes repression)	Revolutionary Armed Task Force (RATF) Symbionese Liberation Army (SLA) George Jackson Brigade Narodnaya Volya Uruguayan Tuparamos Muslim Brotherhood
Disintegration through burnout	Weather Underground Front de libération du Québec (FLQ) Red Brigades
Loss of leaders	Shining Path Real Irish Republican Army (Real IRA) Aum Shinrikyo
Unsuccessful generational transition	Red Brigades The Second June Movement The Japanese Red Army Weather Underground Symbionese Liberation Army Baader-Meinhof group (Red Army Faction)
Loss of popular or external support	Weather Underground Front de libération du Québec Real IRA Red Brigades Shining Path Armenian Secret Army for the Liberation of Armenia (ASALA)
Emerence of new alternatives to terrorism	Front de libération du Québec Provisional IRA Palestinian Liberation Organization Revolutionary Armed Forces of Colombia Khmer Rouge Armed Islamic Group Nepalese Maoists Guatemalan Labor Party/Guatemalan National Revolutionary Unit

NOTES: The cases are not mutually exclusive, as several dominant factors may exist. The historical examples are illustrative, not a comprehensive list. The Nepalese Maoists are formally the Communist Party of Nepal (Maoist) (CPN(M)).

The assessments in Table 7.1 (most of them from Cronin, 2006) are all controversial because of subtleties in definition and interpretation, the on-again off-again nature of some conflicts, and uncertainties as to how influential the violent activities were on results. For example, the *original* IRA, which fought for Irish independence in 1919–1921, did not defeat the British Army but had made governance difficult without military force, thereby pressuring the British (Laqueur, 1976; Hopkinson, 2004). A splinter group (also called the IRA) continued the battle after the 1921 Anglo-Irish treaty but lost that war two years later. Versions of the IRA have continued to exist and, as discussed recently by Alonso (2004), the 1998 Good Friday Agreement between British and Irish governments fails to qualify as a success of the then-current version of the IRA in achieving its main objective—a unified Ireland.

Despite these caveats, the next section discusses the modes separately.

Processes, Developments, and Factors Contributing to the End of Terrorism

This section discusses each of the modes of terrorism decline separately, identifying related processes, developments, and factors. Each subsection also includes some discussion of when the headlined element has not correlated (or need not correlate) with the end of terrorism.

Substantial Success

If success is defined as the achievement of the terrorists' original objectives, then available historical evidence suggests three factors that have been critical to such success (separately or together) in the past:

1. The terrorist organization has narrow, clearly defined, and attainable goals.
2. Powerful outside powers assist the terrorist organization.

3. The organization is waging an anticolonial campaign for independence against a power unwilling to bear the costs of continued struggle (state survival is not at stake and where continued struggle would imperil other state objectives).

Achieving success has often required more than just terrorism, for example, activities such as less violent political action or more traditional warfare tactics (Cronin, 2006; Laqueur, 1977).

Cautions. The success of a terrorist organization need not logically bring an end to terrorism, although the literature is quiet on this matter. A sufficient condition seems logically to be that the terrorist group and its members have both a limited and fixed rationale for using terrorism—a rationale that no longer exists after success is achieved. The victorious group may still choose to preserve its identity and continue to exist in a different capacity, including as a political force. The examples of successful groups include the original IRA after World War I, Croatian Ustasha during World War II, the ANC in South Africa, Irgun in Israel, EOKA in Cyprus, and the Nepalese Maoists.

Partial Success

A terrorist organization often abandons terrorism after achieving public recognition for its organization and its causes, realizing that further violence may alienate supporters, sponsors, and key third-party actors.[2] The result is a transition to a legitimate political process. The PLO and PIRA are two examples of groups that have made this transition.

Some authors have suggested preconditions for a lasting successful outcome (Alonso, 2004; Cronin, 2006; Sederberg, 1990). According to their suggestions, the government must be ready to provide pathways out of violence by offering amnesty and accepting the organization as a legitimate negotiating partner, and the terrorists must agree to compromise on their goals on issues critical to the government—for example, they must abandon aims for the destruction of the state or the regime, or renounce full independence, in exchange for autonomy. In addition, the terrorist organization must possess a strong command and control structure so that a decision to desist from terrorism will be honored, which is more likely when the group's governance is hierarchical and it has a strong leader.

Cautions. Although the "preconditions" mentioned may sometimes be necessary, they are not logically required. For example, a state might cease the activities that caused political turmoil and might even change policies and make de facto concessions without ever negotiating with the terrorist organization. Also, political negotiations with terrorist groups do not necessarily bring an end to terrorism over the long term. This is often the case if the group is a part of a wider resistance movement with a popular cause and strong support base. The violence often *escalates* in the short term as the rival groups step up attacks in an attempt to derail the peace process and enhance their own profile relative to that of competitors (Bloom, 2004; Kydd and Walter, 2002, 2006, pp. 72–78). Likewise, internal divisions within the government, and hostility from the government's own support base toward the government's peace initiatives, may also undermine the peace process (Alterman, 1999; Cronin, 2006).

The evidence seems to indicate that partial success leads to the end of terrorism when it combines with other processes discussed below. Thus, partial success appears to be contributory but is not necessarily sufficient.

Direct State Action (Including Repression)

The government's direct measures against the terrorist group, when they succeed in significantly undermining terrorists' coercive capabilities, can be a major contributor to the end of terrorism. Such measures are targeted against the terrorist organization itself, its people, its resources, and its motivations.[3] A recent RAND study (Jones and Libicki, 2008) concludes that the more important measures tend to be associated with law enforcement and intelligence, often clandestine, rather than traditional military operations.

In some instances, it at first appears that direct state action was successful by itself. Such cases include the success of counterterrorism strategy against the American RATF, SLA, the George Jackson Brigade, Narodnaya Volya, and various anarchist groups in Russia. All involved punitive actions and sometimes preemption against a very small fringe group (usually fewer than 50 members) with little external support. Another example was Egypt's crackdown against the Muslim Brotherhood in the early 1980s. Of course, that was a large-scale effort.

Even in these cases, there were a number of contributing factors, such as the terrorists' failure to sustain popular support (Ross and Gurr, 1989) and strategic miscalculations. In most of the cases mentioned above, the tactic of terrorism was used specifically to provoke government repression that would incite the masses against the government and lead to radical transformation of the existing political and social order. When this tactic failed, the particular terrorists escalated their destructiveness, which in turn alienated the sympathizers, outraged the public, and made it easier for government to destroy the group.[4]

More generally, it is evident that other factors must contribute. Crenshaw (1987) has specified three classes of contributing factors that enhance the effectiveness of direct government actions: (1) the presence of internal disunity within the terrorist organization, (2) strategic miscalculations by the terrorists, and (3) strategic reversals (such as withdrawal of state support, loss of sanctuaries, more attractive alternatives, and collective perception of failures). Again, these factors are not independent variables for the policymaker but rather developments that occurred and that might be intermediate objectives of policy actions.

All of the other elements described in the other subsections of this section have been identified by various authors as contributors in cases characterized by direct state action. Indeed, it is not a matter of there being other contributors. They are often necessary.

In particular, repressive measures are often insufficient when a large population supports the terrorists' cause (Alterman, 1999). Punitive actions can certainly erode support in some respects but cannot fully do so in the absence of policies aimed at addressing popular grievances and splitting off pragmatists from radical rejectionists (for example, through accommodation, cooptation, and amnesty or repentance legislation). Such policies tend to isolate the terrorists and diminish public support for their organization, denying them a strong base from which they operate (Alterman, 1999; Crenshaw, 1991; Gurr, 1990; Sederberg, 1990). As one would expect, the size and the nature of public support seem to determine whether some sort of conciliation strategy is necessary (Sederberg, 1990).

Other factors that affect whether direct state actions are effective include the motivation of terrorists (ideological versus ethnical), their relative isolation or representativeness, the nature and the role of (population and external) support, the nature of the stakes involved, characteristics of their organizations and their internal dynamics, the role of terrorism in the overall strategy, the nature of the state's political system, and the strength of its political foundations and government. These factors appear to be interrelated in complex ways.[5] What matters for the success of counterterrorism is whether they combine in a way that gives the state both the physical and political means necessary to destroy the organization while preventing the emergence of new recruits and new sources of support.[6] What is "enough," though, appears to be different for different groups and in different circumstances: That is, more than one pattern of factors can lead to the successful outcome.

Currently, the social-science literature on terrorism does not offer a satisfactory account of the various circumstances that can lead to the success of direct actions. But it does suggest some interesting regularities. Strong authoritarian states appear to have little problem in suppressing terrorist organizations.[7] Laqueur (1977) has shown that terrorism has succeeded in transforming many liberal societies into authoritarian ones, which then have little trouble controlling it (Hamilton, 1978).[8] However, it should be noted that repression by an authoritarian government can eliminate terrorism in the country in question but may push it elsewhere (Muslim Brotherhood is a case in point) (Lizardo, 2007, p. 159; Pluchinsky, 1987).

Ethnically based terrorism appears harder to end decisively—except in cases of successful anticolonial struggles against a retreating empire—not in the least without government addressing important underlying grievances and offering new alternatives through accommodation and cooptation. Broad-based, mostly rural ethnic/separatist or politically motivated insurgencies, when they do not succeed, might come to an end through some combination of changes in leadership (Shining Path), loss of popular or external support, collapse of internal cohesion, or appearance of new alternatives enabling transition to a legitimate political process or other modus operandi (full insurgency or criminality). Small radical leftist groups in Western Europe (such as the Baader-Meinhof group, Red Brigades, Action Directe, 17 November), despite their unrealistic, vague, and inconsistent demands, had sprung out of larger protest movements (della Porta and Tarrow, 1986; Merkl, 1995; Wieviorka, 1993) and enjoyed a measurable support among the leftist groups—students, workers, and intellectuals—and thus proved hard to repress. Their terrorism subsided because at least one of the following dynamics played a part: the declining commitment of members to the organization caused by fatigue and government amnesty or repentance legislation, loss of support, and unsuccessful generational transition as a result of public backlash and waning fortunes of the Marxist-Leninist ideology after the end of the Cold War. These developments, in turn, made it far easier for the government to hunt down the remaining active group members.

Cautions. Repression of a terrorist group is sometimes counterproductive—a fact well known to counterterrorism scholars and practitioners (Crelinsten and Schmid 1993; Simon, 1987; Sederberg 1995). This is more likely in cases where neither the requisite degree of repression nor effective government control on the spread of information is achievable due to physical or political constraints on government action. Democracies and weak authoritarian regimes are especially vulnerable in this regard, whereas strong authoritarian regimes appear to be immune from the threat (for the relevant arguments see Crenshaw, 1981; Eubank and Weinberg, 1994, 2001; Kydd and Walter, 2006; Li, 2005; Piazza, 2007; Schmid, 1992; Wilkinson, 2006).[9]

Subject to this caveat, repression is more likely to be counterproductive if it is indiscriminate (Art and Richardson, 2007; Crelinsten and Schmid, 1993; Sederberg, 1995). Discriminate use of force is difficult when a significant portion of the population sympathizes with the terrorists and their cause: The terrorists can blend in with the population and good intelligence is hard to come by. In those cases, what are or are perceived to be indiscriminate actions undermine the legitimacy of the state and generate public backlash, which is amplified by terrorist propaganda. Publicity is the lifeblood of the terrorist (see Schmid,

1989; Schmid and de Graaf, 1982; Wilkinson, 1997). In advanced democracies with civil liberties, free press, and an emphasis on human rights, repression gone awry can enable even marginal groups to score a strategic success. In fact, provoking indiscriminate repression is a part and parcel of strategic logic of terrorism (Kydd and Walter, 2006).[10] The potential for negative fallout from repression is even greater with modern communications and increased transborder population movement that allow the terrorists to reach out to the remotest parts of the world and discover new sources of support.

Disintegration Through Burnout

The end of terrorism often occurs as a result of organizational disintegration for reasons internal to the organization (Crenshaw, 1987). The notable examples include the Weather Underground, FLQ, and Red Brigades. Collapse of the organization, in these cases, is thought to have stemmed from intra-organizational dynamics channeled through the psychological process of burnout—loss of the members' commitment to the organizations and its goals (Ross and Gurr, 1989). Burnout often manifests itself through the collapse of organizational cohesion, resulting in risk-avoidance and loss of discipline, factionalization, desertions, and defections (Ross and Gurr, 1989). Several factors work to undermine organizational unity, such as prior cleavages among supporters (ethnic groups or states), disagreement over strategy, ideological differences, power struggles among generational leaders, and personality clashes (Alterman, 1999; Crenshaw, 1987).

However, organizational disunity does not necessarily imply declining commitment to terrorism and, hence, does not always signal the end of terrorism. A common response to disunity is punishment of dissenters and purges. When effective, this strategy enables the organization to retain its identity and restore cohesion. Sometimes the organization splits into several parts as the defectors join a rival organization or establish new ones. Splits and mergers can in fact result in more, rather then less, terrorism. The violence often escalates as the moderates leave the organization, defecting radicals join more militant groups, and the rival groups compete for support by trying to outbid each other

(Bloom, 2004; Crenshaw, 1991; Jaeger and Paserman, 2006; Kydd and Walter, 2006; Toft, 2007).

The circumstances that set the stage for the collapse of internal cohesion and end of terrorism are unclear. The analysis is made difficult by the fact that the literature on the end of terrorism focuses on individual organizations and does not address the reasons for the end of terrorism beyond the particular organization's lifetime. One thing that seems clear is that the terrorism is likely to continue, or even increase in intensity, at least over the short term, when the organization splits into rival factions. A workable hypothesis would be that organizational disintegration does not put an end to terrorism when a sizable demand for terrorism remains (that is, sympathy exists for an action agenda that may employ terrorism as a tactic or even as an end in itself, such as revenge).

A combination of several processes may erode commitment to the organization and lead to organizational disintegration and, possibly, the end of terrorism: fatigue and collective perceptions of failure, backlash from increased destructiveness, loss of support, and the emergence of attractive paths out of violence. A government amnesty or repentance legislation may precipitate widespread defections when the organization faces defeat and support dwindles but seem to be of little effect otherwise (Cronin, 2006). The unraveling process often works through a virtuous circle: The availability of new alternatives or a collective sense of failure resulting from setbacks generates internal disagreements and factionalization; this in turn leads to an escalation in destructiveness followed by backlash from alienated or frightened sympathizers and aggressive government intervention, generating further internal disagreements.

Loss of Leaders

There is some evidence that the loss of leaders through death or imprisonment may in some cases inflict a decisive blow to the terrorism movement, precipitating a significant decline of the group and its activities and ultimately leading to the end of terrorism. However, the evidence is uncertain. Shining Path, PKK, the Real IRA, and Aum Shinrikyo are often cited as examples of cases when loss of the leaders severely disrupted an organization (Cronin, 2006; Hosmer, 2001), but the validity of the claim in each case is still being debated.[11]

It is reasonable to assume that the loss of a leader must be disruptive in one or more ways by producing a break in ideology, strategy, command and control, or competence. Nonetheless, for a decapitation to be successful, good leaders within the terrorist groups must be a scarce resource making it hard for terrorist groups to find replacements (Byman, 2006; Freeman and McCormick, 2007).[12] The research base on such matters is spotty, mainly to the lack of hard data about the internal workings of terrorist organizations.

Several plausible categories of factors with some basis in the literature can be identified and it is often the interaction of factors across these classes that appears to be critical. These include the role of the leader in a group, the character of the leader (for example, a leadership foundation that is not "routinized," as in the case of a cult of personality) (Cronin, 2006; Eisenstadt, 2001; Jenkins, 1987; Jordan, 2004), the structure of the organization (hierarchical and centralized or distributed), the nature of internal disagreements over ideology and strategy (for example, whether the views of the leadership are widely shared and understood in the organization), the size of the organization (Byman, 2006; Eisenstadt, 2001; Gazit, 2002), and the strengths of its support base.

It is now widely believed that the distributed terrorist network, such as the one practicing "leaderless resistance," is less susceptible to disruption through removal of leaders. It is certainly reasonable to assume that when the leadership is "singular" and centralized, removal of the leader is more likely to be effective than when it is abundant or diffuse (Byman, 2006; Freeman and McCormick, 2007).[13] It is notable that the United States claims to have killed an inordinate number of very high-level al-Qaeda leaders, with seemingly no dramatic consequences. Replacements have come forth.

There are some logical issues with conventional wisdom on these matters, however. First, unless "distributed network" is defined in a way that makes the conclusion correct by circular argument, the conclusion about decapitation being ineffective in distributed networks is not obviously valid. Not all nodes of a network are equal and a network's ability to replace a lost node depends on what that node contributed. For example, a charismatic or intellectual leader might be essential to the coherence and motivation of the organization, even though he had little direct administrative or operational control. He might be very difficult to replace. Further, as suggested by correlational evidence, distinctions should be made between simply removing "the" leader and removing most of the entire upper echelon of an organization (Jordan, 2004).

In our view, the fact is that we do not know whether elimination of Osama bin Laden and Ayman al-Zawahiri would have modest, intermediate, or profound effects. We are skeptical of social-science claims on the matter.

We should also note that the effectiveness of "targeted killing" in the context of Israel's continuing battles with Palestinian terrorists remains controversial. It seems that some recent discussions tend to disparage effectiveness (see Gazit, 2002; Hosmer, 2001, pp. 24–27; Kaplan et al., 2006; Kaplan et al., 2005; Stein, 2001). However, other authors (and Israeli officials and officers) disagree (see Byman, 2006; David, 2002; Eisenstadt, 2001; Ganor, 2005; Gazit, 2002; Jaeger and Paserman, 2005; Lotrionte, 2003; Luft, 2003; Zussman and Zussman, 2006), and it is difficult to imagine that a strong conclusion could be valid: Surely, details must matter.

Cautions. Whether or not efforts to remove leaders may sometimes be effective, it is clear that a policy of killing or arresting prominent terrorist leaders may backfire, at least in the short run. Decapitation, especially that involving extrajudicial actions and unintended casualties, may create increased publicity for the terrorist cause and weaken the state's claim on legitimacy in the eyes of its citizens and the international community, inspire and mobilize terrorist sympathizers, enhance the organization's standing in the population, radicalize the moderate faction, promote unity among different terrorist factions, and enable even more-radical leaders to rise to the top. Targeted killing often has the effect of worsening the medium-term prospects of reaching a negotiated settlement.[14] (See Byman, 2006; Cronin, 2002; David, 2002; Ganor, 2005; Jenkins, 1987; Luft, 2003.)

Some authors suggest that killing political leaders is more likely to backfire than killing operational leaders, as the latter are usually less known to the public and their deaths tend to get far less attention (see David, 2002; Ganor, 2005; Gazit, 2002; Zussman and Zussman, 2006). However, the distinction is sometimes difficult to make in practice. In this regard, capturing "the" leader rather than assassinating him may be more effective in damaging the group, provided that the counterterrorists manage to undermine his credibility and cut off inflammatory communications (Cronin, 2006). At the same time, imprisonment is likely to cause further violence as the release of imprisoned terrorists is an important motivation for terrorist attacks (Jenkins, 1987).

It should also be noted that the authors who argue for the effectiveness of the policy of targeted killings tend to do so primarily because they believe in the long-tem cumulative effect—via degrading organizational capability, disrupting organizational routine, and undermining morale—of the sustained campaign against the leaders on the ability and willingness of the terrorist organization to wage campaign of terror (see Byman, 2006; David, 2002; Ganor, 2005; Luft, 2003). These authors usually view, not without certain caveats, the "boomerang effect" (Ganor, 2005)—backlash to targeted killings—more as a short-lived reaction than as a significant contributor to the "cycle of violence."

Unsuccessful Generational Transition
The failures to pass a cause to the next generation seems to have been associated with several interacting and compounding factors: a radical ideology, typically left wing/anarchist,[15] coupled with unclear, inconsistent, and increasingly unrealistic demands for social change (Cronin, 2006); erosion of public support and subsequent marginalization and isolation (often against the background of repression, accommodation, and cooptation by the state); and a large-scale socioeconomic and political change that diminishes the attractiveness of the terrorist ideology. Examples include Red Brigades, the Second June Movement, the Japanese Red Army, Weather Underground, the Symbionese Liberation Army, and the Baader-Meinhof group (Cronin, 2006; della Porta,

1995a; della Porta and Tarrow, 1986; Merkl, 1995; Ross and Gurr, 1989)

Loss of Popular or External Support

The U.S. *National Military Plan for the War on Terrorism*, reflecting a good deal of accumulated social-science lore, considers popular and organizational (state and nonstate supporters) support to be a center of gravity of the terrorist enemy (Chairman of the Joint Chiefs of Staff, 2006, p. 14). Historical evidence suggests that loss of support can indeed spell the end of terrorism by catalyzing the group demise or abandonment of the strategy. It is thought to have been a critical factor in the collapse of terrorism by such organizations as Weather Underground in the United States, FLQ in Quebec, the Real IRA, Red Brigades, Shining Path, and ASALA (Crenshaw, 1991; Cronin, 2006; Ross and Gurr, 1989).

Loss of support adversely affects the capabilities of the terrorist organization by reducing the amount of its resources—materials, funding, recruitment, sanctuaries, and expertise. Cooperation or acquiescence from terrorists' erstwhile supporters gives the state better control over the operational terrain through improved access and intelligence, leading to greater terrorist attrition and disrupting their operations and organization. These setbacks put further pressure on the organization and drive it into hiding. They reinforce a sense of failure among the terrorists and their remaining supporters and bring about fatigue and disenchantment, spurring internal dissent and defections, which in turn lead to further group isolation and marginalization. This process is often self-reinforcing, thrusting the organization into a downward spiral (Crenshaw, 1991; Ross and Gurr, 1989). Loss of qualified personnel, internal dissent, and isolation lead to poor decisionmaking and strategic miscalculations.[16] In an attempt to reverse their waning fortunes, terrorists often escalate the violence, embarking on ever-riskier missions and taking on targets considered illegitimate by supporters and the sympathetic public; descend into internecine, interfactional, or intergroup warfare; enter unpopular alliances; change their claims to fit the need of the moment; and advance a more radical, hence, less relevant and realistic agenda in search of ever more-elusive support (Crenshaw, 1991). What ensues is the series of cascading effects of growing operational and organizational failures; backlash and dwindling support from active supporters, sympathizers, and the international audience; increased calls for tougher measures from citizens; and more intrusive and effective government action. As a result, the organization is physically destroyed or else the terrorists opt to abandon terrorism—either individually or as a group—realizing the futility of their efforts (Crenshaw, 1991; Ross and Gurr, 1989).

A number of developments can lead to diminished popular support. Strategic miscalculations can generate a backlash from actual or potential constituencies. Government repression that raises costs of tolerating terrorists for the sympathetic public often leads to exhaustion and loss of support. Political and socioeconomic reforms targeting the underlying grievances of passive and active supporters also work for the same effect (Cronin, 2006; Ross and Gurr, 1989). In places where neither government nor markets function well, groups that are engaged in terrorism can take up provision of social welfare for the vulnerable groups in exchange for their support (Hamas, Hizballah, and Taliban). When this is the case, policy measures aimed at improving social welfare and building efficient markets can undercut incentives to provide support for the terrorists (Berman, 2003; Burgoon, 2006).[17]

There is some evidence that offering amnesty or reduced prison terms to individual members in exchange for repentance or collaboration contributes to the erosion of support by bringing about desertions, defections, and internal disagreements. However, such a policy appears to be effective only when support is already low or waning and cannot generate decline on its own. (Crenshaw, 1991; della Porta, 1995a; della Porta and Tarrow, 1986).

Success of both repressive and conciliatory measures in undermining population support seems to depend on a wider political context (Alterman, 1999; Sederberg, 1995). The key is whether the government measures enjoy support, or at least acquiescence, from major political forces as well as important segments of the population: Failure by the state and its constituencies to present a unified front against a dissident movement reduces the range of options available to the government and undermines its efforts. The requisite degree of internal

unity is more likely to be attained in strong consolidated democracies and strong authoritarian states than in states with weak political
foundations, democratic or otherwise. However, democratic societies
are unlikely to reach consensus on using repression against the population sympathetic to terrorism (Charters, 1994; Eyerman, 1998; Piazza,
2007; Schmid, 1992). The timing of government initiatives can make a
difference, as repression is more likely to succeed when the government
is in a strong position—for example, when outrage against terrorist
atrocities rallies the public behind it—and the dissident movement is
in disarray. Strong authoritarian states, in contrast, face no problem of
unity. Some initially semidemocratic states under sizable pressure from
terrorists enjoyed considerable success when they scrapped democratic
institutions and relied heavily on repression against the population in
their fight against terrorism.[18]

A better security environment is another factor that can chip away
population support for the terrorist organization. Counterterrorism
and counterinsurgency research suggests that population support is
not always entirely voluntary: When an adequate security environment
is lacking, intimidation and coercion of the population by the terrorists, or their taking over the task of security provision from the state
(such as in a protection racket), is often an important source of support
(Adams, 1986; Horgan and Taylor, 1999; Nagl, 2005; O'Neill, 1990;
Shafer, 1988). This implies that the government's success in improving
security on the ground can contribute to a significant erosion of support in cases where a sizable portion of the population is vulnerable to
depredation and retribution.

State propaganda helps impair support but only in conjunction
with other factors. Furthermore, support can be significantly undermined if the terrorist group overly depends on external support and
the state is able to persuade or compel various state or nonstate external
supporters to cut their aid through provision of politico-diplomatic,
military, and economic incentives and disincentives, such as diplomatic
pressure, economic sanctions, and the use or threat of use of military
force. Finally, support for terrorist groups and their causes can decline
as a result of a large-scale social and political change that renders their
ideology and objectives irrelevant or a dramatic shift in international

order or regional balance of power that wipes out the payoffs from supporting the terrorists (Cronin, 2006; Ross and Gurr, 1989).[19]

Cautions. Sometimes, the loss of support does not lead to the end of terrorism, although the literature on the end of terrorism does not specifically address this issue. Here, it is important to distinguish among support for specific group, support for terrorism as a method, and support for the terrorism cause. One plausible scenario occurs in a situation when significant popular support for the terrorist cause remains as support for their tactics evaporates. This is most likely to be the case when the dissident movement has ethnic or separatist, rather than ideological, goals or at least involves a significant ethic or nationalist component (see Alterman, 1999; Cronin, 2002, p. 1, 2006, p. 13).[20] In this case, it is often the government's excessive and indiscriminate use of force or its failure to provide security for the population from government loyalists, terrorists, or common criminals that rekindles support for the terrorist organization; the group (or its parts) is able to continue its existence by reason of the government's incompetence in fighting the organization, peace overtures by the organization (or its parts), state amnesty, and so forth. Such missteps often provide an opportunity for terrorists to score an unlikely strategic success by escalating the violence. Indeed, terrorist often step up attacks precisely with the goal of provoking government repression and recapturing support.[21] A reversal of the declining fortunes of terrorists may also occur as a result of dramatic exogenous changes in the general political climate or international situation.

A different dynamic takes place when the terrorist group loses support but support for terrorism as a method continues. In this case, the loss of support by a group may lead to group splintering and creation of new groups and may spur competition for resources between rival groups. Violence escalates as the each side tries to outbid the others as they compete for support (Bloom, 2004; Kydd and Walter, 2006).

Loss of support from some sources may have little effect on the organization if support from other sources is available or even strengthens. For instance, cutting external support may have little effect on groups that enjoy significant autonomy enabled by a strong popular support (Sederberg, 1990). In contrast, the loss of domestic support may have little effect if a powerful external supporter is able to take up the burden. New sources of support may also be discovered over time as the group finds new allies or shifts to illegal entrepreneurial activities, such as the drug trade (Cronin, 2006).

Emergence of New Alternatives to Terrorism

Emergence of new alternatives is often a catalyst of the end of terrorism. In the near term, this appears to be true with respect to a particular terrorist group, not necessarily with respect to the larger terrorist movement the group is a part of. New alternatives often impel terrorist groups to shift away from terrorism toward other modus operandi, such as legitimate political processes, more traditional ways of warfare and revolution, or organized criminal behavior, but they can also lead to group breakdown (Alterman, 1999; Cronin, 2006; Ross and Gurr, 1989). The examples that are often cited include FLQ, PIRA, the PLO, FARC, Khmer Rouge, Armed Islamic Group, CPN(M), and the Guatemalan Labor Party/Guatemalan National Revolutionary Unit.

New alternatives often arise from government's conciliatory actions. Political and economic reforms, aimed at groups on which terrorists rely for support, are measures that target the underlying causes of grievances; such reforms reduce support for terrorism by preventing indifferent supporters from becoming sympathetic and separate passive supporters from becoming active supporters. A different type of concession, such as amnesty programs and repentance legislation, targets the challenger organization itself; these policies undermine organizational cohesion and spur defection of activists. Finally, concessions may involve political recognition of some causes: Government accommodation opens up alternative means to the attainment of group goals and spurs defections of a moderate group of leaders (Alterman, 1999; Sederberg, 1995)

None of the aforementioned policy initiatives—reform, amnesty, repentance legislation, accommodation, and cooptation—is sufficient in itself to put an end to terrorism over the long run in the absence of favorable conditions: partial success, collapse of cohesiveness, and loss of population or external support, as discussed elsewhere in this paper.

Political negotiations aimed at achieving some accommodation may fail to end terrorism but can split pragmatists from radicals and help swing public support away from terrorism and toward a peaceful solution (Alterman, 1999; Sederberg, 1995).[22] A wide range of factors can affect the outcome of negotiations. Peace initiatives may fail to bear fruit, at least in the short term, and may even spur an escalation of violence when at least one side is not a cohesive organization (Cronin, 2006; Sederberg, 1995).[23] Thus, peaceful settlement is usually not possible when the challenger movement lacks cohesion[24] and internal control is weak, which is more likely to occur when the organization is hierarchical and hs a strong leader (Cronin, 2006). When the movement comprises various independent factions, the politics of terrorism—a strategic interaction between the moderate and extremist factions and moderate faction and the government—provides mechanisms that not only sustain violence following the government's peace overtures but can lead to increased levels of militancy.[25] Likewise, the peace overtures of a weak governments appear to cause internal divisions and hostility from the government's support base (Alterman, 1999; Cronin, 2006). Furthermore, the terrorist organization must have a strong incentive to transition to legitimate political processes: It must already have some gains that it wants to preserve and must come to recognize that it stands to lose from continuing the struggle (Alterman, 1999).[26] Finally, for the negotiations to succeed, there must be negotiable aims, which in turn requires that the stakes are divisible—a situation that is more likely to exist with territorially based groups than with the fringe left-wing, right-wing, or religious or spiritualist groups (Cronin, 2006).

Terrorist groups may also transition to more classic forms of warfare and revolution. This usually occurs when the groups gain strength against the state as a result of dramatic growth in popular or external state support. Transition to a full-blown insurgency is especially common among ethnonationalist and separatist groups representing a wider territorially based popular movement. Transitions to conventional warfare usually require interference by powerful outside actors.

Finally, terrorist groups can move from the use of terrorism toward lucrative criminal activities, such as racketeering, kidnapping for ransom, and illegal trafficking of drugs, weapons, and so on. The transition toward criminality implies a shift in motivation away from resource collection as a means to political ends and toward material gains as an end in itself (Cronin, 2006). The circumstances of such a shift can be diverse, but they appear to fall within one or more of the different dynamics associated with, for example, loss of the leader and support, organizational breakdown, and generational transition.

Appendix: The Literature and Its Limitations

Although the social-science literature pertaining to the end of terrorism is rich in some respects, yields numerous insights, and suggests the type of theoretical structure described above, it must be acknowledged that the knowledge described is still maturing and lacks "rigor." The following material suggests future directions.

The literature directly addressing the end of terrorism consists largely of "soft theory" (intended to help structure thinking and identify key issues but without pretense of comprehensiveness or logical tightness) with relatively weak empirical foundations (more suggestive than controlled). The empirical work is primarily of the type that seeks to build theory based on observations from historical case studies. Such work can be very insightful but lacks confirmation. Such work may begin by postulating some principal factors and then use historical cases to illustrate the points and, rhetorically, to justify the "theory." However, the theory is then not tested against a wider set of cases, and insufficient distinctions are made among classes of cases. Such problems create doubts about both internal and external validity.

Further, the various authors start with different postulated structures and do not make comparisons among the competing structures; they then iterate to generate increasingly comprehensive structures. They also do little to discuss the relationships among factors (that is, how they interact). None of this is surprising. Rather, it merely indicates that there remains much to be done, building on the very stimulating and suggestive past research. To some extent, this paper (and the larger monograph of which it is part) is an attempt to compare, integrate, and take next steps. However, it is only a first step.

The literature on the end of terrorism also suffers from a near-exclusive preoccupation with terrorist organizations (groups) as the unit of analysis. A consequence is that such literature often equates decline of an organization with the end of terrorism, which need not be the case, as described above (see also Cronin, 2006).

As described in companion papers (Davis and Cragin, 2009), the authors are concerned with building "causal models." However, the knowledge base in the literature is of only limited value in informing that effort. One reason is that the papers drawn on often mix factors that should—from a causal-modeling perspective (as distinct from the authors' original perspectives)—be kept separate. This may even involve mixing what a causal modeler would regard as inputs and outputs. For example, factors identified as contributing to the end of terrorism may include the defeat of the organization (Crenshaw, 1987) or transition to a legitimate political process (Cronin, 2006), both of which might be regarded more as end-state descriptors (outputs). Some of the confusion is arguably more apparent than real, as discussed above. Other aspects of the confusion are more profound.

A related issue is the inclusion of different policy measures among high-level factors contributing to the end of terrorism. However, causal modeling requires that we differentiate between the two, since policies typically affect multiple factors working at many different levels; policies contribute only in the sense that they have direct or mediated effects on high-level factors that are critical to sustaining terrorist activity. For causal modeling, therefore, policy measures can be best represented as a separate set of variables that modify the values of the critical variables in the model.

Despite its weaknesses, the literature on the end of terrorism can be useful for constructing causal models of terrorism in that it points to a range of potential factors that help sustain terrorism as well as to a set of policies that might be used to undermine it. In addition, this literature presents important insights into the processes that lead to terrorism decline. Analysis of such processes can contribute to our understanding of how different factors and policies relate and interact to produce a successful outcome.

Other types of terrorism literature are also useful for providing candidates for contributing policies and factors. They too present specific problems that limit their usefulness for studying the decline of terrorism. The most important of those are discussed below.

The literature on *root causes and origins* has limited value for analyzing terrorism decline. Since the literature on causes does not differentiate between the causes of "outbreak" and those that sustain terrorism, the relevant factors cannot be readily inferred from the "causes of terrorism" unless we are willing to assume that the same variables that "cause" terrorism are implicated in its decline. Yet, terrorism has often ended even when some important hypothesized "causes" remained and vice versa.[27]

Historical studies of the evolution of terrorist groups are rich in contextual details, but they are largely descriptive. Analysis of critical factors is notoriously prone to subjective interpretations, placing a heavy burden on the researcher to correctly identify and extract important factors. They usually display conceptual poverty, advance particularist explanations, and are hard to generalize.

Existing *comparative case studies* primarily look either at the attributes of groups or counterterrorism policies of the state and rarely consider both equally well (Cronin, 2006). Counterterrorism case studies are usually interested in the effect of state counterterrorism policies over the life span of each group but rarely examine dynamic interaction between policies and context (Art and Richardson, 2007, p. 2; Ranstorp, 2006, p. 14). As a result, they tend to have a strong bias toward tying the decline of such groups to specific government policies or the terrorists' failure to achieve objectives and publicity and fail to appreciate the degree to which terrorist groups evolve independent of government action (Cronin, 2006). In addition, much of the comparative counterterrorism research comes as edited collections of individual case studies by different regional experts; they often do not follow a central theme and fail to control relevant variables. The weaknesses in research design make generalizations across different cultural, historical, and political contexts suspect; the findings from this type of counterterrorism research are only suggestive, not definitive.

Organizational perspectives that focus on organizational structures, needs, and dynamics and their influence on terrorist decisionmaking (Crenshaw, 1985, 1992, 2001; McCormick, 2003; Oots, 1989) can be useful for identifying organizational processes that can potentially contribute to the decline—for example, those involving specific vulnerabilities that might lead a group to unravel—but do not specifically address the wider context that determines the values of the critical internal variables. Nonetheless, these perspectives are useful supplementary sources of potential factors and relationships.

The literature investigating the strategic logic behind terrorist actions, much like the organizational perspectives on terrorist decisionmaking, is an important supplementary source of knowledge relevant to the question of decline. This literature includes empirical studies on the effects of terrorism as well as game-theoretic analyses of the strategic interactions between terrorists and governments. Although this literature is sizable, only some of the studies are relevant to the question of terrorism decline. The studies in this category are useful insofar as they provide insights into possible antecedents of terrorists' decisions to abstain from violence as well as their possible reaction to changing circumstances, a government's policy measures, and the factors shaping the response. Particularly helpful is the part of this literature that analyzes the implications of the presence of different types of actors and their divergent interests on the outcome of the peace process. The same holds true for the general literature on conflict resolution and civil war termination, and there is some natural overlap between these types of literature and the literature on strategic interaction between terrorists and governments.

Studies on the evolution of terrorist groups as part of social movements (della Porta, 1992a, 1992b, 1995a, 1995b; della Porta and Tarrow, 1986; Gurr, 1990; Wieviorka, 1993) give more insights into the origins of terrorist groups than to their decline. However, by elucidating the relationship between terrorist groups and their potential base of support, they do suggest conditions and events likely to undermine support for the group.

The last group of works that caught our attention consists of studies that analyze global temporal patterns of terrorism. Some studies in this group (Enders and Sandler, 1999, 2000, 2005a, 2005b; Engene, 1998, 2004) employ large-N time-series analysis of terrorism trends across countries in an effort to identify and account for global patterns of terrorist activities and make predictions about terrorist attacks worldwide. Although global statistical patterns are interesting, they provide no insight into the declines of specific terrorist groups and campaigns (see this argument in Cronin, 2006, p. 14).

Other authors employ historical analysis to develop an argument about the existence global "life-cycles" of terrorism (Rapoport, 2001, 2004; Sedgwick, 2007).[28] This brand of research seems to be motivated by the observation that terrorist campaigns in different places tend to cluster together in time and share certain common characteristics, leading us to suspect that at least some of the origins of terrorism must be global rather than local (see Sedgwick, 2007, pp. 97–98). This type of analysis is concerned with linking together the outbreaks, evolution, and decline of many different terrorist groups and campaigns over time into a succession of identifiable patterns (or "waves") and exploring momentous precipitating events or large-scale changes in conditions that seem to be responsible for those patterns: new ideological/cultural influences, technological transformation and the spread of an innovative terrorism know-how enabled by such transformation, the success of a particular resistance movement serving as an inspiration for similar movements in other parts of the world, and so on. The "cyclical" hypotheses, with their emphasis on exogenous precipitating events or large-scale social change, although perhaps illuminating some important global dynamics at play, are so general that their bearing on the decline of specific groups or campaigns is remote and their policy relevance is limited.[29] In addition, many of the same plausible "global" reasons for terrorism decline (for example, the decline of a particular ideology worldwide) are also discussed in the literature on the end of terrorism.

Bibliography

Abedie, Alberto, *Poverty, Political Freedom, and the Roots of Terrorism,* Cambridge, Mass.: National Bureau of Economic Research, Working Paper No. 10859, October 2004. As of October 2, 2007:
http://papers.nber.org/papers/w10859.pdf

Adams, James, *The Financing of Terror: Behind the PLO, IRA, Red Brigades, and M-19 Stand the Paymasters: How the Groups That Are Terrorizing the World Get the Money to Do It,* New York: Simon and Schuster, 1986.

Alonso, Rogelio, "Pathways Out of Terrorism in Northern Ireland and the Basque Country: The Misrepresentation of the Irish Model," *Terrorism and Political Violence,* Vol. 16, No. 4, 2004, pp. 695–713.

Alterman, Jon B., *How Terrorism Ends,* Washington, D.C.: United States Institute of Peace, Special Report, 1999. As of August 2007:
http://www.usip.org/pubs/specialreports/sr990525.pdf

Art, Robert J., and Louise Richardson, eds., *Democracy and Counterterrorism: Lessons from the Past,* Washington, D.C.: United States Institute of Peace Press, 2007.

Berman, Eli, "Sect, Subsidy, and Sacrifice: An Economist's View of Ultra-Orthodox Jews," *The Quarterly Journal of Economics,* Vol. 115, No. 3, 2000, pp. 905–953.

———, *Hamas, Taliban and the Jewish Underground: An Economist's View of Radical Religious Militias,* Cambridge, Mass.: National Bureau of Economic Research, Working Paper No. 10004, August 2003. As of January 12, 2009:
http://www.nber.org/papers/w10004.pdf

Berrebi, Claude, "The Economics of Terrorism and Counterterrorism: What Matters and Is Rational-Choice Theory Helpful?" in Paul K. Davis and Kim Cragin, eds., *Social Science for Counterterrorism: Putting the Pieces Together,* Santa Monica, Calif.: RAND Corporation, 2009: As of January 29, 2009:
http://www.rand.org/pubs/monographs/MG849/

Berrebi, Claude, and Esteban F. Klor, "On Terrorism and Electoral Outcomes: Theory and Evidence from the Israeli-Palestinian Conflict," *Journal of Conflict Resolution,* Vol. 50, No. 6, December 1, 2006, pp. 899–925.

———, "Are Voters Sensitive to Terrorism? Direct Evidence from the Israeli Electorate," *American Political Science Review,* Vol. 102, No. 3, August 2008, pp. 279–301.

Bloom, Mia M., "Palestinian Suicide Bombing: Public Support, Market Share, and Outbidding," *Political Science Quarterly,* Vol. 119, No. 1, 2004, pp. 61–88.

Bueno de Mesquita, Ethan, "Conciliation, Counterterrorism, and Patterns of Terrorist Violence," *International Organization,* Vol. 59, No. 1, 2005, pp. 145–176.

Bueno de Mesquita, Ethan, and Eric S. Dickson, "The Propaganda of the Deed: Terrorism, Counterterrorism, and Mobilization," *American Journal of Political Science,* Vol. 51, No. 2, 2007, pp. 364–381.

Burgoon, Brian, "On Welfare and Terror: Social Welfare Policies and Political-Economic Roots of Terrorism," *Journal of Conflict Resolution,* Vol. 50, No. 2, April 1 2006, pp. 176–203.

Byman, Daniel, "Do Targeted Killings Work?" *Foreign Affairs,* Vol. 85, No. 2, March/April 2006, pp. 95–111.

Chairman of the Joint Chiefs of Staff, *National Military Strategic Plan for the War on Terrorism,* Washington, D.C., 2006.

Charters, David, *The Deadly Sin of Terrorism: Its Effect on Democracy and Civil Liberty in Six Countries,* Westport, Conn.: Greenwood Press, 1994.

Chen, Daniel, *Economic Distress and Religious Intensity: Evidence from Islamic Resurgence during the Indonesian Financial Crisis,* Cambridge, Mass.: Harvard University, PRPES Working Paper No. 39, 2003.

Crelinsten, Ronald D., and Alex P. Schmid, "Western Responses to Terrorism: A Twenty-Five Year Balance Sheet," in Alex P. Schmid and Ronald D. Crelinsten, eds., *Western Responses to Terrorism,* Portland, Ore.: Frank Cass, 1993, pp. 307–340.

Crenshaw, Martha, "The Causes of Terrorism," *Comparative Politics,* Vol. 13, No. 4, July 1981, pp. 379–399.

———, "An Organizational Approach to the Analysis of Political Terrorism," *Orbis,* Vol. 29, No. 3, Fall 1985, pp. 465–489.

———, "How Terrorism Ends," paper presented at the Annual Meeting of the American Political Science Association, Chicago, Ill., September 1987.

———, "How Terrorism Declines," *Terrorism and Political Violence,* Vol. 3, No. 1, Spring 1991, pp. 69–86.

———, "Decisions to Use Terrorism: Psychological Constraints on Instrumental Reasoning," in Donatella della Porta, ed., *Social Movements and Violence: Participation in Underground Organizations,* Greenwich, Conn.: JAI Press, 1992, pp. 29–42.

———, ed., *Terrorism in Context,* University Park, Pa.: Pennsylvania State University Press, 1995.

———, "Theories of Terrorism: Instrumental and Organizational Approaches," in David C. Rapoport, ed., *Inside Terrorist Organizations,* London: Frank Cass, 2001, pp. 13–31.

Cronin, Audrey Kurth, "Behind the Curve: Globalization and International Terrorism," *International Security,* Vol. 27, No. 3, 2002, pp. 30–58.

———, "How al-Qaida Ends: The Decline and Demise of Terrorist Groups," *International Security,* Vol. 31, No. 1, 2006, pp. 7–48.

David, Steven R., *Fatal Choices: Israel's Policy of Targeted Killing,* Ramat Gan, Israel: The Begin-Sadat Center for Strategic Studies, Bar-Ilan University, Mideast Security and Policy Studies No. 51, September 2002. As of March 25, 2008: http://www.biu.ac.il/Besa/david.pdf

Davis, Paul K., and Kim Cragin, "Introduction," in Paul K. Davis and Kim Cragin, eds., *Social Science for Counterterrorism: Putting the Pieces Together,* Santa Monica, Calif.: RAND Corporation, 2009: As of January 29, 2009: http://www.rand.org/pubs/monographs/MG849/

della Porta, Donatella, "Political Socialization in Left-Wing Underground Organizations: Biographies of Italian and German Militants," in Donatella della Porta, ed., *Social Movements and Violence: Participation in Underground Organizations,* Greenwich, Conn.: JAI Press, 1992a, pp. 259–290.

———, ed., *Social Movements and Violence: Participation in Underground Organizations,* Greenwich, Conn.: JAI Press, 1992b.

———, "Left-Wing Terrorism in Italy," in Martha Crenshaw, ed., *Terrorism in Context,* University Park, Pa.: Pennsylvania State University Press, 1995a, pp. 105–159.

———, *Social Movements, Political Violence, and the State: A Comparative Analysis of Italy and Germany,* Cambridge, U.K., and New York: Cambridge University Press, 1995b.

Della Porta, Donatella, and Sidney Tarrow, "Unwanted Children: Political Violence and the Cycle of Protest in Italy, 1966–1973," *European Journal of Political Research,* Vol. 14, No. 5-6, 1986, pp. 607–632.

DeNardo, James, *Power in Numbers: The Political Strategy of Protest and Rebellion,* Princeton, N.J.: Princeton University Press, 1985.

Eisenstadt, Michael, "'Pre-Emptive Targeted Killings' as a Counter-Terror Tool: An Assessment of Israel's Approach," *Peacewatch,* No. 342, August 28 2001. As of April 20, 2007: http://www.washingtoninstitute.org/templateC05.php?CID=2033

Enders, Walter, and Todd Sandler, "Transnational Terrorism in the Post-Cold War Era," *International Studies Quarterly,* Vol. 43, No. 1, March 1999, pp. 145–166.

———, "Is Transnational Terrorism Becoming More Threatening?: A Time-Series Investigation," *Journal of Conflict Resolution,* Vol. 44, No. 3, June 1 2000, pp. 307–332.

———, "After 9/11: Is it All Different Now?" *Journal of Conflict Resolution,* Vol. 49, No. 2, 2005a, pp. 259–276.

————, "Transnational Terrorism 1968–2000: Thresholds, Persistence, and Forecasts," *Southern Economic Journal,* Vol. 71, No. 3, 2005b, p. 466.

Engene, Jan Oskar, *Patterns of Terrorism in Western Europe, 1950–95,* thesis, Bergen, Norway: Department of Comparative Politics, University of Bergen, 1998.

————, *Terrorism in Western Europe: Explaining the Trends Since 1950,* Cheltenham, U.K., and Northhampton, Mass.: Edward Elgar, 2004.

Eubank, William, and Leonard Weinberg, "Does Democracy Encourage Terrorism?" *Terrorism and Political Violence,* Vol. 6, No. 4, 1994, pp. 417–443.

————, "Terrorism and Democracy: Perpetrators and Victims," *Terrorism and Political Violence,* Vol. 13, No. 1, 2001, pp. 155–164.

Eyerman, Joe, "Terrorism and Democratic States: Soft Targets or Accessible Systems," *International Interactions,* Vol. 24, No. 2, 1998, pp. 151–170.

Freeman, Michael, and Gordon McCormick, "Rethinking Decapitation as a Strategy Against Terrorism," paper presented at the Annual Meeting of the International Studies Association 48th Annual Convention, Chicago, Ill., February 28, 2007.

Gambetta, Diego, *The Sicilian Mafia: The Business of Private Protection,* Cambridge, Mass.: Harvard University Press, 1993.

Ganor, Boaz, *The Counter-Terrorism Puzzle: A Guide for Decision Makers,* New Brunswick, N.J.: Transaction Publishers, 2005.

Gazit, Shlomo, "Israel," in Yonah Alexander, ed., *Combating Terrorism: Strategies of Ten Countries,* Ann Arbor, Mich.: University of Michigan Press, 2002.

Gill, Anthony, and Erik Lundsgaarde, "State Welfare Spending and Religiosity: A Cross-National Analysis," *Rationality and Society,* Vol. 16, No. 4, November 2004, pp. 399–436.

Gupta, Dipak K., *Understanding Terrorism and Political Violence: the Life Cycle of Birth, Transformation, and Demise,* New York: Routledge, 2008.

Gurr, Ted Robert, "Terrorism in Democracies: Its Social and Political Bases," in Walter Reich, ed., *Origins of Terrorism: Psychologies, Ideologies, Theologies, States of Mind,* Washington, D.C.: Woodrow Wilson International Center for Scholars, 1990, pp. 86–102.

Hamilton, Lawrence C., *Ecology of Terrorism: A Historical and Statistical Study,* dissertation, University of Colorado, 1978.

Hopkinson, Michael, *The Irish War of Independence,* Montreal, Canada: McGill-Queen's University Press, 2004.

Horgan, John, and Max Taylor, "Playing the 'Green Card'—Financing the Provisional IRA," *Terrorism and Political Violence,* Vol. 11, No. 2, 1999, pp. 1–38.

Hosmer, Stephen T., *Operations Against Enemy Leaders*, Santa Monica, Calif.: RAND Corporation, 2001. As of January 12, 2009: http://www.rand.org/pubs/monograph_reports/MR1385/

Iqbal, Zaryab, and Christopher Zorn, "The Political Economy of Assasination, 1946–2000," paper presented at the Annual Meeting of the Midwest Political Science Association, Chicago, Ill., April 3–6, 2003.

Jaeger, David B., and M. Daniele Paserman, *The Cycle of Violence? An Empirical Analysis of Fatalities in the Palestinian-Israeli Conflict,* London, U.K.: Centre for Economic Policy Research, CEPR Discussion Paper No. 5320, 2005.

—————, "Israel, the Palestinian Factions, and the Cycle of Violence," *The American Economic Review,* Vol. 96, No. 2, 2006, pp. 45–49.

Jenkins, Brian M., *Should Our Arsenal Against Terrorism Include Assassination?* Santa Monica, Calif.: RAND Corporation, P-7303, 1986. As of January 12, 2009: http://www.rand.org/pubs/papers/P7303/

Jones, Seth G., and Martin C. Libicki, *How Terrorist Groups End: Lessons for Countering al Qa'ida,* Santa Monica, Calif.: RAND Corporation, 2008. As of January 12, 2009: http://www.rand.org/pubs/monographs/MG741-1/

Jordan, Jenna E., "Leadership Decapitation of Terrorist Organizations," paper presented at the Annual Meeting of the International Studies Association 48th Annual Convention, Le Centre Sheraton Hotel, Montreal, Quebec, Canada, March 17, 2004.

Kahler, Miles, "Networks and Failed States: September 11 and the Long Twentieth Century," paper presented at the Annual meeting of the American Political Science Association, Boston, Mass., August 28, 2002.

Kaplan, Edward H., Alex Mintz, and Shaul Mishal, "Tactical Prevention of Suicide Bombings in Israel," *Interfaces,* Vol. 36, No. 6, November 2006, pp. 553–561.

Kaplan, Edward H., Alex Mintz, Shaul Mishal, and Claudio Samban, "What Happened to Suicide Bombings in Israel? Insights from a Terror Stock Model," *Studies in Conflict & Terrorism,* Vol. 28, No. 3, 2005, pp. 225–235.

Kaufmann, Chaim, "Possible and Impossible Solutions to Ethnic Civil Wars," *International Security,* Vol. 20, No. 4, Spring 1996, pp. 136–175.

Kydd, Andrew H., and Barbara F. Walter, "Sabotaging the Peace: The Politics of Extremist Violence," *International Organization,* Vol. 56, No. 2, 2002, pp. 263–296.

—————, "The Strategies of Terrorism," *International Security,* Vol. 31, No. 1, 2006, pp. 49–80.

Lake, David B., "Rational Extremism: Understanding Terrorism in the Twenty-First Century," *Dialogue IO,* Vol. 1, No. 01, 2003, pp. 15–29.

Laqueur, Walter, *Terrorism,* 1st ed., Boston, Mass.: Little, Brown & Co., 1976.

Li, Quan, "Does Democracy Promote or Reduce Transnational Terrorist Incidents?" *Journal of Conflict Resolution,* Vol. 49, No. 2, April 1 2005, pp. 278–296.

Lizardo, Omar B., "The Effect of Economic and Cultural Globalization on Anti-U.S. Transnational Terrorism 1971–2000," *Journal of World-Systems Research,* Vol. 12, No. 1, 2007, pp. 149–186.

Lotrionte, Catherine, "When to Target Leaders," *The Washington Quarterly,* Vol. 26, No. 3, 2003, pp. 73–86.

Luft, Gal, "The Logic of Israel's Targeted Killing," *Middle East Quarterly,* Vol. 10, No. 1, 2003, pp. 6–13.

McCallister, William S., "The Iraq Insurgency: Anatomy of a Tribal Rebellion," *First Monday,* Vol. 10, No. 3, March 2005. As of September 5, 2007: http://outreach.lib.uic.edu/www/issues/issue10_3/mac/index.html

McCormick, Gordon H., "Terrorist Decision Making," *Annual Review of Political Science,* Vol. 6, No. 1, 2003, pp. 473–506.

Merkl, Peter H., "West German Left-Wing Terrorism," in Martha Crenshaw, ed., *Terrorism in Context,* University Park, Pa.: Pennsylvania State University Press, 1995, pp. 160–210.

Mitchell, C. R., and Michael Nicholson, "Rational Models and the Ending of Wars," *The Journal of Conflict Resolution,* Vol. 27, No. 3, September 1983, pp. 495–520.

Nagl, John B., *Learning to Eat Soup with a Knife: Counterinsurgency Lessons from Malaya and Vietnam,* Chicago, Ill.: University of Chicago Press, 2005.

O'Neill, Bard E., *Insurgency & Terrorism: Inside Modern Revolutionary Warfare,* Washington, D.C.: Brassey's, 1990.

Oots, Kent Layne, "Organizational Perspectives on the Formation and Disintegration of Terrorist Groups," *Terrorism,* Vol. 12, 1989, pp. 139–152.

Piazza, James B., "Draining the Swamp: Democracy Promotion, State Failure, and Terrorism in 19 Middle Eastern Countries," *Studies in Conflict and Terrorism,* Vol. 30, No. 6, 2007, pp. 521–539.

Pluchinsky, Dennis, "Middle Eastern Terrorist Activity in Western Europe in 1985: A Diagnosis and Prognosis," in Paul Wilkinson and Alasdair M. Stewart, eds., *Contemporary Research on Terrorism,* Aberdeen, U.K.: Aberdeen University Press 1987, pp. 40–78.

Ranstorp, Magnus, ed., *Mapping Terrorism Research: State of the Art, Gaps and Future Direction*, London, U.K.: Routledge, 2006.

Rapoport, David C., "Fear and Trembling: Terrorism in Three Religious Traditions," *The American Political Science Review,* Vol. 78, No. 3, September 1984, pp. 658–676.

———, "The Fourth Wave: September 11 in the History of Terrorism," *Current History,* Vol. 100, No. 650, December 2001, pp. 419–419.

———, "The Four Waves of Modern Terrorism," in Audrey K. Cronin and James M. Ludes, eds., *Attacking Terrorism: Elements of a Grand Strategy,* Washington, D.C.: Georgetown University Press, 2004, pp. 46–74.

Robison, Kristopher K., Edward M. Crenshaw, and J. Craig Jenkins, "Ideologies of Violence: The Social Origins of Islamist and Leftist Transnational Terrorism," *Social Forces,* Vol. 84, No. 4, June 2006, pp. 2009–2026.

Ronfeldt, David F., "Al Qaeda and Its Affiliates: A Global Tribe Waging Segmental Warfare?" *First Monday,* Vol. 10, No. 3, March 2005. As of January 21, 2009:
http://firstmonday.org/htbin/cgiwrap/bin/ojs/index.php/fm/article/view/1214/1134

Ross, Jeffrey Ian, "The Rise and Fall of Quebecois Separatist Terrorism: A Qualitative Application of Factors from Two Models," *Studies in Conflict & Terrorism,* Vol. 18, No. 4, July 1995, pp. 285–296.

Ross, Jeffrey Ian, and Ted Robert Gurr, "Why Terrorism Subsides: A Comparative Study of Canada and the United States," *Comparative Politics,* Vol. 21, No. 4, July 1989, pp. 405–426.

Rotberg, Robert I., "Failed States in a World of Terror," *Foreign Affairs,* Vol. 81, No. 4, 2002a, p. 126. ABI/INFORM Global (Proquest) database.

———, "The New Nature of Nation-State Failure," *The Washington Quarterly,* Vol. 25, No. 3, 2002b, pp. 85–96.

———, ed., *State Failure and State Weakness in a Time of Terror*, Washington, D.C.: Brookings Institution Press, 2003.

Schmid, Alex P., "Terrorism and the Media: The Ethics of Publicity," *Terrorism and Political Violence,* Vol. 1, No. 4, 1989, pp. 539–565.

———, "Terrorism and Democracy," *Terrorism and Political Violence,* Vol. 4, No. 4, 1992, pp. 14–25.

Schmid, Alex P., and Janny de Graaf, *Violence as Communication: Insurgent Terrorism and the Western News Media*, Beverly Hills, Calif.: Sage Publications, 1982.

Sederberg, Peter C., "Responses to Dissident Terrorism: From Myth to Maturity," in Charles W. Kegley, Jr., ed., *International Terrorism: Characteristics, Causes, Controls,* New York: St. Martin's Press, Inc., 1990, pp. 262–280.

———, "Conciliation as Counter-Terrorist Strategy," *Journal of Peace Research,* Vol. 32, No. 3, August 1995, pp. 295–312.

Sedgwick, Mark, "Inspiration and the Origins of Global Waves of Terrorism," *Studies in Conflict & Terrorism,* Vol. 30, No. 2, 2007, pp. 97–112.

Shabad, Goldie, and Francisco José Llera Ramo, "Political Violence in a Democratic State: Basque Terrorism in Spain," in Martha Crenshaw, ed., *Terrorism in Context,* University Park, Pa.: Pennsylvania State University Press, 1995, pp. 410–469.

Shafer, D. Michael, "The Unlearned Lessons of Counterinsurgency," *Political Science Quarterly,* Vol. 103, No. 1, Spring 1988, pp. 57–80.

Simon, Jeffrey D., *Misperceiving the Terrorism Threat,* Santa Monica, Calif.: RAND Corporation, 1987. As of January 21, 2009: http://www.rand.org/pubs/reports/R3423/

Stein, Yael, *Israel's Assassination Policy: Extra-Judicial Executions,* position paper, translated by Maya Johnston, Jerusalem: B'Tselem, The Israeli Information Center for Human Rights in the Occupied Territories, January 2001. As of April 10, 2008 http://www.btselem.org/Download/200101_Extrajudicial_Killings_Eng.doc

Takeyh, Ray, and Nikolas K. Gvosdev, "Do Terrorist Networks Need a Home?" *The Washington Quarterly,* Vol. 25, No. 3, 2002, pp. 97–108.

Toft, Monica Duffy, "Getting Religion? The Puzzling Case of Islam and Civil War," *International Security,* Vol. 31, No. 4, 2007, pp. 97–131.

Wade, Sara Jackson, and Dan Reiter, "Does Democracy Matter? Regime Type and Suicide Terrorism," *Journal of Conflict Resolution,* Vol. 51, No. 2, April 2007, pp. 329–348.

Walter, Barbara F., "The Critical Barrier to Civil War Settlement," *International Organization,* Vol. 51, No. 3, Summer 1997, pp. 335–364.

———, *Committing to Peace: The Successful Settlement of Civil Wars,* Princeton, N.J.: Princeton University Press, 2002.

Weinberg, Leonard, "Turning to Terror: The Conditions Under Which Political Parties Turn to Terrorist Activities," *Comparative Politics,* Vol. 23, No. 4, July 1991, pp. 423–438.

Wieviorka, Michel, *The Making of Terrorism,* Chicago, Ill.: University of Chicago Press, 1993.

Wilkinson, Paul, "Three Questions on Terrorism," *Government and Opposition,* Vol. 8, No. 3, 1973, pp. 290–312.

————, "The Media and Terrorism: A Reasessment," *Terrorism and Political Violence,* Vol. 9, No. 2, Summer 1997, pp. 51–64

————, *Terrorism Versus Democracy: The Liberal State Response,* 2nd ed., London, U.K., and New York: Routledge, 2006.

Zussman, Asaf, and Noam Zussman, "Assassinations: Evaluating the Effectiveness of an Israeli Counterterrorism Policy Using Stock Market Data," *The Journal of Economic Perspectives,* Vol. 20, 2006, pp. 193–206.

Endnotes

[1] See also Guptka (2008). Guptka's Chapter Eight reviews the demise of terror and highlights three class of outcome: a degree of political success, defeat by the state, or loss of ideological orientation and movement into criminality.

[2] This process is facilitated by the tendency of terrorist organizations to engender strong group dependency that results over time in organization imperatives—for example, survival and viability—replacing political purpose as the dominant incentive for the members (see Crenshaw, 1985, 2001).

[3] They include killing and imprisoning the terrorists, using force to gain intelligence about attacks and organization, destroying and confiscating weapons and materials, denying the terrorists' contact with the sources of aid, disrupting their coordination and communication, isolating them from their audiences, harassing supporters, retaliating in kind, and so on. The instruments include more vigorous law enforcement and criminal prosecution; military strikes involving conventional actions, covert operations, and mirror response; legislative changes resulting in longer prison terms; expanded definition of the culpability for terrorism; a more-restrictive detention and confinement regime; reduced protections against search and arrests; more-invasive search, surveillance and interrogation procedures; relaxed standards of acceptable evidence; special courts and jurisdictions; restrictions on free movement and speech; and other curtailments of civil liberties. Repressive measures undermine terrorists' capabilities by disrupting operations, degrading organization, causing large-scale attrition, denying resources such as weapons and materials, recruitment, funding, and safe havens; and deterring attacks and undermining the terrorists' will to fight.

[4] The anarchist groups were especially vulnerable to repression, as their ideology turned out to be a liability for building a strong, tight-knit organization (see McCormick, 2003). In contrast, similar small radical left-wing groups in Europe enjoyed a measurable and more-enduring, support, albeit limited. As a result, governments found that some conciliatory measures were indispensable.

[5] For example, group organization often depends on the availability of support base: Groups lacking a social base appear to have a simpler structure (see Crenshaw, 1985). It may also depend on the ideological goals of the organization (see McCormick, 2003, pp. 478–480).

[6] For instance, the success of direct actions requires, among other things, that the state have the ability to access the territory from which the organization operates, or to effectively seal off the territory where potential targets are located from terrorists' home base. (This is particularly true in the case of transnational terrorism.)

[7] Strong authoritarian states—which include absolutist or totalitarian states (Schmid, 1992, p. 16)—have better intelligence and access to the group. They also have a near-complete physical control over the population and monopoly over the flow of information. Thus, they face few constraints on their behavior and frequently engage in repressive terror (Wilkinson, 1973). Total government control explains in large measure why little sustained terrorist campaigns are known to have occurred in such totalitarian states as the Soviet Union and China (Crenshaw, 1981, p. 283). Without freedom of information, there is little strategic incentive to engage in terrorism against an overwhelming state power and the groups that do try their luck can be easily suppressed by the government's repressive apparatus.

[8] The histories of the Uruguayan Tuparamos and Chechen groups in Russia suggest that, in weak democracies, the public backlash caused by the destructiveness of terrorists may empower the government or a part of the ruling elite enough to enable it to crack down on democratic institutions and engage in outright repression of the terrorists' support base, resulting in internal splits among terrorists, collapse of public support, and an ultimate defeat of the terrorist movement.

[9] This statement closely parallels the argument found in the literature on root causes of terrorism that points to nonmonotonic relationship between democracy and political freedoms, on the one hand, and terrorism (Abedie, 2004; Eyerman, 1998; Iqbal and Zorn, 2003; Piazza, 2007). The general argument is that the states with intermediate levels of freedoms—new or weak democracies and weakly authoritarian states—are more vulnerable to terrorism than either established democracies or highly authoritarian states: Unlike highly authoritarian states, they have political constraints on their freedom of action in countering the terrorist threat; at the same time, unlike established democracies, they lack strong and durable political institutions to provide terrorists and their supporters with alternatives means to achieve their goals. On the mixed relationship between democracy and terrorism see also Li (2005) and Wade and Reiter (2007).

At the same time, Sederberg (1995, p. 310) suggests that the key variable might involve "less the form of the regime than the essential strength of its political foundations," where "weak states, whether formally democratic or authoritarian, will be both more vulnerable to terrorism and vacillating in their responses." Supplementary insights come from the research on the relationship between terrorism and failed states. The literature maintains that the lack of legitimacy and ability to project power internally make failed states especially vulnerable to terrorist attacks (see Kahler, 2002; Piazza, 2007; Rotberg, 2002a, 2002b, 2003; Takeyh and Gvosdev, 2002).

[10] Significant theoretical and empirical support for this argument comes from the literature on strategic interaction between a terrorist group and a government. This literature suggests that terrorist groups act strategically when using terrorist tactics and examines the implications of the terrorists' strategic interaction with the government (Bueno de Mesquita and Dickson, 2007; Kydd and Walter, 2006; Lake, 2003). The strategic logic behind terrorist actions is based on the calculation that provocation increases the likelihood of repression and repression in turn increases support for the extremists. There is strong empirical evidence justifying this rationalist logic. Using data on the variation of attacks and electoral outcomes across time and space in Israel, Berrebi and Klor (2006, 2008) investigated the interaction between attacks, on the one hand, and electoral preferences and outcomes, on the other, and found that terrorist attacks increase support for the right-wing politicians to the extent that they are sufficient to determine the electoral outcomes. This shift in electoral preferences in turn increases the likelihood of strong government countermeasures (Kydd and Walter, 2006, p. 71). Bueno de Mesquita and Dickson (2007) rely on a game-theoretic model to demonstrate that repression can, under certain circumstances, radicalize the population and increase support for extremists through either economic damage or by revealing information about the government's motivation. Jaeger and Paserman (2005), studying the Israeli experience, found empirical support for the link between repression and radicalization.

[11] Even though its number of members plummeted after the capture of its leader Guzman, Shining Path continued to carry out terrorist activities, albeit at a much reduced level. PKK did shift toward political activities after the capture of Ocalan, declaring a "cease-fire," but has maintained its armed wing and recently resumed attacks against Turkish interests. Aum was, arguably, more a cult practicing terrorism than a political organization.

[12] Israeli experience provides some evidence in favor of this claim. David (2002) argues that targeted assassinations have all but destroyed or substantially reduced the effectiveness of terrorist organizations in cases where leadership, planning, and tactical skills were scarce and limited to a few individuals in the organization (Black September, Islamic Jihad, senior Egyptian intelligence officers organizing terrorist infiltration from Egypt, German scientist developing long-range missiles in Egypt).

[13] Freeman and McCormick also argue that "sustained" decapitation—one practiced repeatedly—may be effective against the organizations with centralized but scarce leadership (see Freeman and McCormick, 2007).

[14] Israel—the country that has practiced the policy of targeted killings for many years—is a case in point. The policy has in the past produced worldwide condemnation (including from such quarters as the United Nations, the European Union, and

even Israel's main ally—the United States) and is thought to have contributed to Israel's international isolation. Becoming a victim of Israeli targeted killing became a badge of honor among Palestinians and unwittingly enhanced the popularity of the organization to which the victim belonged, as evidenced by public opinion polls. The targeted killing of the Hamas bombmaker Yehiya Ayash and the leader of the Popular Front for the Liberation of Palestine (PFLP) Mustafa Zibri is claimed to have undermined temporary cease-fire negotiations and resulted in greater terrorist action. Assassination in January 2002 of Tanzim leader Raed al-Karmi ended a cease-fire declared by Yasir Arafat and caused the relatively moderate Al-Aqsa Brigades affiliated with Fatah to engage in suicide bombings for the first time; it is also suggested to have led to the appearance of women suicide bombers and unprecedented casualties among Israelis. The July 2002 killing of Hamas leader Sheikh Salah Shehada that claimed the lives of 14 innocent bystanders derailed what many believed to be promising negotiations. That year also saw a significant increase in the number of suicide bombers. Some argued that Israeli actions to assassinate Palestinian terrorists who were thought to be pragmatists and might have proven to be useful political negotiators (such as PLO second-in-command Abu-Jihad) reduced the chances for reaching agreement with the Palestinians. The Israeli policy of targeted assassination might as well have led to enhanced cooperation between Islamic Jihad, Hamas, and Palestinian Authority. Finally, when the Israelis assassinated the PFLP's theater director, Mohammad Boudia, he was replaced by an even more notorious terrorist—Carlos "the Jackal." (See Byman, 2006; David, 2002; Ganor, 2005; Jenkins, 1987; Luft, 2003.)

[15] The evolution of right-wing groups does not conform to the pattern identified here despite some similarities. Even though many right-wing groups appeared to have a difficulty persisting over generations, this is likely to reflect the challenges of tracking them over time (see Crenshaw, 1991; Cronin, 2006).

[16] Several authors use the organizational dynamics perspective to reflect on the causes and outcomes of less-than-perfect decisionmaking in terrorist organizations. Gordon McCormick discusses the following features of imperfect decisionmaking driven by organizational/group dynamic factors: isolation and loss of a sense of reality, bias toward action, use of preexisting "scripts," interorganizational competition, "group-think" and excessive risk-taking as a result of "self-censorship," consensus-building, suppression of dissent, and strong group dependency. Crenshaw instead focuses on the organizational needs and effects on decisionmaking of terrorist organizations. She points out that strong interest in group preservation over time can supplant the ends as survival of the group becomes the end in itself; terrorist behavior becomes self-sustaining over time regardless of changing conditions and political results (see Crenshaw, 1985, 2001; McCormick, 2003).

[17] Berman (2003) argues that, in cases where social services are inadequate, the terrorists can seize the opportunity to win over supporters and that voluntary religious groups have a distinct advantage. His rational-choice model demonstrates that, in places where the government is a poor provider of local public goods, such

as health care, education, and public safety, and the market is an inefficient provider of income and insurance, voluntary religious organizations act as "mutual insurance clubs" where the membership provides the access to these services in exchange for costly commitment signals in the form of sacrifices demanded by the membership. (For economic explanations of the influence of religious-political organizations, see also Berman, 2000; Chen, 2003; Gill and Lundsgaarde, 2004.) The latter feature also makes these groups excellent potential militias—as demonstrated by the experience of such groups as Hamas, Hizballah, and the Taliban—as it ensures the commitment of members that is crucial for effective functioning of militias (see Berman, 2003). The implication is that an efficient market economy and a functioning secular state providing basic public goods pose a threat to these affiliations as they reduce the need for their services and raise the opportunity cost of membership. However, to be effective, social services must be provided both to members and nonmembers without discrimination, as any exclusion would create incentives for membership (Berman, 2003). Although this analysis suggests the potential effectiveness of social-service provision by the government in undercutting support for such groups, it also implies that the latter are likely to resist fiercely any such effort by the state. In fact, terrorism may not be the original or even primary purpose for the existence of these groups and may be employed precisely to ward off such threats to the groups' existence. Ronfeldt (2005) and McCallister (2005) point to the existence of very similar incentive structures in traditional tribal societies that seem to be fueling terrorism and insurgency by al-Qaeda and their affiliates. (For additional relevant insights, see also Diego Gambetta's excellent analysis of the Sicilian Mafia's protection business [Gambetta, 1993].)

There is an argument that government's social policies, as they involve transfer of resources, may play into the hands of terrorists by increasing their capacity to organize terror. A recent pooled time-series analysis provides evidence that, on balance, social-welfare policies reduce international and domestic terrorism (Burgoon, 2006). Although the study does not directly address the link between these policies and public support, it does find that welfare policies are associated with less poverty, income inequality, economic insecurity, and religious extremism—all thought to be the correlates of public support. However, it seems logically necessary to us that, to effectively undermine support for the terrorist organization, social-services provision by governments must compete with that by the terrorists—that is, the terrorists should not be allowed to take credit for those services.

[18] Such was the case in Uruguay when the military toppled the liberal democratic government, crushed the Tuparamos, and remained in power. Russia's brutal war in Chechnya is, arguably, another example of the success of repression strategy by an increasingly strong and autocratic state. This comparative advantage of strong autocracies over democratic polities seems to explain why oppositional terrorism rarely takes root in strong authoritarian societies.

[19] The end of the Cold War is a case in point. The failure of the Socialist bloc not only led to curtailment of external aid to terrorist groups worldwide from the Soviet

Union and its satellites, but (perhaps more important) also undercut the attractiveness of the driving ideology of many terrorist movements. Similarly, Ross and Gurr (1989) argue that the change in the general political climate that occurred in United States in 1970s, as manifested in the shift in public interest from public purposes and social change to pursuit of private interests, made it increasingly difficult for militants in the 1970s and 1980s to establish constituencies in support of radical change.

[20] Some religious terrorist organizations can be usefully grouped with ethnic and nationalist groups, as religion is employed as a marker of identity in contraposition to a rival identity; others look more like ideologically motivated groups. Religious groups are often similar to, and sometimes overlap, some ethnic and nationalist groups in that they fight the state dominated by a group representing a different religion or ethnicity or a foreign power. Such groups may demand greater autonomy or independence; they may also vie, as some ethnic groups do, for control of the state or a greater representation, especially when the dissident group represents a repressed majority. As distinct from these groups, other religious groups have clear ideological aims of transforming the established social order and are fighting to take over the state from their more-secular (or apostate) coreligionists. The distinction is often blurred in practice, however, as many dissident movements have both elements, as is the case of many terrorist groups in the Arab Middle East. It also seems to be the case that the large size of the religious terrorist group's following often signals the presence of an identity-based component also present in ethnic and nationalist groups. On the other hand, some authors seem to favor the explanation that strong support exists for the religious terrorist groups because of the inherent staying power of sacred and spiritual motivations when they argue that the latter accounts for the relative longevity of religious groups (see Cronin, 2006, p. 13; Rapoport, 1984).

[21] See footnote 9 for the relevant details.

[22] DeNardo (1985) has demonstrated how government concessions can split "pragmatists" and "purists" in revolutionary movements, although he did not specifically address terrorism.

[23] See the earlier discussion on the effects of organizational disunity on terrorist decisionmaking.

[24] It appears that negotiations are more likely to undermine the cohesiveness of the terrorist movement and result in splits when the constituencies with strong preferences support the terrorists' cause (see Cronin, 2006, p. 26).

[25] The literature on strategic interaction between terrorists and the government offers interesting insights into the mechanisms responsible for sustaining violence following the government's peace initiatives. For one, terrorist extremists have plenty to gain from continued violence. Terrorist attacks often increase support for radical measures in the government's support base and provoke government retaliation, which in turn radicalizes the terrorists' support base and increases support

for extremist factions at the expense of the moderates (see footnote 9 for details). Violence offers extremist factions a real opportunity to thwart moderates' acceptance of concessions that they consider insufficient. Kydd and Walter (2002) suggest that terror attacks by extremist factions can sabotage the peace process when they manage to undermine government's trust of the moderates' willingness and ability to abide by the terms of agreement in situations when the consequence of abortive settlement for the government is resumption of hostilities on more disadvantageous terms (as in the case of Israel handing a portion of West Bank back to the Palestinians). (On "spoiling" as one of the principal strategic logics behind terrorist campaigns, see also Kydd and Walter, 2006.) Moderate factions themselves might have incentives that can contribute to continued violence. For instance, while the negotiations are still under way, moderates have an incentive to appear more radical to secure better terms of settlement; however, this behavior often has an unintended consequence of radicalizing supporters to the point where they are no longer willing to endorse moderate positions. After a peace agreement has been reached, moderates may have a stake in the continued existence of the extremist faction to secure the government's commitment to the terms of settlement (Bueno de Mesquita, 2005). The government might itself have incentives to aid extremist challengers to moderate groups when the extremists are still the weaker side; they might choose to do so (as in the case of Israel's support for Hamas and Islamic Jihad) as a way to put pressure on the moderates to accept concessions while the latter are still in charge and at the same time minimizing the level of concessions made (see Bueno de Mesquita, 2005, p. 164).

Not only can terrorism attacks continue following the government's peace initiatives, but the level of violence can increase as well. The "increased militancy" argument suggests that, with extremists' having a continued stake in violence, moderates' acceptance of the government's conciliatory offer frees the movement from the restraining influence of the moderates (see Bueno de Mesquita, 2005).

The literature on termination of civil wars and conflict resolution also discusses how the problems with agreeing on and ensuring commitment to the terms of settlement can complicate achievement of an enduring peaceful outcome (see Kaufmann, 1996; Mitchell and Nicholson, 1983; Walter, 1997, 2002).

[26] Nonrationalist explanations of continued violence following government concessions have also been advanced in the literature. For instance, Shabad and Llera Ramo (1995) argue that the increase in terror after the granting of Basque autonomy was due to the "culture of violence" that prevented the bask terrorist group Euskadi Ta Askatasuna (ETA) from abandoning violent tactics. Other similar explanations—such as desire for revenge or religious zeal—are extrapolations from nonrationalist explanations advanced in the literature on individual motivations for terrorism. These nonrationalist incentives for terrorism are discussed in detail in Berrebi (2009). We note here only that the whole literature on the general causes of terrorist behavior is relevant to this discussion insofar as those causes point to enduring sources of the decision to engage in terrorism. "Irrational" incentives, for one,

exemplify rigidities in terrorist motivation that cannot be affected by government concessions.

[27] Some authors working from organizational perspectives have suggested that terrorism might develop its own momentum and become self-sustaining regardless of changing conditions and results (see Crenshaw, 1985, p. 473, 1995, 2001, p. 21). This would indicate that the way terrorism "goes up" is different from the way it "goes down," implying somewhat different processes and factors.

[28] See also the statistical study by Robison, Crenshaw, and Jenkins (2006) for an attempt at empirical evaluation of the argument.

[29] The value of this research is that it suggests that terrorist organizations may have a limited lifetime. However, the problem with this type of work is that different historians may, and do, disagree on both the identification of individual waves and the reasons for their origins and decline. Sedgwick (2007) demonstrates how a different conceptualization of terrorism can lead to a very different characterization of the waves. More important, although these works are articulate about the causes for the outbreak of a wave, they are less so for its decline and for a good reason. Specifically, for a succession of waves to exist, no other reason for the decline of each preceding wave is logically required than the arrival of the next wave, as individual terrorist groups that formed the part of the old wave are being caught up in the new one or fading away into irrelevance. Additionally, the hypotheses about "waves" are concerned with explaining a wave, not individual cases of terrorism within a wave: A decline of the wave is by no means thought to imply a decline of every case of terrorism that forms a part of the wave.

CHAPTER EIGHT

Disengagement and Deradicalization: Processes and Programs

Darcy M.E. Noricks

Introduction

The question of why individuals move away from terrorism has been much less studied than how terrorism arises or even how it ends. This paper reviews the available literature, much of which is quite recent. A key theme, reflected in the paper's title, is the need to address both disengagement and deradicalization because they turn out to be quite different. I first discuss what is known about these processes and then describe programs that have been or are being used to encourage disengagement or deradicalization.

Processes

Despite a number of deradicalization programs targeting individuals arrested on terrorism charges, which have been implemented in such countries as Saudi Arabia, Singapore, Yemen, Egypt, and the United Kingdom in recent years,[1] the issues of deradicalization and disengagement from a terrorist group are two topics about which we know relatively little. These processes are some of the most undertheorized concepts in the terrorism literature, but not because they are unimportant. A closer look suggests that the lack of theory is particularly problematic, since deradicalization is not merely the radicalization process in reverse. Although there are some similarities, disengagement from terrorism appears to be "as complex a process as that which helps us understand initial involvement in the first place" (Horgan, 2007a, p. 117). It may be that the weak theory in this area is a result of the particularly idiosyncratic nature of deradicalization and disengagement processes. But as Horgan (2007b) emphasizes, *it is in the disengagement phase that practical counterterrorism initiatives might best be applied.*

Deradicalization can be either ideological or behavioral. Omar Ashour (2008a) defines the concept as a process that leads an individual (or group) to change his attitudes about violence—specifically about the appropriateness of violence against civilians. Ideological deradicalization results from a change in beliefs, whereas behavioral deradicalization emphasizes changes in actions. Horgan (2008) also emphasizes the need for clarity in distinguishing deradicalization (attitudinal modification) from disengagement (behavioral modification). He underscores the fact that, from a counterterrorism perspective, disengagement is more important than deradicalization, since the former can occur without the latter. Ashour (2008a) also postulates a third category, organizational deradicalization, which is a group-level phenomenon that, if successful, would move the entire group away from terrorism—ideally this would occur without the group spinning off violent splinter groups. Examples of organizational deradicalization include former terrorist groups (for example, the Palestine Liberation Organization or South Africa's African National Congress) and militia groups (for example, Amal in Lebanon).

Although the vacuum may be partly filled with the forthcoming publication of Horgan's book, *Walking Away from Terrorism,* the terrorism literature is currently lacking any detailed debate over theories of deradicalization, although there is a small literature on how and why terrorism ends—both at the group level and as a larger phenomenon.[2] Conference proceedings from a RAND-sponsored conference in 2005 represent some of the first efforts to explicitly address the issue of deradicalization in the context of contemporary Islamist radicalization and recruitment (Benard, 2005). Discussion has occurred in a number of other fields, however, about the factors that might lead an individual to turn away from violence, leave a particular group, or halt participation in a particular activity or movement. Renee Garfinkel interviewed individuals who were former activists, terrorists, and political and military leaders about their experiences in turning away from violent strategies and organizations and toward ones that espoused a nonviolent or cooperative perspective. Tore Bjorgo's work on disengagement from extremist groups that embrace racist ideologies provides a discussion of relevant "push" and "pull" factors. Scott Decker and Barrik Van Winkle illuminate some of the reasons that gang members leave gangs, and Galanter provides details about both voluntary and forced departure from religious cults and sects. John Horgan (2005, 2006, 2007a, 2007b, 2008, and forthcoming), Tore Bjorgo (2006), Bjorgo and Horgan (2009), and Omar Ashour (2007, 2008a, 2008b, and forthcoming) are three authors who have dealt with deradicalization in the most theoretical fashion.

Renee Garfinkel's (2007) review of seven deradicalization cases—Muslim, Jewish, and Christian—leads her to conclude that deradicalization can be as much of a spiritual experience, similar to a religious conversion, as the initial radicalization may have been.[3] In contrast to the radicalization experience, however, the deradicalized individuals in her small study did not adopt their new ideology as a function of their participation in a supportive peer group. The decision to deradicalize was most often an individual decision, which subsequently isolated that person from his or her existing social group (p. 11). Relationships with role models were cited as important in making the move away from radical beliefs, however. One commonality with radicalization pathways was the experience of trauma preceding the decision or process of deradicalization. Trauma acted as a precipitating event for the transformation of personal beliefs.[4] In many cases, trauma coincided with the unexpected experience of compassion from those previously identified as enemies or "other." Individuals who turned away from radical ideologies first experienced a perceived failure of their existing values and beliefs (p. 12), which was similar to their experience of radicalization. Even then, recognition of the shared humanity of the "enemy" was a difficult step that had to be repeated over and over throughout the lengthy move from violence to nonviolence (p. 14).

Writing about defection from right-wing groups, Tore Bjorgo (2006) distinguishes between push and pull factors that affect an activist's decision to leave the group. Push factors are negative circumstances or social forces that make it unattractive to continue membership in a particular organization. These factors might include criminal prosecution, parental or social disapproval, or counterviolence from oppositional groups. Alternatively, as the movement evolves, "the terrorist may find that some of his or her most deeply held political ideals—the ones that led them to become involved in the movement in the first place—are being compromised as a result of some new stifling organizational 'climate' within the group or through the role of certain individuals within it" (Horgan, 2005, p. 147). Additionally, one may discover that the original ideological impetus for radicalization no longer resonates with the individual.

Pull factors are opportunities or social forces that attract an individual to a more promising alternative. These might include "longing for the freedoms of a normal life," new employment or educational prospects that could be undermined if an individual's group membership were known, or the desire to establish a family and take on parental and spousal roles—one of the strongest motives for leaving a militant group (Bjorgo, 2006, pp. 11–12). Bjorgo emphasizes that the effect of push factors can be difficult to determine in advance. Negative sanctions may lead more recent members to leave the group, but those same sanctions could also increase members' solidarity within the group as the group bands together to meet the outside threat. The latter is particularly a risk when negative sanctions are not matched with positive incentives.

One of the most common reasons for staying in the group is that the activist has nowhere to go, because of the nature of the relationships he or she destroyed or abandoned when joining the group in the first place. The defector "risks ending up in a social vacuum," isolated, alone, and lonely (Bjorgo, 2006, p. 14)—a likely outcome that Garfinkel's evidence supports. Pull factors represent the shifting of a militant's priorities. But these factors are often challenged by high barriers to exit, including concerns about the sunk costs of time and effort already invested in the group, fear about reprisals from the group, and lack of protection against former enemies. Moreover, even if the activist no longer believes in the group's ideology or political goals, leaving the group is akin to leaving a family, a community, and an identity. Table

8.1 lists a number of factors that fall under the categories of push and pull factors.

A search for identity and the reward of "belonging" have been identified as major influences that motivate radical behavior and continuing participation in radical groups (Victoroff, 2005). The effect of these factors on deradicalization has been explored in such historical case studies as the Red Brigades (Jamieson, 1990a), the Baader-Meinhof Group (Post, 1987), and the Irish Republican Army (O'Callaghan, 1998). These studies reflect the internal pressures to stay competing with the external pressures to go—focusing specifically on psychological pressures and the "spiraling of commitment" that often keeps members within the group (Taylor, 1988, p. 168). Writing about deradicalization in Southeast Asia, Zachary Abuza (forthcoming) notes that

Table 8.1
Sample Push and Pull Factors

Class of Factor	Examples
Push factors	Criminal prosecution
	Parental or social disapproval
	Counterviolence from oppositional groups
	Loss of faith in ideology or politics of group
	Discomfort with group's violent activities
	Disillusionment with group's leadership
	Loss of confidence, status or position in group
	Ejection from the group
	Exhaustion from tension and uncertainty as a member of a targeted group
	Increased activity in a "competing role," for example, political activity that displaces the violent role
Pull factors	Desire for a normal life
	Desire to establish a family and take on parental and spousal roles
	Other changing priorities
	New employment or educational opportunities that could be undermined if group membership were known
	New role model or social group
	New, more compelling ideology or belief structure

the success of such an effort "is driven in large part by societal attitudes: will former terrorists be welcomed back into society, or will they be treated as outcasts?"

Decker and Van Winkle's (1996, pp. 262–264) assessment of participation in, and attrition from, street gangs picks up on issues also identified by Horgan, Garfinkel, and Bjorgo. Decker and Van Winkle describe the difficulty in defining the act of "leaving the gang." In some cases, leaving means refraining from participation in illegal activities but maintaining friendships with current gang members—particularly when these friendships predate gang membership. However, they note that even after leaving the gang, prior antagonisms may continue to be played out in resident neighborhoods, drawing the "ex-" gang member back into the group.

The most common reason for leaving a gang is the personal or indirect experience of violence by the gang member. This is also a common reason for leaving a right-wing group, according to Bjorgo. Decker and Van Winkle specify *that it is the period immediately following a violent confrontation between gangs that is most ripe for intervention*—but that this intervention must take place before the gang can reframe the violent confrontation as something that increases solidarity (p. 270). This experience may be the same as the "trauma" experienced by Garfinkel's interviewees. Gang scholars do not agree over the role of positive inducements to encourage defection from gangs. Klein's (1971) assessment of gang intervention programs led him to conclude that social-service programs that targeted gangs as a unit increased the solidarity of the group, which eventually led to increased violence rather than gang attrition (see also Short and Strodtbeck, 1965). But more recent studies suggest that positive inducements are useful if targeted at members on the fringe of the group and the most recent members to join the group. If this conclusion is correct, it could have significant implications for intervention efforts in counterterrorism.

The importance of solidarity and the specific organizational composition of the group suggested by Klein is picked up in Abuza's (forthcoming) discussion of Jemaah Islamiyyah (JI). Abuza notes that this is a highly interconnected group with friendship and kinship ties reinforced through strategic marriages. He suggests that "the high level of inter-connectedness of members probably affects the rate of rehabilitation," and notes that JI remained cohesive even after the loss of their leaders and the restructuring of the organization following a series of widespread arrests.

Galanter's (1989) study of both voluntary and involuntary defections from the Unification Church is also revealing. He observed, as did both Garfinkel and Bjorgo, that those who left the sect voluntarily usually moved toward this end over a long period of time and after disillusionment with the internal management of the organization or a loss of commitment to the organization's values—because of a perceived failure of those values and beliefs. Forced departure was often initiated by families, with the help of deprogrammers or organizations representing the deprogrammers. Deprogramming "refers to the use of physical restraint by a family or its representatives in an attempt to dislodge a member" from a cult or sect (p. 166). Deprogramming always involves a counterideological component, a reeducation component, and usually a period of isolation not only from the sect but also from family and friends. In a study of 66 former Unification Church members, those who were forcibly removed from the church had more negative views on the church in later years and "showed a greater alienation from the church, scoring lower on loyalty toward the members they knew best and on their relative acceptance of the church creed" (p. 175). In this case, disengagement was more closely linked to deradicalization when members were forcibly removed from the church rather than when they left of their own accord. When members left of their own accord, they indicated positive feelings toward existing members of the church and even continued to accept some specific church tenets. However, at least half of the former members surveyed felt that "current members should leave the Unification Church" (p. 174).

In each of these cases, departure from the group was made more difficult by the fact that membership in the radical group entailed substantial isolation from other social networks and from potentially countervailing influences. In contrast to the experience of joining the group, with all of the concomitant benefits of fraternity, acceptance, purpose, identity, and even status, individuals who try to leave the group are faced with the prospect of trying to repair the mended relationships they left behind in addition to trying to figure out who they are and what they believe, once they take the step of rejecting their adopted identity. Because of this dynamic, individuals who may question their allegiance to the group are often faced with twice as many reasons to stay as to leave. Leaving brings condemnation from the group but also fails to provide a context in which approbation for the decision to leave is provided on the other end.

Nor can we afford to forget that leaving an ideologically based terrorist group may not be the same as leaving a nonideologically driven group such as a gang. Mark Juergensmeyer's (2001) study of terrorism in five religious traditions concluded that although religion—by itself—did not generally lead to violence, religion did in many cases provide the ideological foundation, motivation, and organizational structure of the terrorist group as well as fostering group cohesion. Leaving a radical Islamist group implies a rejection of the radical ideology espoused by the group. Hence, even if a militant experiences both push and pull factors that lead him to consider leaving the group, the articulation—by credible religious authorities—of "theologically grounded imperatives for renouncing violence could be an important factor in catalyzing the decision to leave the radical group."[5] Garfinkel and, to a lesser extent, Galanter emphasize the importance of countering the radical ideology, but neither provide lessons about how best to do so. Counterideological education is the foundation of most of the existing state-run deradicalization programs, discussed in the next section.

Programs

Structured, state-sponsored deradicalization programs have been attempted in Egypt, Yemen, Saudi Arabia, Jordan, Algeria, Tajikistan, Malaysia, Indonesia, Singapore, and the United Kingdom in recent years. We currently know the most about programs in Indonesia, Saudi Arabia, and Egypt. Most reeducation and rehabilitation programs have an ideological foundation—reinterpretation of theological arguments to "de-legitimize the use of violence against the state, the society and the 'other'" (Ashour, 2008a, p. 11). These programs typically include other social-service and individual counseling components as well, making it difficult to determine which aspects of the program are more or less effective in achieving deradicalization. What follows draws most heavily on the Saudi experience as currently reported.

The case for the reinterpretation of theological arguments in favor of jihad is often made in small group settings that bring together religious authorities and radicalized individuals. Recent counterideological efforts in Salafism have been assisted by the publishing of Sayyid Imam al-Sharif's *Document for Guiding Jihad in Egypt and in the World*. Al-Sharif, also known as Abd al-Qadir Ibn Abd al-Aziz, was a former al-Qaeda ideologue and the Emir of al-Jihad in Egypt from 1987 to 1993. He was replaced in 1993 by Ayman al-Zawahiri, now considered Osama bin Laden's right-hand man.

In addition to the ideological component, there is often a counseling or psychological component, as well as a social services component targeting both the family of the detained as well as the detainee, after his release. The Saudi government's program, begun in 2004, is based around a counseling program that includes detainee participation in religious debates, as well as participation in psychological counseling. The goal of the program is for individuals to "repent and abandon terrorist ideologies" (Boucek, 2007a). Six week courses include sessions on loyalty, allegiance, terrorism and even self-esteem. Religious dialogue focuses on the idea that prisoners were tricked into believing a false interpretation of Islam; the correct interpretation is then provided.[6] This process is facilitated both by the participation of former militants in the Advisory Committee, and by the religious authority of the Saudi state (Boucek, 2007a).

In addition to religious dialogue, the Psychological and Social Subcommittee of the Saudi program evaluates each prisoner's social status, psychological problems, and types of social assistance the prisoner and his family will need during the detention period. Families are provided with schooling, health care, and financial assistance to offset the loss of income during incarceration. On release, job assistance programs and government stipends for cars and apartments are provided to those who successfully complete the program and "repent." Single men are encouraged to marry and have children. Finally, the Saudi program makes it clear to the prisoner's wider familial network that they will be held partly responsible for his behavior after his release (Boucek, 2007a).[7]

Participants in the Saudi program are limited to those arrested for minor infractions or for sympathy with radical ideologies or groups (for example, individuals may have been caught with jihadi literature). Those who have committed more serious acts of terrorism are not part of the group targeted for deradicalization. In some ways, then, this is preemptive deradicalization.

Although every country with a deradicalization program reports success, there are limited data available to confirm either the degree of success or the reasons for it (that is, how important is the counterideological component compared with the financial incentives). Moreover, any accurate measure of recidivism relies on tracking and reporting rearrests. Ashour (2008a, p. 11) claims that the successful 1997 effort to deradicalize the Egyptian Islamic Group (IG) "removed more than 15,000 IG militants from the Salafi-Jihadi camp led currently by al-Qaeda." Moreover, he notes that deradicalization may have a "domino effect" when one group influences another—as was the case with the small, violent Takfiri and Salafi-jihadi groups that joined al-Jihad's re-education efforts in 2007 (Ashour, 2007, pp. 596–597). These numbers seem severely inflated when compared with the number of individuals interviewed by Horgan for his forthcoming book, *Walking Away from Terrorism*. Horgan says, "In the sample of individuals I interviewed from 2006–2008, while almost all of the interviewees could be described as disengaged, not a single one of them could be said to be 'de-radicalized'" (Horgan, 2008, p. 6). The Saudi effort has reportedly resulted in the release of 700 of the 2,000 prisoners who participated in the deradicalization program. Saudi authorities claim that only nine individuals have been rearrested, a 1–2 percent recidivism rate (Boucek, 2007a).

A closer look at the Yemeni deradicalization program revealed that a number of the program's supposed graduates were fighting in Iraq. The BBC interviewed a graduate of the program in 2005, who said, "We understood what the judge wanted and he understood what we wanted from him. The Yemeni Mujahideen in prison know Hitar (head of the Religious Dialogue Council) is the way for them to get released, so they ingratiate themselves with him. There was no long or complex dialogue" (Whewell, 2005). This is anecdotal evidence, of course, but it is an important caution nonetheless in a new research area for which little reliable information exists.

In Singapore, the discovery of a Jemaah Islamiyyah cell in 2003 led to the establishment of the Religious Rehabilitation Group (RRG) initially focused on individuals detained for terrorism offenses. The RRG later expanded its focus to include detainee family members on a voluntary basis and then education of the broader Muslim public. Local Muslim scholars help the RRG to address JI's misinterpretation of Islam, publish and distribute moderate Islamic tracts (such as *Unlicensed to Kill: Countering Imam Samudra's Justification for the Bali Bombing* [Hassan, 2006]),[8] and provide education on moderate Islam to both the detainees and the broader community. In addition, a number of loosely affiliated local Muslim groups provide financial and psychological support to the families of the detained. As with the programs mentioned above, few data are available about the efficacy of ideological intervention on deradicalization. However, 19 out of 51 detainees were released after an average of three years detention between 2001 and the end of 2007 (Hassan, 2007, p. 8).

The International Crisis Group (2007) recently warned that the issue of deradicalization in Indonesia is inextricably linked to the issue of prison reform. This is not only because corruption in the prison system reinforces the idea of the government as un-Islamic, but also because jihadi solidarity is reinforced by the need to band together for protection against dangerous prison gangs (p. 5). Existing deradicalization programs are similar to those in Saudi Arabia and Singapore, emphasizing the involvement of former JI militants who have renounced their actions—albeit not always their ideology. In addition to the ideological component, the Indonesian programs also emphasize the need to meet the economic needs of imprisoned radicals' families. About two dozen former JI members and several members of other jihadi organizations have agreed to cooperate with Indonesian police (p. i).

A contrast between notable JI figures participating in the program is striking. Nasir Abas, a former JI leader who split from the group early on over disagreements about the use of violence, takes a traditional counterideological perspective when working with prisoners or speaking in public—emphasizing "right" and "wrong" interpretations of Islam. Ali Imron takes a very different tack. Imron was one of the key JI members involved in the 2002 Bali bombings. He apologized for his role in the bombings during his trial and is now working with police on deradicalization. Rather than emphasizing a wrongheaded interpretation of Islam, Imron says that JI's understanding of jihad was correct and that attacks on the Indonesian state were justified at a time when Muslims were being killed in regional conflicts and the government's failure to fully implement sharia had allowed "deviant teachings, secularism and idolatry to flourish, immorality to rise, splits among Muslims to surface and the gap between rich and poor to widen" (p. 12). Where JI went wrong, according to Imron, was in acting without the support of the Muslim community, in failing to attempt to persuade by other (nonviolent) means those they targeted, and by acting precipitously, before the group had the strength to fulfill their lofty goals (p. 13). This is precisely the problem with conceptualizing deradicalization as equivalent to disengagement. Horgan (2008, p. 5) notes "Often there can be physical disengagement from terrorist activity, but no concomitant change or reduction in ideological support, or indeed, the social and psychological control that the particular ideology exerts on the individual."

Conclusions

If we think about radicalization as a staircase (Moghaddam, 2005), with each step up constituting deeper commitment to the group, then deradicalization would likely begin at the point when the investment of time and resources begins to outweigh the material, psychological, and communal benefits of belonging to the group. Abandonment of the group is most likely if the individual believes that increased commitment will fail to produce a more desirable outcome in the future.

The time at which this decision point occurs will differ in accordance with each individual's calculation of his own investment and rewards. This calculation is also likely to differ in accordance with the roles and responsibilities of the individual within the group. Jeff Victoroff (2005, p. 33) has observed that leaders are psychologically distinct from followers. This may be reflected in the fact that group leadership is generally more stable than group membership. Other scholars have observed that leaders and followers have different levels of commitment, different interests, and even different goals (Crenshaw, 1981; Chai, 1993). In addition, terrorism is a group phenomenon and an attempt to understand the process of deradicalization from only the individual perspective is one-sided. Further, just as the individual's decisions are nested in a group context, the radical group is nested within a specific political, economic, and cultural context. Hence, just as radicalization pathways are somewhat context-specific, so too are deradicalization pathways likely to be affected by the political-economic and sociocultural context in which the individual and group are nested.

Despite the myriad possibilities for variation, several common themes with potential implications for counterterrorism stand out in the existing literature. Given the cited importance of relationships with role models in the decision to reject violence, Garfinkel's (2007) observation that the decision to deradicalize was often an individual one is more likely attributed to necessity than to choice. The "reeducation" efforts of state deradicalization programs as well as their focus on the detainees' families seem more in line with our understanding of the important role that social and familial ties have on identity, values, and beliefs, as well as on an individual's extracurricular activities. If any area of terrorism studies can be said to have reached a level of consensus, it is the role of social networks in contributing to both recruitment and radicalization (Sageman, 2004, 2008; Bakker, 2006; Hegghammer, 2006a, 2006b; and Cragin, Chalk, Grant, Helmus, Temple and Wheeler, 2006). It is therefore extremely likely that this particular factor will also play a key role in deradicalization. Bjorgo (2006) emphasizes the double-bind of having to leave one's new social group behind while having no new social ties to sustain the decision to disengage at the other end. Decker and Van Winkle (1996) underscored that the lack of a "receiving group" was an important hurdle not only for reasons of identity and belonging but also for protection from both random and reprisal-related violence.

These findings suggest that the existence of alternative social networks could be a critical pull factor that is currently missing in discussions of deradicalization from terrorist groups. Bjorgo (2006) reports that parental network groups in Norway successfully intervened to help youth disengage from neo-Nazi and other racist groups. Between 1995 and mid-2000, some 130 parents representing 100 youths participated in parental network groups targeting disengagement. By the end of that period, 90 percent of the youths were no longer involved in a right-wing group. Bjorgo reports that "parental involvement played a decisive role in many cases," although numerous other factors were also important in the decision to leave the group (p. 27).[9] Also of note, post-release detainees from the Saudi program usually continue to meet regularly with the same religious study group and imam to whom they were assigned while in the detention facility. If the Saudis' reported recidivism rates are even partially correct, this factor might account for some of their success.

A study of Germany's deradicalization programs for right-wing radicals found that the majority of those who became involved with the programs voluntarily contacted a deradicalization organization looking for help (Grunenberg and van Donselaar, 2006). This is also a common experience for those working to rehabilitate former gang members, which suggests the need to establish and publicize the availability of deradicalization assistance and resources.

The authors writing about deradicalization also cited the importance of a traumatic or precipitating event that immediately preceded the decision to reject violence or the beginning of the process of disengagement. Remarkably, the majority of those interviewed—from religious groups, gangs, and right-wing groups alike—cited an experience of violence or trauma or an event that forced them to question their existing values and beliefs and led them to consider disengagement or deradicalization. Decker and Van Winkle noted the criticality of timing; inducements to leave the group grew weaker as time passed following the incident's occurrence. Just as radicalization is a long evolutionary process for some and a rapid "snap" for others, so, too, is the process of deradicalization. However, the similarity of the experiences cited above, across all different types of extreme groups, suggests implications for the timing of deradicalization initiatives.

Positive incentives are reportedly more durable than negative incentives in the larger picture. Negative sanctions were just as likely to increase group solidarity as to lead members to leave the group. Positive incentives included new employment or educational prospects as well as the possibility of taking on parental and spousal responsibilities. These incentives were relevant specifically when ties to the violent organization would inhibit access to these opportunities. Boucek told a reporter from the *Boston Globe* that the Saudi government even found wives for released detainees in a few cases, hoping to "insulate them from the predominantly male world of aspiring jihadis."[10]

The role of positive incentives is complex. As mentioned in other papers, economists have found no correlation between income or gross domestic product and participation in terrorist activities. However, a 2003 RAND project that compared social and economic programs in three countries targeted at reducing terrorism found that social and economic development policies did weaken local support for terrorist activities. This was true, for example, in cases where social and economic development policies led to the expansion of a new middle class in communities that traditionally supported terrorist groups (Cragin and Chalk, 2003, p. x).

Recent RAND research focusing on the detainee population in Iraq found that large numbers of those involved in preparing and placing improvised explosive devices were not ideologically motivated. Instead, they had been recruited by local insurgents and lured by the promise of financial compensation (O'Connell and Benard, 2006). One U.S. postdetention program took this factor into account when designing a six-month followup program for released detainees. Detainees check in with the command each month for six months following their release and are paid a small fee for their continued cooperation and for remaining disengaged from the insurgency (Bowman, 2008).

The few scholars commenting on deradicalization programs are uniformly skeptical about the effectiveness of deradicalization from an ideological perspective (Bennett, 2008). But this skepticism reinforces Horgan's point about the need to distinguish between deradicalization and disengagement. If we assume that groups of individuals in essentially every country hold ideas that are contrary, or even abhorrent, to those held by the majority (for example, the Ku Klux Klan in contemporary America), one possible course of action is to target not the ideological orientation of radicalized individuals but their action orientation. In this case, the goal is not to change an individual's worldview but to get him to stop engaging in terrorism.

In addition to gaining a better understanding of the potentially divergent deradicalization pathways of terrorist group leaders and terrorist group followers, of how members of different types of radical groups (for example, nationalist, religious, or single-issue) might experience deradicalization incentives differently, and of how the different social, economic, and political conditions of the state affect the likelihood of deradicalization, a future deradicalization research agenda should[11]

1. examine the effect of temporary cease-fires or other cessations of terrorist activity on group members and any steps taken to maintain group cohesion

2. examine the extent to which former militants express remorse and the ways they make amends or take action to alleviate the associated psychological stress

3. follow up on Gallanter's findings with respect to the difference between voluntary and forced disengagement and compare the experiences of imprisoned terrorists with those of voluntarily disengaged former radicals

4. expand the comparison of deradicalization experiences across different roles and functions within terrorist organizations (for example, do fundraisers have different attrition rates from gunmen?) (Horgan, 2007b, p. 124)

5. obtain a better understanding of what produces dissension and internal fragmentation of groups and what factors reduce popular support for extremist groups (Horgan, 2007b, p. 120)

6. compare the successes, failures, and lessons learned of past and present deradicalization programs worldwide; identify any unique sociocultural, political, or economic factors that seem to influence the success or failure of the programs

7. explore the potential role of the Muslim community in encouraging deradicalization of Islamist extremists and in reinforcing the effects of both push and pull factors.[12]

Bibliography

Abuza, Zachary, "The Disengagement and Rehabilitation of Jemaah Islamiyah Detainees in Southeast Asia: A Preliminary Assessment," in John Horgan and Tore Bjørgo, eds., *Leaving Terrorism Behind: Individual and Collective Disengagement,* New York: Routledge, forthcoming.

Al-Sharif, Sayyed Imam, *Wathiqat Tarshid Al-'Aml Al-Jihadi fi Misr w'Al-'Alam* (Document for Guiding Jihad in Egypt and in the World), Kuwait City: Al-Jarida, Cairo: Al-Masri Al-Yawm, 2007 (in Arabic), cited in Diaa Rashwan, "Egypt's Contrite Commander," *Foreign Policy,* March/April 2008.

Ashour, Omar, "Lions Tamed? An Inquiry Into the Causes of De-Radicalization of the Egyptian Islamic Group," *Middle East Journal,* Vol. 61, No. 4, Autumn 2007, pp. 596–597.

———, "De-Radicalization of Jihad? The Impact of Egyptian Islamist Revisionists on Al-Qaeda," *Perspectives on Terrorism* II/5, Vol. 11, No. 14, March 2008a.

———, "Islamist De-Radicalization in Algeria: Successes and Failures," *The Middle East Institute Policy Brief,* No. 21, November 2008b.

———, *The Deradicalization of Jihadists: Transforming Armed Islamist Movements,* New York and London: Routledge, forthcoming, April 2009.

Bakker, Edwin, *Jihadi Terrorists in Europe,* Clingendael, Netherlands: Netherlands Institute of International Relations, December 2006. As of January 24, 2009: http://www.clingendael.nl/publications/2006/20061200_cscp_csp_bakker.pdf

BBC News, "Yemeni Anti-Terror Scheme in Doubt," BBC News Web site, October 11, 2005. As of January 21, 2009: http://news.bbc.co.uk/2/hi/programmes/crossing_continents/4328894.stm

Benard, Cheryl, *A Future for the Young: Options for Helping Middle Eastern Youth Escape the Trap of Radicalization,* Santa Monica, Calif.: RAND Corporation, WR-354, 2006. As of January 21, 2009:
http://www.rand.org/pubs/working_papers/WR354/

Bennett, Drake, "How to Defuse a Human Bomb," *The Boston Globe,* April 13, 2008. As of January 21, 2009:
http://www.boston.com/bostonglobe/ideas/articles/2008/04/13/
how_to_defuse_a_human_bomb/?page=1

Bjorgo, Tore, "Reducing Recruitment and Promoting Disengagement from Extremist Groups: The Case of Racist Sub-Cultures," in Cheryl Benard, ed., *A Future for the Young: Options for Helping Middle Eastern Youth Escape the Trap of Radicalization,* Santa Monica, Calif.: RAND Corporation, WR-354, 2006. As of January 21, 2009:
http://www.rand.org/pubs/working_papers/WR354/

Bjorgo, Tore, and John Horgan, eds., *Leaving Terrorism Behind: Disengagement from Political Violence,* New York: Routledge, 2009.

Boucek, Christopher, "Extremist Reeducation and Rehabilitation in Saudi Arabia," *Terrorism Monitor,* Vol. 5, No. 16, August 16, 2007a.

———, "The Saudi Process of Repatriating and Reintegrating Guantanamo Returnees," *CTC Sentinel,* Vol. 1, No. 1, December 2007b, pp. 10–12.

———, "Jailing Jihadis: Saudi Arabia's Special Terrorist Prisons," *Terrorism Monitor,* Vol. 6, No. 2, January 24, 2008, pp. 4–6.

Boucek, Christopher, Shazadi Beg, and John Horgan, "Opening the Jihadi Debate: Yemen's Committee for Dialogue," in Tore Bjørgo and John Horgan, eds., *Leaving Terrorism Behind: Disengagement from Political Violence,* New York: Routledge, forthcoming.

Bowman, Tom, "U.S. Offers Training, Pay as It Frees Iraqi Detainees," Morning Edition, NPR News, May 16, 2008. As of January 21, 2009:
http://www.npr.org/templates/story/story.php?storyId=90506939

Brachman, Jarret, "Leading Egyptian Jihadist Sayyid Imam Renounces Violence," *CTC Sentinel,* Vol. 1, No. 1, December 2007, pp. 12–14.

Brandon, James, "The UK's Experience in Counter-Radicalization," *CTC Sentinel,* Vol. 1, No. 5, April 2008, pp. 10–12.

Chai, Sun-Ki, "An Organizational Economics Theory of Antigovernment Violence," *Comparative Politics,* Vol. 26, No. 1, October 1993.

Cragin, Kim, and Peter Chalk, *Terrorism & Development: Using Social and Economic Development to Inhibit a Resurgence of Terrorism,* Santa Monica, Calif.: RAND Corporation, 2003. As of July 16, 2008:
http://www.rand.org/pubs/monograph_reports/MR1630/

Cragin, Kim, Peter Chalk, Audra Grant, Todd C. Helmus, Donald Temple, and Matt Wheeler, "Curbing Militant Recruitment in Southeast Asia," Santa Monica, Calif.: RAND Corporation, unpublished, 2006.

Crenshaw, Martha, "The Causes of Terrorism," *Comparative Politics,* Vol. 13, No. 4,1981, pp. 379–399.

Darwish, Nonie, *Now They Call Me Infidel: Why I Renounced Jihad for America, Israel, and the War on Terror,* New York: Penguin, 2006.

Decker, Scott H., and Barrik Van Winkle, *Life in the Gang: Family, Friends, and Violence,* Cambridge, U.K.: Cambridge University Press, 1996.

Ebaugh, Helen Rose Fuchs, *Becoming an Ex: The Process of Role Exit*, Chicago, Ill.: University of Chicago Press, 1988.

Fink, Naureen Chowdhury, and Ellie B. Hearne, *Beyond Terrorism: Deradicalization and Disengagement from Violent Extremism,* New York: International Peace Institute, October 2008.

Galanter, Marc, *Cults: Faith, Healing and Coercion,* New York: Oxford University Press, 1989.

Garfinkel, Renee, "Personal Transformations: Moving from Violence to Peace," *United States Institute of Peace Special Report,* Vol. 186, April 2007.

Grunenberg, Sara, and Jaap van Donselaar, "Deradicalisation: Lessons from Germany, options for the Netherlands?" in Jaap van Donselaar and Peter R. Rodrigues, eds., *Racism & Extremism Monitor: Seventh Report,* Department of Public Administration, Leiden University, 2006.

Gvineria, Gaga, "How Does Terrorism End?" in Paul K. Davis and Kim Cragin, eds., *Social Science for Counterterrorism: Putting the Pieces Together,* Santa Monica, Calif.: RAND Corporation, 2009. As of January 21, 2009:
http://www.rand.org/pubs/monographs/MG849/

Hannah, Greg, Lindsay Clutterbuck, and Jennifer Rubin, *Radicalization or Rehabilitation: Understanding the Challenge of Extremist and Radical Prisoners,* Santa Monica, Calif.: RAND Corporation, TR-571-RC, 2008. As of January 21, 2009:
http://www.rand.org/pubs/technical_reports/TR571/

Hassan, Muhammad Haniff, *Unlicensed to Kill: Countering Imam Samudra's Justification for the Bali Bombing,* Peace Matters, Singapore, 2006.

―――, "Singapore's Muslim Community-Based Initiatives Against JI," *Perspectives on Terrorism,* Vol. 1, No. 5, December 2007, pp. 3–8.

Hegghammer, Thomas, *Saudi Militants in Iraq: Backgrounds and Recruitment Patterns,* Norwegian Defence Research Establishment (FFI), February 5, 2006a. As of January 21, 2009:
http://rapporter.ffi.no/rapporter/2006/03875.pdf

————, *Terrorist Recruitment and Radicalization in Saudi Arabia,* Middle East Policy Council, Vol. 13, No. 4, 2006b, pp. 39–60.

Homeland Security Policy Institute, *Out of the Shadows: Getting Ahead of Prisoner Radicalization,* special report, Critical Incident Analysis Group, George Washington University, University of Virginia, 2007.

Horgan, John, *The Psychology of Terrorism,* London: Routledge, 2005.

————, "Psychological Factors Related to Disengaging from Terrorism: Some Preliminary Assumptions and Assertions," in Cheryl Benard, ed., *A Future for the Young: Options for Helping Middle Eastern Youth Escape the Trap of Radicalization,* Santa Monica, Calif.: RAND Corporation, WR-354, 2006. As of January 21, 2009:
http://www.rand.org/pubs/working_papers/WR354/

————, "Understanding Terrorist Motivation: A Socio-Psychological Perspective," in Magnus Ranstorp, ed., *Mapping Terrorism Research: State of the Art, Gaps, and Future Discussion,* New York: Routledge, 2007a, pp. 106–126.

————, "From Profiles to Pathways: The Road to Recruitment," *Foreign Policy Agenda,* an EJournal of the US Department of State/Bureau of International Information Programs, Vol. 12, No. 5, May 2007b.

————, "Deradicalization or Disengagement? A Process in Need of Clarity and a Counterterrorism Initiative in Need of Evaluation," *Perspectives on Terrorism,* Vol. 2, No. 4, February 2008, pp. 3–8.

————, *Walking Away from Terrorism: Accounts of Disengagement from Radical and Extremist Movements,* London: Routledge, forthcoming.

Horowitz, Donald L., *The Deadly Ethnic Riot,* Berkeley, Calif.: University of California Press, 2001.

International Crisis Group, "'Deradicalisation' and Indonesian Prisons," *Asia Report,* No. 142, November 19, 2007.

Jamieson, Allison, *The Heart Attacked: Terrorism and Conflict in the Italian State,* London: Marian Boyays, 1989.

————, "Entry, Discipline and Exit in the Italian Red Brigades," *Terrorism and Political Violence,* Vol. 2, No. 1, Spring 1990a, pp. 1–20.

————, "Identity and Morality in the Italian Red Brigades," *Terrorism and Political Violence,* Vol. 2, No. 4, Winter 1990b, pp. 508–520.

Jones, Seth G., and Martin C. Libicki, *How Terrorist Groups End: Lessons for Countering al Qa'ida,* Santa Monica, Calif.: RAND Corporation, 2008. As of January 21, 2009:
http://www.rand.org/pubs/monographs/MG741-1/

Juergensmeyer, Mark, *Terror in the Mind of God: The Global Rise of Religious Violence,* Berkeley, Calif.: University of California Press, 2001.

Klein, Malcolm W., *Street Gangs and Street Workers,* Englewood Cliffs, N.J.: Prentice-Hall, 1971.

Mason, David T., Martha Crenshaw, Cynthia McClintock, and Barbara Walter, "How Political Violence Ends: Paths to Conflict Deescalation and Termination," presented at the 2007 Meeting of the American Political Science Association, Chicago, Ill., August 22, 2007.

Moghaddam, Fathali, "The Staircase to Terrorism, A Psychological Exploration," *American Psychologist,* Vol. 60, No. 2, 2005, p. 161.

Newman, Edward, "Exploring the "Root Causes" of Terrorism," *Studies in Conflict & Terrorism,* Vol. 29, 2006, pp. 749–772.

O'Callaghan, Siobhan, *The Informer,* London: Granta, 1998.

O'Connell, Ed, and Cheryl Benard, "The Myth of Martyrdom: Young People and the Insurgency in Iraq," in Cheryl Benard, ed., *A Future for the Young: Options for Helping Middle Eastern Youth Escape the Trap of Radicalization,* Santa Monica, Calif.: RAND Corporation, WR-354, 2006. As of January 21, 2009: http://www.rand.org/pubs/working_papers/WR354/

Post, Jerrold M., "Group and Organisational Dynamics of Political Terrorism," in Paul Wilkinson and A. M. Stewart, eds., *Contemporary Research on Terrorism,* Aberdeen: Aberdeen University Press, 1987.

Rabasa, Angel, Cheryl Benard, Lowell H. Schwartz, and Peter Sickle, *Building Moderate Muslim Networks,* Santa Monica, Calif.: RAND Corporation, 2007. As of January 21, 2009: http://www.rand.org/pubs/monographs/MG574/

Sageman, Marc, *Understanding Terror Networks,* Philadelphia: University of Pennsylvania Press, 2004.

———, *Leaderless Jihad: Terror Networks in the Twenty-First Century,* Philadelphia: University of Pennsylvania Press, 2008.

Sharif, Sayyid Imam, *Rationalizations on Jihad in Egypt and the World,* serialized in the Egyptian daily newspaper *Al-Masry al-Youm,* November–December 2007.

Short, J. F., and F. L. Strodtbeck, *Group Processes and Gang Delinquency,* Chicago, Ill.: The University of Chicago Press, 1965.

Taarnby, Michael, "Yemen's Committee for Dialogue: The Relativity of a Counter Terrorism Success," in Cheryl Benard, ed., *A Future for the Young: Options for Helping Middle Eastern Youth Escape the Trap of Radicalization,* Santa Monica, Calif.: RAND Corporation, WR-354, 2006. As of January 21, 2009: http://www.rand.org/pubs/working_papers/WR354/

Taylor, Max, *The Terrorist,* London, U.K.: Brassey's, 1988.

Victoroff, Jeff, "The Mind of the Terrorist: A Review and Critique of Psychological Approaches," *Journal of Conflict Resolution,* Vol. 49, No. 1, 2005.

Watkins, Eric, "Yemen's Innovative Approach to the War on Terror," *Terrorism Monitor,* February 24, 2005.

Whewell, Tim, "Yemeni Anti-Terror Scheme in Doubt," BBC, October 11, 2005. As of January 24, 2009:
http://news.bbc.co.uk/2/hi/programmes/crossing_continents/4328894.stm

Zannoni, Marco, "Deradicalization in the Netherlands," presentation at "Radicalization in Europe: A Post-9/11 Perspective," conference hosted by the Danish Institute for International Studies, Copenhagen, Denmark, August 2007.

Endnotes

[1] Other programs have been undertaken in Algeria, Jordan, Tajikistan, and Malaysia.

[2] For additional discussions of organizational disengagement, see also Jones and Libicki (2008). For additional discussion of how political violence, more broadly interpreted, ends, see Mason et al. (2007).

[3] Garfinkel's account is more a collection of anecdotes than a methodologically rigorous study; however, the insights she derives from the deeply personal experiences of ideological deradicalization are all the more powerful for the similarities across anecdotes.

[4] Garfinkel cites some long-term trauma studies, which found that post-traumatic stress syndrome was sometimes displaced by post-traumatic growth. Some of the personality traits that determine which path is more likely include optimism and the type of coping strategies the individual embraces.

[5] Author's discussion with RAND colleague Angel Rabasa, senior political scientist, August 2008.

[6] Whether the allegedly "correct interpretation" being taught in Saudi Arabia is something that a Westerner or a moderate Muslim would regard as such is not currently known, at least to this author.

[7] Additional details about the Saudi program are provided in Boucek (2007a, 2007b, and 2008).

[8] Samudra was the mastermind behind the 2002 Bali bombings, among other JI attacks.

[9] However, disengagement in Colombia is reported to have been more successful when individuals made the decision to disengage (for example, members of the Fuerzas Armadas Revolucionarias de Colombia) than when a collective disengagement agreement was made by a militant group (for example, the United Self-Defense Forces of Colombia) (Fink and Hearne, 2008).

[10] See various interviews included in Bennett (2008).

[11] Several of the following are adapted from Horgan (2007b, pp. 120, 123–124).

[12] See for example, Rabasa, Benard, Schwartz, and Sickle (2007).

CHAPTER NINE

Social-Science Foundations for Strategic Communications in the Global War on Terrorism

Michael Egner

Introduction

One of the major themes emerging from terrorism research is the need to reduce public support for terrorist organizations. That, in turn, has highlighted the potential significance of "strategic communications," a subject that has long been fraught with controversy. This paper reviews the social science that can inform both discussion and execution of strategic communications. Interestingly, many of the insights seem strikingly obvious until it is recognized how often the principles they suggest are violated in practice.

Background

Strategic communications, as a broad and multipronged policy endeavor, has roots in many of the social sciences. A recent report (Defense Science Board [DSB], 2008) defines strategic communications as integrating "the development, implementation, assessment, and evolution of public actions and messages in support of America's interests at home and abroad," using a mix of methods that

> includes but goes beyond media affairs and short-term news streams to focus on mid-range and long-term objectives that require multi-disciplinary capabilities, engaging in a dialogue of

ideas, and durable partnerships with civil society organizations. (pp. 1–2)

Strategic communications can be used to attack terrorism or insurgency at a number of different critical points. A review of factors identified in companion papers suggests many potential roles for strategic communications, including

- easing short-term frustrations among vulnerable populations before they harden into long-term grievances
- reducing the perceived rewards and increasing the perceived risks of joining a terrorist group or engaging in particular tactics
- fostering resistance to terrorist and insurgent recruitment messages
- reducing terrorist group cohesion and recruiting key allies
- reducing social pressures to aid terrorists, the perceived legitimacy of terrorism, and social acceptance of violence.

Different social-science disciplines offer unique contributions in the pursuit of these objectives. A psychologist might focus on the developmental processes driving attitudes toward terrorism and the points along this process at which targeted communications might have an effect. An anthropologist, on the other hand, might study the culturally specific images and symbols that could subtly enhance or poison counterterrorism (CT) or counterinsurgency (COIN) communication efforts. Other potential areas of social-science work include the strategies necessary to fight the deterioration of communities affected by terrorism (sociology), patterns of success and failure in previous communications campaigns (history), the interaction between communications and behavioral incentives (economics), and predicting the behavior of key stakeholders in response to a hypothetical communications campaign (political science and simulation).

Many writings have aggregated these contributions into broad, philosophical recommendations for the future direction of strategic communications. However, less common are micro-level discussions of pragmatic (and, in particular, empirically based) prescriptions that explicitly lay out which communications strategies tend to work and which do not. This paper aims to extract such lessons from social-science research, supplementing the primarily observational lessons from terrorism and insurgency communications with empirical research from analogous policy efforts to prevent or deter high-risk or illegal behaviors (for example, in health policy, education, and criminology) and lessons from the worlds of public relations, government-sponsored broadcasting, and political communications. Of course, the findings of such studies are not statistically generalizable to the problems of terrorism or insurgency. Nevertheless, they are useful in suggesting generally promising approaches, noting specific cross-cultural mistakes, warning of other potential pitfalls, and highlighting research questions for future examination.

It can be difficult to talk about "what works" in strategic communications because the practice is often confused with careless or heavy-handed propaganda. A more subtle problem is that strategic communications operates on hugely varying time scales; what might move the needle in daily public opinion polls might not have any effect on long-term strategic relationships, and what might be a useful long-term strategy, such as education, may be very hard to justify using short-term metrics.

As a result, the remainder of this paper divides the discussion into three sections, corresponding with the three time frames of strategic communications identified by the DSB: "short-term news streams," "medium-range campaigns on high-value policies," and "long-term engagement" through relationships and dialogue (DSB, 2008). Generally speaking, the first section explores lessons in media relations and crisis communications; the second reviews lessons for designing, launching, and evaluating a communications campaign; and the third reviews literature on relationship-building and long-term changes in community norms. The discussion is not meant to exhaustively cover all of the relevant social science but rather to highlight potentially useful theory and practice for CT/COIN efforts.

Short-Term Communications

In countering terrorism or insurgency, short-term communications efforts—such as crisis response or rapid message adaptation in the face of changing events—can affect many of the key factors identified in companion papers. For instance, the communications response to news of collateral deaths during a security operation might influence perceived grievances among the general population, whereas the response to a particularly violent terrorist attack might affect the perceived legitimacy of terrorism. Most of the prescriptive literature in this area focuses on the former type of events, that is, crises or news events that pose a threat to the reputation or legitimacy of those doing the communicating.

Planning for Short-Term Crises

One common prescription for short-term communications is that organizations should plan for crises and other surprises ahead of time. Within the literature on crisis communications is a generally agreed-on list of preparatory actions an organization should take, including writing a crisis response plan, assigning crisis spokespersons and teams, training staff for crises, identifying communications risks and vulnerabilities, maintaining media contact lists, and monitoring the media for key events (Borda and Mackey-Kallis, 2003; Cloudman and Hallahan, 2006; Gainey, 2006). There has been some empirical research on which organizational factors are associated with successful completion of the above tasks; among these factors are an autonomous communications department, an environment encouraging the delegation of authority, a philosophy that is process-oriented (as opposed to, for instance, outcome-oriented), and past experience with crises (Cloudman and Hallahan, 2006; Guth, 1995).

Study of whether or not these tasks actually predict crisis communications "success" has been primarily observational. Although the bulk of reviews conclude that planning is effective, others find that crisis plans are weaker predictors of crisis outcomes than simply having a proactive, aggressive, "tell our side of the story" organizational culture (Marra, 1998) or following a consistent code of values (Fitzpatrick,

1995). In an attempt to distinguish why some plans work and some do not, one reviewer (Fearn-Banks, 2007) argued that communications planning is ineffective when there has been a failure of imagination (the particular type of crisis was not anticipated) or a failure to regularly update the plan.

Methods of Response

Within the communications literature, some attempt at short-term response to crises and rumors—particularly those that threaten an organization's reputation or legitimacy—is, not surprisingly, considered essential (Borda and Mackey-Kallis, 2003; Gainey, 2006; Maynard, 1993). This observational lesson has theoretical roots in the fundamental attribution error (Jones and Nisbett, 1972), which states that when an observer watches an actor commit a negative behavior—but does not know the full context behind the behavior—the observer will too easily attribute the behavior to negative traits of the actor. In other words, a civilian who notices longer wait times at a security checkpoint—but is not aware that this is in response to a recent foiled attack—is likely to attribute the change to simple malice or insensitivity among the security forces. Therefore, one role of short-term communications is to provide the public with the contextual information behind a crisis response so that the response is not simply attributed to ill will. Two other roles, detailed below, are in responding to fact-based crises (that is, crises rooted in a mistake, controversial action, or unexpected event) and in responding to controversies based in rumor and disinformation.

Response to Fact-Based Crises. In the terrorism and insurgency contexts, one example of this crisis might be the public outcry after a security operation results in collateral civilian injuries. Public relations literature has addressed this topic with such theories as situational crisis communication theory (Coombs and Holladay, 2002) and the communicative response model (Bradford and Garrett, 1995), which match different response types to different crisis types. For instance, the latter model lays out four crisis responses and matches each with appropriate crisis conditions: denial (when actors can show they did not commit the offense), excuses (when actors clearly committed the offense but can show they were not in control of the situation), justi-fication (when actors were clearly in control of the situation but there were no clear standards of right and wrong), and concession (when clear standards were violated).

Empirical lessons from an initial quantitative study of this frame-work were that failure to give any response at all had a clear negative effect on the organization's image, and, contrary to theory, conces-sion was an optimal response for maintaining one's image, even where theory implied that concession was not appropriate (for example, when there was evidence that could have been used to make a denial) (Brad-ford and Garrett, 1995). In dealing with the culturally sensitive crises of terrorism and insurgency, the power of concession and apology might be even greater. However, later research in this same vein found that using the situation-appropriate crisis response as predicted by theory was correlated with positive subsequent media coverage of the crisis (Huang, 2006).

Taking a slightly different perspective, other studies (Coombs, 1999; Coombs and Holladay, 2006) found that an offending organiza-tion's offer of sympathy and compassion (as opposed to instructional messages, formal apologies, or victim compensation) led to a more sym-pathetic and less-angry audience and that apologies delivered by print media were received slightly better than identically worded apologies delivered by video. Truth and completeness of the communications response, of course, is also strongly recommended in the literature; sur-veys have found that 95 percent of individuals are more angered by lying about a crisis than they are by the actions precipitating the crisis itself (Maynard, 1993). Audience reaction to these messages can also be affected by such factors as spokesperson attractiveness and ethnic-ity; the more that audiences judge a crisis spokesperson to be similar to themselves, the more likely they are to perceive the crisis communica-tion as credible (Arpan, 2002).

A final dimension to crisis response is the speed with which it is delivered. A recent review of crisis research concluded that rapid response is a necessary but insufficient condition for successful crisis communications (Borda and Mackey-Kallis, 2003). Although some studies have quantified the required response speed—concluding, for instance, that an organization has up to 12 hours after a story breaks to gain control of the message (Gainey, 2006; Small, 1991)—most literature discusses the benefits of speed in relative terms, that is, getting the organization's "story" out faster than competing narratives. However, greater response speed may have drawbacks. A recent experiment involving journalists (Arpan and Pompper, 2003) found that if an organization rushes to break a story to the media before the media discovers the story from a third party, the media will rate the organization as more credible, but the media will also be more interested in the story than it would have been otherwise.

Response to Rumors and Disinformation. This type of response has always been a necessary part of wartime strategic communications; for instance, the failure to rebut rumors of U.S. germ warfare during the Korean War (in the mistaken belief that the rumors would disappear on their own) has been considered a clear mistake (Shaw, 1999). Coalition forces in Iraq were the subject of potentially damaging rumors (for example, that U.S. soldiers distribute pornography to children) as early as the spring of 2003 (Hendon and Holton, 2003).

The literature on rumor psychology, and more specifically rumor management, offers a number of useful insights. A meta-analysis of prior theory and empirical research (Rosnow, 1991) identified three necessary factors for the dissemination of rumor: general uncertainty about the rumored issue, high personal anxiety levels, and the believability of the particular rumor. Consequently, rumors and enemy disinformation can be attacked by weakening any of these three factors (for example, offering greater information to the public to quell uncertainty, or attacking the credibility of a rumor source). Not surprisingly, spokespersons perceived as honest are most effective in reducing rumor anxiety and believability; although a more knowledgeable or high-status spokesperson will further enhance this effectiveness, knowledge and status alone are insignificant in the absence of perceived honesty (Bordia, DiFonzo and Schulz, 2000).

Other survey-based research (DiFonzo and Bordia, 2000) concluded, among other findings, that two strategies—rumor-control hotlines and denial of rumors by trusted third parties—are generally underused relative to their perceived effectiveness. Vigorous refutations of rumors may be most effective when the source of the rumor or disinformation is portrayed as having something to gain from the rumor and when those doing the refuting end their message on a conciliatory note (Iyer and Debevec, 1991). As for how best to transmit these refutations, U.S. sources on the ground in Iraq have argued that face-to-face rumor control is much more effective than print or radio efforts (Steele, 2003). However, close monitoring and message adaptation is essential to ensure that the refutation does not backfire; recent political-science research has suggested that, among committed ideologues, the attempted refutation of a rumor only strengthens the underlying misperceptions (Nyhan and Reifler, 2008).

Mid-Range Communications Campaigns

Mid-range communications consist of preplanned information campaigns designed to change public attitudes in support of CT/COIN policies (that is, attitudes toward cooperating with security forces). Campaigns can also change behaviors: Recent research on the U.S. public has found that the more that individuals consider terrorism to be a serious social problem in which they are personally involved, the more they will seek out information about it, which in turn predicts a greater likelihood of performing recommended protective behaviors (Lee and Rodriguez, 2008). In general, social-science contributions for communications campaigns can be divided into three areas: formative evaluation (that is, what to ask before launching a campaign), message content and delivery, and summative evaluation (that is, what to ask after the campaign ends to measure effects and outcomes).

Formative Evaluation

In the social sciences, formative evaluation has been defined simply as "the collection of information that helps to shape the campaign" (Coffman, 2002). The commonsense proposition that this process benefits subsequent program success has been empirically validated (Brown and Kiernan, 2001), but detailed lessons on precisely how best to run this process are, again, primarily observational in nature. The key contribution of social science in this area has been in providing a list of common questions a communications officer may wish to answer before launching a campaign. A nonexhaustive selection of useful questions is listed below, as drawn from a variety of evaluation literature (Atkin and Freimuth, 2001; Barthe, 2006; Coffman, 2002; Valente, 2001; Wixon, 1998). Evidence supporting the value of particular questions is included where applicable.

Defining a Campaign "Theory of Action." One formative evaluation lesson confirmed in practice has been the key role of a theoretical underpinning in predicting campaign success (O'Keefe and Reid, 1990). This goes beyond borrowing from existing theory and includes the creation of a program-specific "theory of action," explicitly laying out each hypothesized logical relationship linking program design to eventual program effects (for example, a particular message will increase awareness of issue X, which will reduce behavior Y among the public, which will reduce the ability of terrorist groups to do Z).

One useful theoretical backbone for such an exercise is information processing theory (McGuire, 1978), which posits a pathway of message effectiveness through (1) audience exposure (to the message), (2) audience attention, (3) comprehension of the message, (4) agreement with the message, (5) retention, and (6) behavior change. Because every step is necessary for success, this type of layout is useful in pinpointing precisely where an ineffective campaign has broken down (Coffman, 2002; Wixon, 1998). It also highlights trade-offs in communications strategy; for instance, a message may be broadly comprehensible at the expense of persuasiveness.

Basic questions to ask at this stage include

- What is the specific knowledge, attitude, or behavior that the campaign is trying to change?
- What is the logical or theoretical relationship between the communications effort and the desired change?
- What assumptions or intermediate steps are required for the campaign to perform as predicted?

- What are the alternative messages, themes, tactics, and strategies available to the campaign?
- How long will the campaign last?

Defining and Segmenting the Target Audience. Communications research generally confirms the benefits of audience segmentation. Audiences prefer messages that are personally relevant to them and will judge broadly targeted messages as relatively ineffective (Fishbein et al., 2002). Conversely, national communications campaigns are improved when they include local tags or references (O'Keefe and Reid, 1990). In general, although segmenting by simple demographics or geography is easiest, a campaign may be most effective when segmenting along those factors (for example, attitudes toward suicide bombing) that underlie the behavior to be changed (Slater, 1995). Failure to segment can be dangerous: Different audiences can differ in their basic interpretation of the same message (Barthe, 2006), and a particular message can be well-received by one segment but judged as not credible by a different segment (Skinner and Slater, 1993). However, such intergroup differences are not always as large as might be expected (Borzekowski and Poussaint, 2000).

General segmentation questions to ask include

- Will the campaign target segments of the general public, current or potential terrorists/insurgents, their current or potential supporters, their social companions, or others?
- What factors underlie the behavior to be changed? Can the audience be segmented along those factors? If not, what alternative segmentation strategy will be used?

Understanding Audience Segments. Segmentation is useful only when the individual audience segments are studied and understood. For campaigns that aim to affect behavioral decisions (such as whether or not to join a terrorist group or report suspicious activity to the police), stages-of-change models (DiClemente and Prochaska, 1985) provide a useful theoretical structure. The theory categorizes audiences into five stages depending on their relationship with key behavior at issue: precontemplation (the behavior is not personally relevant to them), contemplation (the behavior is relevant, but they are not prepared to perform it), preparation (they are prepared to perform it, but have not actually done so), action (they have performed it, but not maintained the behavior over time), or, ultimately, maintenance of the behavior. This theory highlights the importance of fit between message and stage (that is, individuals thinking about engaging in terrorism or insurgency will likely require a very different message from those already practicing it).

Another useful way to strengthen campaign design through audience research is in isolating the target audience's "hot button" and "cold button" values. This method, which has already been studied in the context of terrorist deterrence (MacNulty, 2008), entails profiling the audience to determine which hot button values are important to them (such as material wealth or religion) and which cold button values are not (for example, loyalty or freedom of thought). Communications campaigns can then focus on the right levers to influence attitudes or behavior. This approach can also help avoid counterproductive messaging. For example, consider differences in the degree to which individuals value the ability to think for themselves: An information campaign highlighting the inconsistency of joining a terrorist group and being able to think for oneself may be effective for people who hold this as an important value, ineffective for people who do not care about this value, and counterproductive for people who, for whatever reason, prefer to simply follow the decisions of others.

Audience research to consider at this stage includes

- What are each audience segment's values, beliefs, and attitudes toward terrorism/insurgency and CT/COIN policy?
- How important are different values and desires in driving a target group's terrorism- or insurgency-related behaviors?
- How salient a problem is violence to each segment?
- What are the skills and self-efficacy beliefs of each segment regarding terrorism- or insurgency-related behaviors (such as joining a group or reporting suspicious activity)?
- Who are the most credible messengers for each segment?

Understanding the Program Environment. Formative research can also include analysis of on-the-ground factors that may help or hurt the campaign, such as cultural or language barriers, relationships with community leaders, political constraints, and staff skill levels. To gauge how these factors might change over the lifetime of the campaign, a number of increasingly sophisticated computer simulation and prediction tools, such as the Senturion program (Abdollahian et al., 2006), allow for the forecast of political dynamics over a mid-range time frame (up to two years). Such programs can also strengthen formative evaluation by simulating the future effects of different versions of the communications campaign.

In general, researchers may wish to study the following items:

- What resources (for example, information, staff skills, local community support) are currently adequate and which are still needed or otherwise unsatisfactory?
- What factors in the target community will be obstacles to program success, and what factors will be helpful?
- What do local stakeholders think about the campaign, and how will they react once the campaign begins?

Understanding the Media Environment. The issue of media environment goes beyond merely determining which channels the audience is watching. However, these mundane issues should not be overlooked; for instance, the BBC's anticommunist broadcasts to Albania in the early Cold War were played 90 minutes before the electricity was switched on in Albanian towns (Defty, 2002).

Media theories, such as priming theory (Iyengar and Kinder, 1987) and the agenda-setting model (McCombs and Shaw, 1973), point to the fundamental role of the media in determining which issues the public deems to be important (issue salience). Increased issue salience, in turn, leads to increased public knowledge about the issue, stronger opinions on it, and a greater likelihood of participating in political actions related to the issue (Weaver, 1991). Not surprisingly, then, public service announcements are generally more effective when dealing with topics already covered extensively by the news media (O'Keefe and Reid, 1990). A better understanding of current local media coverage of terrorism or insurgency, therefore, will help better estimate the effectiveness of a potential campaign.

Questions to ask include

- What media channels and communication methods are most frequently used by each audience segment?
- When (that is, time of day) are these channels or formats most frequently used by each audience segment?
- How do local media sources currently depict issues of terrorism and insurgency?

Message Testing and Feedback. Another common recommendation in the literature is that a campaign be subjected to audience testing before a full launch. At the most basic level, this is necessary to make sure that the message matches the cognitive capabilities of the audience. Otherwise, overly simple communications may be judged by the audience as superficial and condescending (as happened with U.S.-sponsored documentaries screened in Asia, Latin America, and the Middle East during the Cold War) (Green, 1988); on the other hand, overly complex communications may actually decrease audience understanding of the issue at hand (Mitchell, 1973).

Message-testing can also reveal the mistaken cultural assumptions that lead to counterproductive communications outcomes. One common assumption is that an audience is hostile merely because they do not know enough about the messenger or its objectives; yet providing a hostile audience with more information may only arm them with more reasons to continue their hostility. Another dangerous assumption is that the audience shares the messenger's lifestyle preferences; an interesting example of this mistake comes from the Cold War, when an anticommunist campaign in Iraq screened the film *Ninotchka*, which unflatteringly compared the dour Soviet lifestyle with the excitement of Paris. Despite the Western slant of the movie, Iraqi audiences preferred the "somber" life in the USSR to the "immoral" way of life in France, shifting their attitudes in the opposite direction than was intended (Battle, 2002).

Audience feedback does seem to be an effective tool for strengthening campaigns. One recent study in the education literature (Brown and Gerhardt, 2002) distilled five key lessons confirmed by available empirical data on feedback: (1) expert feedback should be collected from individuals with different types of expertise, (2) audience feedback should be collected and used to revise messages, (3) audience feedback should be collected in small groups with active discussion of problems and suggestions, (4) all feedback should be summarized in specific and directive ways, and (5) message designers should be held accountable for incorporating feedback. Even though these lessons seem merely logical, they often are not followed. For example, a 16-state study of evaluations in the education field found that, contrary to theory, many stakeholders, particularly those with larger administrative fiefdoms, had neither the time nor the desire to be involved in the planning or implementation of evaluations (Smith, 1980).

Many surveys of successful programs also point to the importance of community involvement even after the campaign is launched. Examples of such involvement include fine-tuning messages for a better local fit (Knight et al., 1998); fostering audience reply and debate through phone hotlines, followup groups, or putting people in touch with those in similar circumstances (Weiss and Tschirhart, 1994); and supplementing traditional media with targeted publications for key stakeholders and efforts to persuade citizens to persuade each other (O'Keefe and Reid, 1990).

Basic testing and feedback questions include

- What types of pretesting strategies are available (for example, interviews, surveys, and day-after message recall)?
- Which messages or arguments work best with each segment?
- Do the members of each audience segment adequately comprehend the communications directed toward them?
- How will future feedback and community involvement be incorporated into the campaign?

Message Content and Delivery

The literature on how to best design an effective message is quite large, spanning many disciplines and policy fields, and with many prescriptive lessons based in empirical research. This section will extract highlights from this research, taking care to note (where applicable) how the optimal content choice varies by context.

One-Sided Versus Two-Sided Messages. One major line of theory and research has studied whether, and under what conditions, the effectiveness of a message is bolstered by including opposing arguments. In general, two-sided communications are most effective when the audience is relatively knowledgeable about the issue or is currently opposed to the position being advocated by the messenger (Fisher and Misovich, 1990). Opposing arguments can increase perceptions of messenger credibility but may begin to decrease persuasiveness if their volume exceeds about 40 percent of the total message (Crowley and Hoyer, 1994; Golden and Alpert, 1987). A useful meta-analysis of existing research (Allen, 1991) concluded that a two-sided message without refutation (of the opposing argument) is less persuasive than a one-sided argument, which in turn is less persuasive than a two sided message with refutation.

A specific two-sided approach with potential use for the global war on terrorism is behavioral inoculation (Lumsdaine and Janis, 1953; McGuire, 1970), in which individuals at risk of accepting arguments in favor of a negative behavior (such as joining a terrorist or insurgent organization) are proactively presented with those arguments in an attempt to "inoculate" them. Effective inoculation should include both threat messages (realistically explaining to the audience that they will be pressured to commence the negative behavior) as well as refutational preemption (refuting the specific arguments they will hear from the other side, starting with the most common) (Pfau, 1995). Effective inoculation should also be tailored to the way the audience thinks about the issue and where there is no clear winner between using peer or more authoritative inoculators (Best et al., 1988). Inoculation has been shown to be effective in such diverse contexts as inoculating adolescents against prosmoking arguments (Banerjee and Greene, 2006) and inoculating supporters of political candidates against the attack ads of opposing candidates (Pfau and Burgoon, 1988). Although a recent attempt at inoculating individuals against a propagandistic YouTube video failed to show an effect, this was likely because of limitations in the design of the study (Lim and Ki, 2007).

A final topic to note here is whether or not a one-sided argument should be softened by using such qualifiers as "perhaps" or "maybe." This technique has been criticized as detrimental to the attention-getting power of the message (Wiener and Mehrabian, 1968), but others (Browne, 1983) have attributed the high credibility of the BBC to contributors who use this type of qualified language.

Gain Framing Versus Loss Framing. In making choices, individuals take greater risks when faced with a dilemma framed in terms of losses (such as discussing "lives lost") than in dilemmas framed in terms of gains (such as discussing "lives saved"), even if the two dilemmas are otherwise identical (Tversky and Kahneman, 1981). This effect has been evaluated in the context of message design. Research has found that loss-framed messages are deemed more persuasive by highly authoritative individuals (that is, those aggressive toward outsiders and submissive toward authority), whereas gain-framed messages are deemed better by those scoring low on authoritarianism (Lavine et al., 1999). As might also be expected, loss frames are more effective when persuading individuals to perform behaviors perceived as risky, whereas gain frames are more effective for encouraging behaviors perceived as safe (Rothman and Salovey, 1997). In terrorism and insurgency, of course, many actions—such as the decision to provide information on suspicious activity to security forces—may be perceived as anything from very risky to moderately safe, depending on an individual's personality and the local security environment.

An extension of loss framing that may be useful for CT/COIN communications is perceptual deterrence (Gibbs, 1975; Zimring and Hawkins, 1973), in which potential offenders are informed of the risk of arrest or punishment to deter behavior. For deterring criminal behavior, the *probability* of arrest is more influential than the *consequences* of arrest (Barthe, 2006; Burkett and Ward, 1993); indeed, increased publicity surrounding executions may actually lead to more homicides (Bowers and Pierce, 1975). The danger of merely sensationalizing wrongdoing by focusing on violent punishment would seem doubly true in the martyrdom-seeking cultures of certain terrorist or insurgent organizations.

Nevertheless, one highly effective use of perceptual deterrence against gang violence (Braga et al., 2001) explicitly spelled out the law enforcement consequences of violent acts to violent gang members; the program was not trying to destroy the gang but simply trying to encourage it to end violence. Another perceptual deterrence strategy involving gun crime was not able to increase knowledge of the deterrent penalties, possibly because of a failure to properly segment the audience (Haas and Turley, 2007).

Positive Emotions Versus Negative Emotions. A cousin of the gain-framed versus loss-framed debate is whether messages are more effective when they appeal to positive emotions (such as pride or hope) or to negative emotions (such as fear or anger). On the one hand, positive antiviolence messages with realistic characters and situations score highest on attention and interest as well as on understandability, credibility, and effectiveness (Borzekowski and Poussaint, 1999; 2000). Appeals to positive emotions are effective for audiences unfamiliar with the message and help them make faster decisions and be more compliant (Isen, 1987). In a message, the use of positive emotions increases the effectiveness of logically weak arguments but decreases the effectiveness of logically strong arguments (Batra and Stayman, 1990). However, a positive tone is generally ineffective if the audience is already predisposed to disagree with the message (Monahan, 1995). This limitation may be critical in CT/COIN contexts, where predispositions against the communications arm of security forces may be relatively high.

Negative emotions (particularly fear) do seem to affect personal risk-avoidance behaviors but have an ambiguous record on both personal preventive behaviors and collective crime prevention (Rosenbaum and Heath, 1990). There is also the danger that too many appeals to fear can actually lead to accelerated neighborhood decline (Skogan, 1986). This risk would seem to be especially acute in precisely those areas most affected by terrorism or insurgency. Fear-inducing messages are less effective when the audience is forced to watch them instead of choosing to watch (Horowitz, 1969), when the audience already has high baseline levels of anxiety, or when targeting younger audiences (Boster and Mongeau, 1984). The fearful message should be supplemented by communicating personal efficacy (you can do something to avoid the fearful outcome) and response efficacy (it will actually work) (Hale and Dillard, 1995). As for persuasive messages using negative emotions besides fear—particularly sadness and anger—recent empirical work (DeSteno et al., 2004; Rucker and Petty, 2004) emphasizes the importance of matching the tone of the message with the audience's current mood.

Truth Versus Untruth. In strategic communications, appearances of inauthenticity can severely harm a campaign; for instance, an expensive, three-year project to film U.S. progress in Vietnam was shelved after it was discovered that some battle scenes had been staged (Dizard, 2004). Another major risk of exaggeration or manipulation in strategic communications is the risk to overall source credibility. The link between source credibility and message persuasiveness is quite strong, although the effect may dissipate over time as people remember the message but forget who said it (known as the "sleeper effect") (Hovland and Weiss, 1951; Kumkale and Albarracin, 2004). However, source credibility on its own cannot save a weak or misleading argument; if even a highly credible speaker is relying on irrelevant evidence or evidence from a poor-quality source, audience attitudes shift away from the position the speaker is advocating (Luchok and McCroskey, 1978). Audiences have demonstrated resistance to persuasion based on spurious evidence, particularly when the claims are tangible or when alternative sources of information on the subject are available (Reinard, 1988).

Attention-Grabbing Versus Understandable Messages. A message can effectively grab audience attention by triggering the audience's active thought processes. This can be done by simply instructing the audience to pay attention; or by using novel, unexpected, or discrepant messages and media (Louis and Sutton, 1991; Parrott, 1995). However, this novelty can reduce audience comprehension of the message if taken too far. Music-themed antiviolence messages can be attention-getting but are less likely to be understood; special effects–based or abstract messages may also generate interest but falter on understandability and credibility. Simple visuals (particularly graphic visuals) may represent an effective combination of vividness and persuasiveness (Borzekowski and Poussaint, 1999, 2000).

Peer Versus Nonpeer Spokespersons. A final issue spanning both mid-term and long-term objectives is whether, and how, to deliver messages to an audience through their peers. Given the cultural divide between the U.S. and foreign strategic publics, this issue is of particular importance for CT/COIN efforts. There is a diverse theoretical basis justifying the use of peers in persuasion. For instance, when encouraging positive behaviors among an audience, stages of change theories (DiClemente and Prochaska, 1985) suggest that the most effective role models are just one stage beyond that of the target audience and therefore in a position to explain how they were able to take the next incremental step toward the desired behavior (for example, transitioning from merely thinking about leaving a terrorist support network to actually doing so) (Maibach and Cotton, 1995).

The general benefit of using peer spokespersons has been confirmed in many empirical studies. As discussed above, crisis spokespersons are more effective when they resemble their audience (Arpan, 2002). Early persuasion research (Brock, 1965) also found that messengers who are similar to members of the target audience are more persuasive than dissimilar messengers, even if the latter are perceived as more knowledgeable. For antiviolence messages, spokespersons similar to the audience also score higher on attention, interest, and credibility than even well-known authority figures (Borzekowski and Poussaint, 1999, 2000). The limited research on antigang advertising has also found that antigang messages delivered by actual and former gang members were judged the most influential by both current gang members and individuals at high risk of joining gangs (Chapel, Peterson, and Joseph, 1999; Lafontaine, Ferguson, and Wormith, 2005).

However, peer spokespersons may not always be preferable. Additional research on this topic has found that messages delivered by similar sources may be best at reinforcing normative or value judgments (for example, that terrorism is wrong), whereas messages delivered by dissimilar sources are best at reinforcing factual beliefs (Goethals and

Nelson, 1973). Therefore, one potential role of nonpeer spokespersons may be to confirm widely held factual beliefs that have recently come under attack (for instance, in debunking strange conspiracy theories).

Longer-term strategies for actually changing social norms about violence—rather than simply choosing the identity of a spokesperson—will be examined below.

Summative Evaluation

Although different researchers employ different terminology and taxonomies, here we equate summative evaluation with "back-end" evaluation (Coffman, 2002), encompassing (1) process evaluation, which measures effort, direct accomplishments, and implementation; (2) outcome evaluation, which measures program effects and direct changes against baseline levels; and (3) impact evaluation, which measures aggregate or long-term results, requiring experimental or quasi-experimental research to determine causality. Summative evaluation frameworks are useful for highlighting tensions (such as the trade-off between using evaluation funds for showing results rather than program improvement [Patton, 1997]) and for identifying key research questions. Below is a selection of key questions drawn from several studies (Barthe, 2006; Coffman, 2002; Weiss and Tschirhart, 1994); they have been reworded to fit within counterterrorism or counterinsurgency contexts, when necessary.

Process Evaluation

- How much communications output was created and distributed?
- Was the audience targeted successfully?
- How much media placement did the campaign receive?
- How many people were reached and for how long?
- Did the audience comprehend and recall the message?

Outcome Evaluation

- How have key audience factors (that is, knowledge, awareness, saliency, attitudes, anxiety, norms, self-efficacy, behavioral intentions, or behaviors) changed from precampaign levels?

- How has the media's framing of the problem changed from before the campaign?
- What is the level of community satisfaction with the communications campaign?

Impact Evaluation

- What are the long-term behavioral changes in the target community or among potential or current terrorists and their supporters?
- What is the long-term effect of the program on the incidence, severity, lethality, cost, or geographic distribution of terrorist and insurgent attacks within the area covered by the communications campaign?
- Have there been any systems-level responses to the campaign, such as the development of new organizations?
- Have there been any significant policy or political responses to the campaign?

There is an enormous body of empirical evidence on how best to run a summative evaluation, covering topics from survey design to advanced statistical methods. A detailed discussion of measuring outcomes appears in a companion paper on metrics (Bahney, 2009), but it is worth noting that these methodological lessons occasionally intersect with some of the approaches discussed above. For instance, when evaluating a perceptual deterrence campaign (that is, deterring terrorists or insurgents by conveying the threat of punishment), effects should be measured separately for those who believe terrorism is morally wrong and those who find it morally neutral or acceptable. Otherwise, the effectiveness of the campaign will be underestimated because those who find terrorism morally wrong show no deterrence effect; they have already decided not to engage in it (Burkett and Ward, 1993).

There are also important theory-based lessons. For example, theory on behavior change highlights the need to measure audience attitudes toward the relevant behavior at the heart of a campaign rather than the relevant outcome (for example, measuring specific attitudes toward reporting suspicious activity rather than attitudes toward terrorism or insurgency in general) (Ajzen and Fishbein, 1980).

Long-Term Srategies

Long-term strategic communications has been defined to include two methods: building partnerships with civil society and engaging in a dialogue of ideas (DSB, 2008). The discussion below slightly broadens this first goal (to include all strategic relationship-building) and narrows the second (to focus on the long-term de-legitimization of terrorist groups, tactics, and ideologies).

Strategic Relationship-Building

Strong strategic relationships can contribute to the long-term success of CT/COIN strategic communications in a number of ways. Good relationships with the public can help create and maintain credibility and trust; good relationships with local leaders can help amplify one's message, maintain a community's willingness to accept a communications intervention, and prevent disputes from escalating into crises; and good relationships with the media can help curtail the public relations fallout from such crises, should they occur.

The first contribution of social science in this effort has been in defining what a good relationship is and how it can be measured. An influential research program (Grunig, 2002; Hon and Grunig, 1999) has isolated and ranked four key elements that reflect a successful relationship. The most important is control mutuality, or the degree to which partners in a relationship accept the current balance of control; this is followed by mutual trust, commitment to the relationship, and relationship satisfaction. Reliable survey tools have been developed to capture and measure these factors (Hon and Grunig, 1999). In the military context, transparency (making all legally releasable information available) has been added to this list, whereas the phrase "control mutuality" (and its potential replacement, "mutual influence") has been viewed somewhat critically in light of government mandates against exerting undue influence over populations (Plowman, 2007).

Apart from developing a measurable definition of relationship, social-science research has examined both the causes and effects of good relationships. On the effects side, it should not be surprising that a great deal of empirical research finds that strong relationships are useful—in boosting an organization's reputation, its perceived performance, and its ability to prevent defections to competitors (Bronn, 2007; Grunig and Hung, 2002; Hagan, 2007; Ledingham and Bruning, 1998; Yang and Grunig, 2005). The remainder of this section will detail lessons on the "causes" side; that is, what an organization can and should do to build and maintain relationships with different strategic groups in the first place.

Relationships with the General Public. In the public relations literature, the last two decades have seen an increased emphasis on "putting relationships back into public relations" (Ledingham and Bruning, 2001; Williams, 1996). Research has found, unsurprisingly, that the strength of an organization's relationship with the public grows with the amount of time in the relationship (Ledingham, Bruning and Wilson, 1999). For very popular or very unpopular organizations, public perceptions of the relationship also depend critically on the handful of behaviors by the organization's leaders that the public can recall (Grunig and Hung, 2002).

One major arm of research in public-organization relationships has been to study the effect of inviting the public to participate in an organization's decisionmaking. There are clear dangers for aloof organizations; if individuals never have direct public experience with an organization, they are left to evaluate it solely by what they have heard or read from others, and they may shun an organization despite having only a superficial knowledge about it (Grunig and Hung, 2002). Of course, this is a very real danger in the hypersaturated, politically charged media environment of the Middle East.

Depending on the study, public involvement and participation have been found to help aspects of relationships (Yang and Grunig, 2005) or to have an insignificant effect (Rawlins, 2007). This discrepancy might be explained by the importance of whether public involvement is truly meaningful. Psychological research has identified a frustration effect, whereby giving people a voice—but not allowing that voice to change circumstances people view as unfair—will make them more dissatisfied than giving them no voice at all (Folger, 1977). Similarly, the public's perceived fairness of a decisionmaking process in which they are involved has some influence on the degree to which they will accept the outcome of that process (McComas et al., 2007). If the public perceives that a real dialogue has taken place—in which the organization shares their goals, listens to them, and empathizes with them—empirical research suggests that they are more likely to believe they have benefited from the relationship, which in turn makes members of the public more likely to support the organization (Bruning, Dials and Shirka, 2008).

Two other recent studies evaluated the efforts an organization might take boost the element of trust in public relationships. The first (Rawlins, 2007) divided "transparency" with the public into four elements—accountability, sharing of information, lack of secrecy, and stakeholder participation—and measured their effects on relational trust. Accountability had the strongest effect on trust, followed by the sharing of information; the other two factors had only weak or insignificant effects. The second study, a university experiment (Baksh-Mohammed, Choi, and Callison, 2007), evaluated an organization's involvement in acts of goodwill (charity) on perceived credibility. Although the charitable action had a positive effect on credibility, this effect disappeared during crisis situations—in other words, when credibility would presumably be needed the most.

In sum, the literature recommends that communication relationships with the public be initiated early, focus on communicating a few key positive behaviors, have meaningful public involvement and dialogue, and exhibit both accountability and information sharing.

Relationships with Strategic Partners and Key Stakeholders. Partners and stakeholders in CT/COIN strategic communications— e.g., community leaders, local security forces, or nongovernmental organizations—can critically affect long-term communications success. Good local partners can provide cultural feedback on the appropriateness of particular messages and secure access to media channels and other resources. With good relationships, community leaders will also express their disapproval privately to community liaisons rather than publicly to political leaders or the media (Ledingham and Bruning, 2001).

There exists a broad area of social-science research on partnerships in general (that is, not necessarily for communications), exploring why organizations enter into partnerships or alliances, why they leave, and what makes the partnership effective (Gulati, 1998). A recent study tying together theory and empirical data on partnerships in social-service delivery (Graddy and Chen, 2006) evaluated how program success is affected by the specific reason for choosing a particular partner. Direct and indirect program goals were helped when partners were chosen on the basis of their ability to help meet resource needs, to help bolster the organization's reputation among key audiences, or their shared philosophical vision; goals were negatively affected when partners were chosen simply to lay the foundation for a future relationship, to continue a previous collaboration, or because no other partner could be found. Given that CT or COIN partnerships contain real risks (particularly when partnering with unstable or otherwise compromised temporary allies), such findings may be useful when weighing the costs and benefits of a particular alliance.

Strategic partnerships have also been evaluated in empirical marketing literature. One key obstacle to the success of marketing alliances, as identified in research, has been asymmetries in the balance of resources and power in the relationship (Bucklin and Sengupta, 1993). Of course, this is an unavoidable element of most partnerships involving the United States, but stresses caused by this asymmetry can be anticipated and monitored. Research has also explored cross-cultural marketing partnerships (Aulakh, Kotabe, and Sahay, 1996), finding a positive empirical link between partnership performance and (1) expectations that the relationship will continue, (2) willingness to be flexible in the relationship, and (3) the use of informal (as opposed to rigid and output-focused) monitoring by the dominant partner.

Relationships with the Media. Although crisis-based interactions with the media are a "short-term" issue, developing relationships with media practitioners is a long-term affair. Close media relationships, developed over time, can help minimize the media fallout from crises and help earn media trust. Social science provides many specific recommendations for how best to build and maintain this relationship. A recent review of the relationship between the Australian military and the press (Hibbert and Simmons, 2006) attributed some of the Australian media's dissatisfaction with the relationship to lack of troop access (for example, through embedding). A review of research on media relationships (Desiere and Sha, 2005) added that organizations should develop a media strategy that is proactive (not waiting for the phone to ring), continuous (not waiting for a crisis to break), open, accurate, based on meaningful personal relationships (knowing the media's needs and providing useful information to assist them quickly), and that emphasizes preparedness before speaking with the media.

Another recent study on this topic from the health field (Cho and Cameron, 2007) looked at the organization-media relationship in terms of five "powers" that public affairs officials believe they have over the media: power to serve as a source of expertise ("expert power"), power to provide additional information to reporters they like ("information reward power"), and power to withhold information, advertising, or influence in response to undesirable press behavior. Communications officers who were personally close to reporters had significantly greater expert power and information reward power. This suggests that if officials are not willing to take the time to develop personal relationships with the press, they are less likely to be recognized as experts or to have collaborative information exchanges with the press.

Delegitimization

Another long-term goal in counterterrorism and counterinsurgency is to fundamentally delegitimize terrorist or insurgent groups, ideologies, and behaviors. The difference between this goal and that of "mid-range" communications (discussed above) is primarily one of scope and degree; in the long term, strategic communications aims to not merely change one or two specific attitudes or behaviors but to prompt a widespread, public dialogue over whether terrorism or violent insurgency is normatively acceptable.

Moral Delegitimization. The most obvious strategy for reducing the legitimacy of terrorism or insurgency is to cast it as immoral. In many countries, this effort would require, first of all, the removal or reduction of institutionalized and often religion-based incitements to violence in the media, in political discourse, and in educational materials. Within the realm of strategic communications, however, methods have included painting terrorism as inconsistent with the teachings of Islam (as done through projects such as the "terrorism has no religion" campaign in the Middle East and the "Islam is peace" campaign in the United Kingdom); equating terrorism with such global scourges as slavery and genocide; and promoting the voices of nonviolent moral and religious authorities.

Recent meta-analyses on the decades-old "hellfire and delinquency" line of social-science research has confirmed that religiosity and moral condemnation are generally associated with lower levels of criminal behavior (Baier and Wright, 2001; Burkett and Ward, 1993; Higgins and Albrecht, 1976; Johnson et al., 2000), particularly in actively religious communities (Stark, 1996) and for offenses not already condemned by all segments of secular society (Burkett and White, 1974). That is, moral deterrence should be especially useful in contexts where secular society provides a mixed message on whether terrorism or insurgency is wrong.

However, the studies above take place in contexts where religion uniformly comes down on the side of condemning violent acts. Many if not most populations vulnerable to terrorism and insurgency are exposed to religious and moral arguments both for and against violent action. Although it does appear that positive moral messages can be effective in a noncompetitive communications environment, more research is necessary to evaluate the effectiveness of moral or religious antiviolence arguments when they compete with moral arguments in *favor* of terrorism or insurgency.

Social Delegitimization. Existing social-science theory and research has more to offer on the strategy of delegitimizing behaviors by casting them as socially unacceptable. Going beyond the use of peer spokespersons (discussed above), social-science theory provides a strong rationale for focusing on the long-term social attractiveness of terrorism and insurgency. For instance, the theory of reasoned action (Ajzen and Fishbein, 1980) views behavioral choices as the product of two beliefs: whether engaging in the behavior will lead to good or bad outcomes, and whether people important to the individual will approve or disapprove of the behavior.

In a very similar fashion, reference group theory (Deutsch and Gerard, 1955) holds that behavior is guided in part by conformity with a reference group consisting of those important to the individual. This conformity can manifest itself in two ways: normative, or conforming to the values of others (for example, that suicide terrorism is wrong); and informational, or conforming to the group's factual beliefs. Different long-term communication strategies might (1) directly communicate with or develop relationships with reference group members, (2) strengthen an individual's positive reference groups, (3) provide individuals with the knowledge or self-efficacy to seek out new positive reference groups, or (4) weaken the forces of conformity in an individual's negative reference groups. In general, individuals have been shown to be less vulnerable to conformity with their reference groups as the group gets smaller, less cohesive, and less unanimous in opinion (Fisher and Misovich, 1990).

Moving from theory to on-the-ground reality, it does seem clear that social context has a potentially important role in battling terrorism and insurgency. Recently, a military spokesman said of foreign fighters in Iraq that:

> Most of these young men wanted to make an impression, but paradoxically they did not tell their families they were going off to Iraq to fight for Al Qaeda out of fear of disapproval (Zavis, 2008).

Social science has studied efforts to harness these social forces in a variety of contexts. Empirical research has confirmed that both the real and perceived levels of social acceptance of many problem behaviors—from homicide and domestic violence to alcohol abuse and underage smoking—influence whether or not an individual will initiate the behavior (Archer, 2006; Eisenberg and Forster, 2003; McAlister, 2006), and that efforts by family members or the media to communicate social disapproval can be effective in preventing the behavior (Thomson et al., 2005; Yanovitzky and Stryker, 2001). One element that might be avoided in these efforts is normative ambiguity (for example, that terrorism is sometimes acceptable, just not here or now), which can reduce the effectiveness of a behavior change program (Yanovitzky and Stryker, 2001).

A promising strategy in conducting social delegitimization for some of the aforementioned problems has been social norms marketing (Perkins, 2002), which directly addresses the perceived acceptance of a negative behavior by explaining to the audience that the negative behavior is in fact not as popular or common as might have been assumed. Such a strategy may be useful in the contexts of terrorism and insurgency, where some individuals may support or tolerate terrorist activity in the mistaken belief that it is widely supported by others in their community.

In its more aggressive form, long-term influence over social norms can involve shifting the public image of terrorism to that of a doomed, pathetic, and deeply shameful act, deserving of scorn and ridicule. A well-known uses of this approach in communications during the global war on terrorism was the 2006 release of outtakes from a video message of Abu Mus'ab al-Zarqawi, showing him unable to fire a machine gun. This approach is quite distinct from moral or political arguments that paint terrorism as evil or highlight the pain terrorists have inflicted on others; certain individuals may be attracted to images of evil power and destruction, and talk of terrorist "masterminds" and the like may only romanticize the tactic and further encourage these individuals to seek out violence. In making this argument, social scientists (Pech, 2003) have highlighted the importance of language, counseling communicators discussing terrorism to avoid such words such "assassin" and "revenge" in favor of such labels as "insecure" and "weak."

However, in general, the body of public relations research would seem to offer two caveats to this strategy: (1) that it be undertaken only in a consistent manner (for example, avoiding the muddled message that terrorists are both pathetic and existentially dangerous) and (2) that it be undertaken only in tandem with efforts to offer a viable and satisfactory alternative to terrorism and violence. British analysts in the early Cold War used the rule of thumb that negative messaging failed unless accompanied by at least as much material promoting positive alternatives (Shaw, 1999). Thus, social delegitimization of terrorism may fail without efforts to promote nonviolence and tolerance in its place. Recent experimental research on teaching tolerance (Lillis and Hayes, 2007) found that acceptance and commitment therapy—in which participants explore their own prejudicial thoughts and emotional responses, accept them as a consequence of learning from others, and learn to control them—increased positive behavioral intentions toward other racial and ethnic groups significantly more than did lecture-based approaches. However, as suggested above, this issue goes beyond strategic communications and requires politically difficult choices about fundamental educational reform.

Conclusion

A review of the contributions of social science to strategic communications provides many useful lessons. Chief among them, as seen elsewhere in this paper, is that *context matters*—a message that effectively changes attitudes in one setting or population subgroup may have no effect, or even a counterproductive effect, in a different culture or for a different demographic. The cross-cultural mistakes in terrorism and insurgency communications discussed in this paper have primarily been anecdotal but still serve as valuable warnings. It is common, but dangerous, to assume that all audiences want the same things, or get their information the same way, or need to be approached using simplistic language or ideas.

As a result, a second important—but often overlooked—lesson is that message construction is best left to those with expert knowledge of a particular audience and its subgroups. This knowledge is not only important at the front end of a communications campaign—when segmenting, researching, and testing messages among audiences—but as the campaign continues, to nurture local relationships and adapt the campaign to changing circumstances. Unexpected changes in such factors as ideological polarization or community anxiety can poison a campaign, and therefore centralized, intuitively constructed messages will often be inferior to messages based on continuous local monitoring and rapid adaptation

Despite these gaps and uncertainties, domestic strategic communications research has identified dozens of extremely useful questions to answer, such as when and how social disapproval of violence can effectively reduce it. Specific findings may not hold up in other cultures—for example, perceptual deterrence and religious condemnation of violence are likely to operate quite differently in many Middle Eastern cultures—but the existing research provides a useful framework for exploring these questions abroad.

In addition, many of the communications best practices identified in the literature should be applicable to, or at the very least adaptable to, the problems of terrorism or insurgency. Below, I summarize a few of the best practices discussed in this paper.

Short-Term Strategic Communications

- Plan for communications crises by creating crisis plans, crisis teams, and training sessions and by monitoring the media. Plans should be updated regularly and should be imaginative, anticipating many different crisis types.
- Respond to fact-based crises truthfully, compassionately, and quickly. Concession is a generally robust response; there may be a trade-off between responses that build credibility and those that minimize interest in the crisis.
- Respond to rumor-based crises by not only debunking the particular rumor and those who spread it but also by providing general information and reducing public anxiety on the rumor-relevant topic.

Mid-Range Strategic Communications

- Plan for a communications campaign by defining a theory of what the campaign will accomplish, segmenting and researching the audience, researching the program and media environment, receiving feedback on the campaign, and keeping the community involved as the campaign proceeds.

- Message comprehensibility and authenticity are key common-sense factors that are sometimes overlooked.
- CT/COIN efforts may benefit by exploring promising communications approaches from other policy areas, such as behavioral inoculation.

Long-Term Strategic Communications

- Build relationships with the public, key partners, and the media by engaging in proactive early outreach, meaningful involvement, accountability, flexibility, informality (if possible), and substantive information-sharing.
- Delegitimize terrorist violence by revealing it to be inconsistent with moral or religious teachings, socially unacceptable, or unattractive. Further research is needed to evaluate the effectiveness of these approaches and others, such as social norms marketing, in the CT context.

In sum, strategic communications benefits most from detailed cultural expertise, empirically based adaptations, and a realistic assessment of its strengths and limitations. Ultimately, actions speak louder than words, and good communications can only partially mitigate the effects of an unpopular policy or action. Indeed, it is paradoxically when words are needed the most—during a crisis or when the credibility of a messenger is on the line—that the effectiveness of mere "spin" drops even further. Yet if good communications cannot whitewash CT or COIN policy crises, bad communications can certainly worsen them; at the very least, therefore, there is clear utility in training communications officers to avoid the mistakes of the past.

Bibliography

Abdollahian, Mark, Michael Baranick, Brian Efird, and Jacek Kugler, *Senturion: A Predictive Political Simulation Model,* Washington, D.C.: Center for Technology and National Security Policy, National Defense University, 2006.

Ajzen, Icek, and Martin Fishbein, *Understanding Attitudes and Predicting Social Behavior,* Upper Saddle River, N.J.: Prentice Hall, 1980.

Allen, Mike, "Meta-Analysis Comparing the Persuasiveness of One-Sided and Two-Sided Messages," *Western Journal of Speech Communication,* Vol. 55, No. 4, 1991, pp. 390–404.

Archer, John, "Cross-Cultural Differences in Physical Aggression Between Partners: A Social-Role Analysis," *Personality and Social Psychology Review,* Vol. 10, No. 2, 2006, pp. 133–153.

Arpan, Laura M., "When in Rome? The Effects of Spokesperson Ethnicity on Audience Evaluation of Crisis Communication," *The Journal of Business Communication,* Vol. 39, No. 3, 2002, pp. 314–339.

Arpan, Laura M., and Donnalyn Pompper, "Stormy Weather: Testing 'Stealing Thunder' as a Crisis Communication Strategy to Improve Communication Flow Between Organizations and Journalists," *Public Relations Review,* Vol. 29, 2003, pp. 291–308.

Atkin, Charles K., and Vicki S. Freimuth, "Formative Evaluation Research in Campaign Design," in Ronald E. Rice and Charles K. Atkin, eds., *Public Communications Campaigns,* Thousand Oaks, Calif.: Sage, 2001, pp. 125–145.

Aulakh, Preet S., Masaaki Kotabe, and Arvind Sahay, "Trust and Performance in Cross-Border Marketing Partnerships: A Behavioral Approach," *Journal of International Business Studies,* Vol. 27, No. 5, 1996, pp. 1005–1032.

Bahney, Benjamin, "Analytic Measures for Counterterrorism and Counterinsurgency," in Paul K. Davis and Kim Cragin, eds., *Social Science for Counterterrorism: Putting the Pieces Together,* Santa Monica, Calif.: RAND Corporation, 2009. As of January 20, 2009:
http://www.rand.org/pubs/monographs/MG849/

Baier, Colin J., and Bradley R. E. Wright, "'If You Love Me, Keep My Commandments': A Meta-Analysis of the Effect of Religion on Crime," *Journal of Research in Crime and Delinquency,* Vol. 38, No. 1, 2001, pp. 3–21.

Baksh-Mohammed, Sufyan, Min-Hwan Choi, and Coy Callison, "Cashing In Goodwill Capital in Times of Crisis: Does Public Awareness of Organizational Charitable Contributions Lead to Leniency When Problems Arise?" paper presented at the 10th International Public Relations Research Conference, Miami, Fla., 2007, pp. 23–33.

Banerjee, Smita C., and Kathryn Greene, "Analysis Versus Production: Adolescent Cognitive and Attitudinal Responses to Antismoking Interventions," *Journal of Communication,* Vol. 56, 2006, pp. 773–794.

Barthe, Emmanuel, "Crime Prevention Publicity Campaigns," Office of Community Oriented Policing Services, Washington, D.C.: U.S. Department of Justice, 2006.

Batra, Rajeev, and Douglas M. Stayman, "The Role of Mood in Advertising Effectiveness," *The Journal of Consumer Research,* Vol. 17, No. 2, 1990, pp. 203–214.

Battle, Joyce, ed., "U.S. Propaganda in the Middle East—The Early Cold War Version," *National Security Archive Electronic Briefing Book No. 78,* 2002. As of January 5, 2009:
http://www.gwu.edu/~nsarchiv/NSAEBB/NSAEBB78/essay.htm

Best, J. Allan, Shirley J. Thomson, Susanne M. Santi, Edward A. Smith, and K. Stephen Brown, "Preventing Cigarette Smoking Among School Children," *Annual Review of Public Health,* Vol. 9, 1988, pp. 161–201.

Borda, Jennifer L., and Susan Mackey-Kallis. "A Model for Crisis Management," in Dan Pyle Millar, and Robert L. Heath, eds., *Responding to Crisis: A Rhetorical Approach to Crisis Communication,* Mahwah, N.J.: Lawrence Erlbaum, 2003, pp. 117–139.

Bordia, Prashant, Nicholas DiFonzo, and Cassandra A. Schulz, "Source Characteristics in Denying Rumors of Organizational Closure: Honesty Is the Best Policy," *Journal of Applied Social Psychology,* Vol. 30, No. 11, 2000, pp. 2309–2321.

Borzekowski, Dina L.G., and Alvin F. Poussaint, "Common Themes from the Extremes: Using Two Methodologies to Examine Adolescents' Perceptions of Anti-Violence Public Service Announcements," *Journal of Adolescent Health,* Vol. 26, 2000, pp. 164–175.

———, "Public Service Announcement Perceptions: A Quantitative Examination of Anti-Violence Messages," *American Journal of Preventive Medicine,* Vol. 17, No. 3, 1999, pp. 181–188.

Boster, Franklin J., and Paul Mongeau, "Fear-Arousing Persuasive Messages," in Robert N. Bostrom, ed., *Communication Yearbook,* Beverly Hills, Calif.: Sage, 1984, pp. 330–375.

Bowers, William, and Glenn Pierce, "The Illusion of Deterrence in Isaac Ehrlich's Research on Capital Punishment," *Yale Law Journal,* Vol. 85, 1975, pp. 187–208.

Bradford, Jeffrey L., and Dennis E. Garrett, "The Effectiveness of Corporate Communicative Responses to Accusations of Unethical Behavior," *Journal of Business Ethics,* Vol. 14, No. 11, 1995, pp. 875–892.

Braga, Anthony A., David M. Kennedy, Elin J. Waring, and Anne Morrison Piehl, "Problem-Oriented Policing, Deterrence, and Youth Violence: An Evaluation of Boston's Operation Ceasefire," *Journal of Research in Crime and Delinquency,* Vol. 38, No. 3, 2001, pp. 195–225.

Brock, Timothy C., "Communicator-Recipient Similarity and Decision Change," *Journal of Personality and Social Psychology,* Vol. 1, No. 6, 1965, pp. 650–654.

Bronn, Peggy Simcic, "Relationship Outcomes as Determinants of Reputation," *Corporate Communications: An International Journal,* Vol. 12, No. 4, 2007, pp. 376–393.

Brown, J. Lynne, and Nancy Ellen Kiernan, "Assessing the Subsequent Effect of a Formative Evaluation on a Program," *Evaluation and Program Planning,* Vol. 24, No. 2, 2001, pp. 129–143.

Brown, Kenneth G., and Megan W. Gerhardt, "Formative Evaluation: An Integrative Practice Model and Case Study," *Personnel Psychology,* Vol. 55, 2002, pp. 951–983.

Browne, Donald R., "The International Newsroom: A Study of Practices at the Voice of America, BBC and Deutsche Welle," *Journal of Broadcasting,* Vol. 27, No. 3, 1983, pp. 205–231.

Bruning, Stephen D., Melissa Dials, and Amanda Shirka, "Using Dialogue to Build Organization-Public Relationships, Engage Publics, and Positively Affect Organizational Outcomes," *Public Relations Review,* Vol. 34, 2008, pp. 25–31.

Bucklin, Louis P., and Sanjit Sengupta, "Organizing Successful Co-Marketing Alliances," *Journal of Marketing,* Vol. 57, No. 2, 1993, pp. 32–46.

Burkett, Steven R., and David A. Ward, "A Note on Perceptual Deterrence, Religiously Based Moral Condemnation, and Social Control," *Criminology,* Vol. 31, No. 1, 1993, pp. 119–134.

Burkett, Steven R., and Mervin White, "Hellfire and Delinquency: Another Look," *Journal for the Scientific Study of Religion,* Vol. 13, No. 4, 1974, pp. 455–462.

Chapel, Gage, Kristin M. Peterson, and Roy Joseph, "Exploring Anti-Gang Advertisements: Focus Group Discussion with Gang Members and At-Risk Youth," *Journal of Applied Communication Research,* Vol. 27, No. 3, 1999, pp. 237–257.

Cho, Sooyoung, and Glen T. Cameron, "Power to the People—Health PR People That Is!" *Public Relations Review,* Vol. 33, 2007, pp. 175–183.

Cloudman, Reghan, and Kirk Hallahan, "Crisis Communications Preparedness Among U.S. Organizations: Activities and Assessments by Public Relations Practitioners," *Public Relations Review,* Vol. 32, 2006, pp. 367–376.

Coffman, Julia, *Public Communication Campaign Evaluation: An Environmental Scan of Challenges, Criticisms, Practice, and Opportunities*, Harvard Family Research Project, 2002. As of January 5, 2009:
http://www.hfrp.org/content/download/1116/48621/file/pcce.pdf

Coombs, W. Timothy, "Information and Compassion in Crisis Responses: A Test of Their Effects," *Journal of Public Relations Research,* Vol. 11, No. 2, 1999, pp. 125–142.

Coombs, W. Timothy, and Sherry J. Holladay, "Effects of Response Strategies and Media on Post-Crisis Perceptions and Intentions," paper presented at the 9th International Public Relations Research Conference, Miami, Fla., 2006, pp. 109–117.

———, "Helping Crisis Managers Protect Reputational Assets: Initial Tests of the Situational Crisis Communication Theory," *Management Communication Quarterly,* Vol. 16, No. 2, 2002, pp. 165–186.

Crowley, Ayn E., and Wayne D. Hoyer, "An Integrative Framework for Understanding Two-Sided Persuasion," *Journal of Consumer Research,* Vol. 20, No. 4, 1994, pp. 561–574.

Defense Science Board, "Report of the Defense Science Board Task Force on Strategic Communication," Washington, D.C.: Defense Science Board, 2008.

Defty, Andrew, "'Close and Continuous Liaison': British Anti-Communist Propaganda and Cooperation with the United States, 1950–51," *Intelligence and National Security,* Vol. 17, No. 4, 2002, pp. 100–130.

Desiere, Scott, and Bey-Ling Sha, "Analyzing Organization-Media Relationships: Exploring the Development of an Organizational Approach to Media Relations," paper presented at the 8th International Public Relations Research Conference, Miami, Fla., 2005, pp. 43–58.

DeSteno, David, Richard E. Petty, Derek D. Rucker, Duane T. Wegener, and Julia Braverman, "Discrete Emotions and Persuasion: The Role of Emotion-Induced Expectancies," *Journal of Personality and Social Psychology,* Vol. 86, No. 1, 2004, pp. 43–56.

Deutsch, Morton, and Harold B. Gerard, "A Study of Normative and Informational Influences upon Individual Judgment," *Journal of Abnormal and Social Psychology,* Vol. 51, 1955, pp. 629–636.

DiClemente, Carlo C., and James O. Prochaska, "Processes and Stages of Change: Coping and Competence in Smoking Behavior Change," in Saul Shiffman and Thomas A. Wills, eds., *Coping and Substance Abuse,* New York: Academic Press, 1985, pp. 319–342.

DiFonzo, Nicholas, and Prashant Bordia, "How Top PR Professionals Handle Hearsay: Corporate Rumors, Their Effects, and Strategies to Manage Them," *Public Relations Review,* Vol. 26, No. 2, 2000, pp. 173–190.

Dizard, Wilson P., *Inventing Public Diplomacy: The Story of the U.S. Information Agency,* Boulder, Colo.: Lynne Rienner, 2004.

DSB—*See* Defense Science Board.

Eisenberg, Marla E., and Jean L. Forster, "Adolescent Smoking Behavior: Measures of Social Norms," *American Journal of Preventive Medicine,* Vol. 25, No. 2, 2003, pp. 122–128.

Fearn-Banks, Kathleen, *Crisis Communications: A Casebook Approach,* 3rd ed., Mahwah, N.J.: Lawrence Erlbaum, 2007.

Fishbein, Martin, Kathleen Hall-Jamieson, Eric Zimmer, Ina von Haeften, and Robin Nabi, "Avoiding the Boomerang: Testing the Relative Effectiveness of Antidrug Public Service Announcements Before a National Campaign," *American Journal of Public Health,* Vol. 92, No. 2, 2002, pp. 238–245.

Fisher, Jeffrey D., and Stephen J. Misovich. "Social Influence and AIDS-Preventive Behavior," in John Edwards, R. Scott Tindale, Linda Heath, and Emil J. Posavac, eds., *Social Influence Processes and Prevention,* New York: Plenum Press, 1990, pp. 39–70.

Fitzpatrick, Kathy R., "Ten Guidelines for Reducing Legal Risks in Crisis Management," *Public Relations Quarterly,* Vol. 40, 1995, pp. 33–38.

Folger, Robert, "Distributive and Procedural Justice: Combined Impact of 'Voice' and Improvement on Experienced Inequity," *Journal of Personality and Social Psychology,* Vol. 35, No. 2, 1977, pp. 108–119.

Gainey, Barbara S., "Crisis Management Best Practices: A Content Analysis of Written Crisis Management Plans," paper presented at the 9th International Public Relations Research Conference, Miami, Fla., 2006, pp. 182–195.

Gibbs, Jack P., *Crime, Punishment, and Deterrence,* New York: Elsevier, 1975.

Goethals, George R., and R. Eric Nelson, "Similarity in the Influence Process: The Belief-Value Distinction," *Journal of Personality and Social Psychology,* Vol. 25, No. 1, 1973, pp. 117–122.

Golden, Linda L., and Mark I. Alpert, "Comparative Analyses of the Relative Effectiveness of One-Sided and Two-Sided Communications for Contrasting Products," *Journal of Advertising,* Vol. 16, No. 1, 1987, pp. 18–28.

Graddy, Elizabeth A., and Bin Chen, *The Consequences of Partner Selection in Service Delivery Collaborations,* Los Angeles, Calif.: USC Bedrosian Center on Governance and the Public Enterprise, 2006.

Green, Fitzhugh, *American Propaganda Abroad,* New York: Hippocrene Books, 1988.

Grunig, James E., *Qualitative Methods for Assessing Relationships Between Organizations and Publics,* Gainesville, Fla.: The Institute for Public Relations, 2002. As of December 30, 2008:
http://www.instituteforpr.org/files/uploads/2002_AssessingRelations.pdf

Grunig, James E., and Chun-ju Flora Hung, "The Effect of Relationships on Reputation and Reputation on Relationships: A Cognitive, Behavioral Study," paper presented at the 5th Annual International, Interdisciplinary Public Relations Research Conference, Miami, Fla.: PRSA Educator's Academy, 2002.

Gulati, Ranjay, "Alliances and Networks," *Strategic Management Journal,* Vol. 19, No. 4, 1998, pp. 293–317.

Guth, David W., "Organizational Crisis Experience and Public Relations Roles," *Public Relations Review,* Vol. 21, 1995, pp. 123–136.

Haas, Stephen M., and Erica Turley, "The Impact of WV's PSN Media Awareness Campaign on Citizen Attitudes Toward Crime and Criminal Justice System Responses," paper presented at the BJS/JRSA Conference, Pittsburgh, Pa.: West Virginia Statistical Analysis Center, 2007.

Hagan, Linda M., "For Reputation's Sake: Managing Crisis Communication," in James E. Grunig, Larissa A. Grunig, and Elizabeth L. Toth, eds., *The Future of Excellence in Public Relations and Communication Management: Challenges for the Next Generation,* Mahwah, N.J.: Lawrence Erlbaum, 2007, pp. 413–440.

Hale, Jerold L., and James Price Dillard, "Fear Appeals in Health Promotion Campaigns: Too Much, Too Little, or Just Right?" in Edward Maibach, and Roxanne Louiselle Parrott, eds., *Designing Health Messages: Approaches from Communication Theory and Public Health Practice,* Thousand Oaks, Calif.: Sage, 1995, pp. 65–80.

Hendon, David W., and Kevin Holton, "Notes on Church-State Affairs," *The Journal of Church and State,* Vol. 45, No. 3, 2003, pp. 619–642.

Hibbert, Zoe, and Peter Simmons, "War Reporting and Australian Defence Public Relations, an Exchange," *Prism,* Vol. 4, No. 2, 2006.

Higgins, Paul C., and Gary L. Albrecht, "Hellfire and Delinquency Revisited," *Social Forces,* Vol. 55, 1976, pp. 952–958.

Hon, Linda Childers, and James E. Grunig, *Guidelines for Measuring Relationships in Public Relations,* Gainesville, Fla.: The Institute for Public Relations, Commission on PR Measurement and Evaluation, 1999.

Horowitz, Irwin A., "Effects of Volunteering, Fear Arousal, and Number of Communications on Attitude Change," *Journal of Personality and Social Psychology,* Vol. 11, 1969, pp. 34–37.

Hovland, Carl I., and Walter Weiss, "The Influence of Source Credibility on Communication Effectiveness," *Public Opinion Quarterly,* Vol. 15, 1951, pp. 635–650.

Huang, Yi-Hui, "Crisis Situations, Communication Strategies, and Media Coverage: A Multicase Study Revisiting the Communicative Response Model," *Communication Research,* Vol. 33, 2006, pp. 180–205.

Isen, Alice M., "Positive Affect, Cognitive Processes and Social Behavior," in Leonard Berkowitz, ed., *Advances in Experimental Social Psychology,* New York: Academic Press, 1987, pp. 203–253.

Iyengar, Shanto, and Donald R. Kinder, *News That Matters: Television and American Opinion,* Chicago, Ill.: University of Chicago Press, 1987.

Iyer, Easwar S., and Kathleen Debevec, "Origin of Rumor and Tone of Message in Rumor Quelling Strategies," *Psychology & Marketing,* Vol. 8, No. 3, 1991, pp. 161–175.

Johnson, Byron, Spencer De Li, David B. Larson, and Michael McCullough, "A Systematic Review of the Religiosity and Delinquency Literature," *Journal of Contemporary Criminal Justice,* Vol. 16, No. 1, 2000, pp. 32–52.

Jones, Edward E., and Richard E. Nisbett, "The Actor and the Observer: Divergent Perceptions of the Causes of Behavior," in Edward E. Jones, David E. Kanhouse, Harold H. Kelley, Richard E. Nisbett, Stuart Valins, and Bernard Weiner, eds., *Attribution: Perceiving the Causes of Behavior,* Morristown, N.J.: General Learning Press, 1972, pp. 79–94.

Knight, Myra Gregory, Karen Kemp, Jane D. Brown, and Frank Biocca, "Stop the Violence: Lessons from Antiviolence Campaigns Using Mass Media," in James Hamilton, ed., *Television Violence and Public Policy,* Ann Arbor, Mich.: University of Michigan Press, 1998, pp. 267–312.

Kumkale, G. Tarcan, and Dolores Albarracin, "The Sleeper Effect in Persuasion: A Meta-Analytic Review," *Psychological Bulletin,* Vol. 130, No. 1, 2004, pp. 143–172.

Lafontaine, Tania, Myles Ferguson, and J. Stephen Wormith, *Street Gangs: A Review of the Empirical Literature on Community and Corrections-Based Prevention, Intervention and Suppression Strategies,* Regina, Saskatchewan: Ministry of Corrections, Public Safety and Policing, 2005.

Lavine, Howard, Diana Burgess, Mark Snyder, John Transue, John L. Sullivan, Beth Haney, and Stephen H. Wagner, "Threat, Authoritarianism, and Voting: An Investigation of Personality and Persuasion," *Personality and Social Psychology Bulletin,* Vol. 25, No. 3, 1999, pp. 337–347.

Ledingham, John A., and Stephen D. Bruning, "Relationship Management in Public Relations: Dimensions of an Organization-Public Relationship," *Public Relations Review,* Vol. 24, No. 1, 1998, pp. 55–65.

———, *Public Relations as Relationship Management: A Relational Approach to the Study and Practice of Public Relations,* LEA's Communication Series, Mahwah, N.J.: Lawrence Erlbaum, 2001.

Ledingham, John A., Stephen D. Bruning, and Laurie J. Wilson, "Time as an Indicator of the Perceptions and Behavior of Members of a Key Public: Monitoring and Predicting Organization-Public Relationships," *Journal of Public Relations Research,* Vol. 11, No. 2, 1999, pp. 167–183.

Lee, Susan, and Lulu Rodriguez, "Four Publics of Anti-Bioterrorism Information Campaigns: A Test of the Situational Theory," *Public Relations Review,* Vol. 34, 2008, pp. 60–62.

Lillis, Jason, and Steven C. Hayes, "Applying Acceptance, Mindfulness, and Values to the Reduction of Prejudice: A Pilot Study," *Behavior Modification,* Vol. 31, No. 4, 2007, pp. 389–411.

Lim, Joon Soo, and Eyun-Jung Ki, "Resistance to Ethically Suspicious Video Spoof on YouTube: A Test of Inoculation Theory," paper presented at the 10th International Public Relations Research Conference, Miami, Fla., 2007, pp. 283–297.

Louis, Meryl R., and Robert I. Sutton, "Switching Cognitive Gears: From Habits of Mind to Active Thinking," *Human Relations,* Vol. 44, 1991, pp. 55–76.

Luchok, Joseph A., and James C. McCroskey, "The Effect of Quality of Evidence on Attitude Change and Source Credibility," *Southern Speech Communication Journal,* Vol. 43, 1978, pp. 371–383.

Lumsdaine, Arthur A., and Irving L. Janis, "Resistance to 'Counterpropaganda' Produced by One-Sided and Two-Sided 'Propaganda' Presentations," *Public Opinion Quarterly,* Vol. 17, No. 3, 1953, pp. 311–318.

MacNulty, Christine A. R., "Values as a Basis for Deterring Terrorists," *2008 Unrestricted Warfare Symposium,* Baltimore, Md.: The Johns Hopkins University, 2008.

Maibach, Edward W., and David Cotton, "Moving People to Behavior Change: A Staged Social Cognitive Approach to Message Design," in Edward Maibach and Roxanne Louiselle Parrott, eds., *Designing Health Messages: Approaches from Communication Theory and Public Health Practice,* Thousand Oaks, Calif.: Sage, 1995, pp. 41–64.

Marra, Francis J., "Crisis Communication Plans: Poor Predictors of Excellent Crisis Public Relations," *Public Relations Review,* Vol. 24, No. 4, 1998, pp. 61–74.

Maynard, Roberta, "Handling a Crisis Effectively: Will You Be Ready if Your Company Comes Under Scrutiny? Here's How to Sharpen Your Public-Relations Skills—Crisis Management," *Nation's Business,* 1993.

McAlister, Alfred L., "Acceptance of Killing and Homicide Rates in Nineteen Nations," *European Journal of Public Health,* Vol. 16, No. 3, 2006, pp. 259–265.

McComas, Katherine, Leah Simon Tuite, Leah Waks, and Linda Ann Sherman, "Predicting Satisfaction and Outcome Acceptance with Advisory Committee Meetings: The Role of Procedural Justice," *Journal of Applied Social Psychology,* Vol. 37, No. 5, 2007, pp. 905–927.

McCombs, Maxwell, and Donald L. Shaw, "The Agenda-Setting Function of the Mass Media," *Public Opinion Quarterly,* Vol. 37, 1973, pp. 62–75.

McGuire, William J., "An Information Processing Model of Advertising Effectiveness," in Harry J. Davis, and Alvin J. Silk, eds., *Behavioral and Management Science in Marketing,* New York: Roland Press, 1978, pp. 156–180.

———, "A Vaccine for Brainwash," *Psychology Today,* Vol. 3, No. 9, 1970, pp. 36–39.

Mitchell, Austin, "The Decline of Current Affairs Television," *The Political Quarterly,* Vol. 44, No. 2, 1973, pp. 127–136.

Monahan, Jennifer L., "Thinking Positively: Using Positive Affect When Designing Health Messages," in Edward Maibach, and Roxanne Louiselle Parrott, eds., *Designing Health Messages: Approaches from Communication Theory and Public Health Practice,* Thousand Oaks, Calif.: Sage, 1995, pp. 81–98.

Nyhan, Brendan, and Jason Reifler, "When Corrections Fail: The Persistence of Political Misperceptions," unpublished manuscript, 2008.

O'Keefe, Garrett J., and Kathaleen Reid, "The Uses and Effects of Public Service Advertising," in James E. Grunig, and Larissa A. Grunig, eds., *Public Relations Research Annual,* Hillsdale, N.J.: Lawrence Erlbaum, 1990, pp. 67–91.

Parrott, Roxanne Louiselle, "Motivation to Attend to Health Messages: Presentation of Content and Linguistic Considerations," in Edward Maibach, and Roxanne Louiselle Parrott, eds., *Designing Health Messages: Approaches from Communication Theory and Public Health Practice,* Thousand Oaks, Calif.: Sage, 1995, pp. 7–23.

Patton, Michael Q., *Utilization-Focused Evaluation,* Thousand Oaks, Calif.: Sage, 1997.

Pech, Richard J., "Inhibiting Imitative Terrorism Through Memetic Engineering," *Journal of Contingencies and Crisis Management,* Vol. 11, No. 2, 2003, pp. 61–66.

Perkins, H. Wesley, "Social Norms and the Prevention of Alcohol Misuse in Collegiate Contexts," *Journal of Studies on Alcohol, Supplement,* Vol. 14, 2002, pp. 164–172.

Pfau, Michael, "Designing Messages for Behavioral Inoculation," in Edward Maibach, and Roxanne Louiselle Parrott, eds., *Designing Health Messages: Approaches from Communication Theory and Public Health Practice,* Thousand Oaks, Calif.: Sage, 1995, pp. 99–113.

Pfau, Michael, and Michael Burgoon, "Inoculation in Political Campaign Communication," *Human Communication Research,* Vol. 15, No. 1, 1988, pp. 91–111.

Plowman, Kenneth D., "Measuring Relationships in Strategic Communications for Public Relations," paper presented at the 10th International Public Relations Research Conference, Miami, Fla., 2007, pp. 407–418.

Rawlins, Brad, "Measuring the Relationship Between Organizational Transparency and Trust," paper presented at the 10th International Public Relations Research Conference, Miami, Fla., 2007, pp. 425–439.

Reinard, John C., "The Empirical Study of the Persuasive Effects of Evidence: The Status After Fifty Years of Research," *Human Communication Research,* Vol. 15, No. 1, 1988, pp. 3–59.

Rosenbaum, Dennis P., and Linda Heath, "The 'Psycho-Logic' of Fear-Reduction and Crime-Prevention Programs," in John Edwards, R. Scott Tindale, Linda Heath, and Emil J. Posavac, eds., *Social Influence Processes and Prevention,* New York: Springer, 1990, pp. 221–229.

Rosnow, Ralph L., "Inside Rumor: A Personal Journey," *American Psychologist,* Vol. 46, No. 5, 1991, pp. 484–496.

Rothman, Alexander J., and Peter Salovey, "Shaping Perceptions to Motivate Healthy Behavior: The Role of Message Framing," *Psychological Bulletin,* Vol. 121, No. 1, 1997, pp. 3–19.

Rucker, Derek D., and Richard E. Petty, "Emotion Specificity and Consumer Behavior: Anger, Sadness, and Preference for Activity," *Motivation and Emotion,* Vol. 28, No. 1, 2004, pp. 3–21.

Shaw, Tony, "The Information Research Department of the British Foreign Office and the Korean War, 1950–53," *Journal of Contemporary History,* Vol. 34, No. 2, 1999, pp. 263–281.

Skinner, Ellen R., and Michael D. Slater, "Family Communication Patterns, Rebelliousness, and Adolescent Reactions to Anti-Drug PSAs," *Annual Meeting of the International Communication Association,* Washington, D.C., 1993.

Skogan, Wesley, "Fear of Crime and Neighborhood Change," *Crime and Justice,* Vol. 8, 1986, pp. 203–229.

Slater, Michael D., "Choosing Audience Segmentation Strategies and Methods for Health Communications," in Edward Maibach and Roxanne Louiselle Parrott, eds., *Designing Health Messages: Approaches from Communication Theory and Public Health Practice,* Thousand Oaks, Calif.: Sage, 1995, pp. 186–198.

Small, William J., "Exxon Valdez: How to Spend Billions and Still Get a Black Eye," *Public Relations Review,* Vol. 17, No. 1, 1991, pp. 9–25.

Smith, Nick L., "Evaluation Utility and Client Involvement in Accreditation Studies," *Education Evaluation and Policy Analysis,* Vol. 2, No. 5, 1980, pp. 57–65.

Stark, Rodney, "Religion as Context: Hellfire and Delinquency One More Time," *Sociology of Religion,* Vol. 57, No. 2, 1996, pp. 163–173.

Steele, Dennis, "Delivering the Message," *Army,* Vol. 53, No. 11, 2003.

Thomson, Carey Conley, Michael Siegel, Jonathan Winickoff, Lois Biener, and Nancy A. Rigotti, "Household Smoking Bans and Adolescents' Perceived Prevalence of Smoking and Social Acceptability of Smoking," *Preventive Medicine,* Vol. 41, No. 2, 2005, pp. 349–356.

Tversky, Amos, and Daniel Kahneman, "The Framing of Decisions and the Psychology of Choice," *Science,* Vol. 211, Issue 4481, 1981, pp. 453–458.

Weaver, David, "Issue Salience and Public Opinion: Are There Consequences of Agenda-Setting?" *International Journal of Public Opinion Research,* Vol. 3, No. 1, 1991, pp. 53–68.

Weiss, Janet A., and Mary Tschirhart, "Public Information Campaigns as Policy Instruments," *Journal of Policy Analysis and Management,* Vol. 13, No. 1, 1994, pp. 82–119.

Wiener, Morton, and Albert Mehrabian, *Language Within Language: Immediacy, a Channel in Verbal Communication,* New York: Appleton Century Crofts, 1968.

Williams, Terrie M., "Putting Relationships Back into Public Relations," *Vernon C. Shranz Distinguished Lectureship in Public Relations,* Muncie, Ind.: Department of Journalism, Ball State University, 1996.

Wixon, Renee V., "Approaches to Youth Violence Prevention: A Program Planning Guide," Minneapolis, Minn.: Minneapolis Department of Health and Family Support, 1998.

Yang, Sung-Un, and James E. Grunig, "Decomposing Organisational Reputation: The Effects of Organisation-Public Relationship Outcomes on Cognitive Representations of Organisations and Evaluations of Organisational Performance," *Journal of Communication Management,* Vol. 9, No. 4, 2005, pp. 305–325.

Yanovitzky, Itzhak, and Jo Stryker, "Mass Media, Social Norms, and Health Promotion Efforts: A Longitudinal Study of Media Effects on Youth Binge Drinking," *Communication Research,* Vol. 28, No. 2, 2001, pp. 208–239.

Zavis, Alexandra, "Foreign Fighters in Iraq Seek Recognition, U.S. Says," *Los Angeles Times,* March 17, 2008.

Zimring, Franklin E., and Gordon J. Hawkins, *Deterrence: The Legal Threat in Crime Control,* Chicago, Ill.: University of Chicago Press, 1973.

CHAPTER TEN

Cross-Cutting Observations and Some Implications for Policymakers

Kim Cragin

Does religious extremism play a significant role in the radicalization and recruitment of new al-Qaeda fighters? Is recruitment even a vulnerability for terrorists? If so, to what degree? To what extent do terrorist groups prioritize public support in their own decisionmaking? Social-science research can provide insight into some of the most pressing issues in U.S. counterterrorism policy. Yet it also has its shortfalls. This paper looks across the earlier papers in this volume by Noricks, Helmus, Paul, Jackson, Gvineria, Berrebi, and Egner to address some cross-cutting observations such as these. The chapter begins with what may be the most interesting, which is a discussion of several tensions that exist not only in the academic literature but also in the various papers of this volume. It then discusses significant points of consensus. It ends with selective observations about possible implications for the use of various instruments of power.

Introduction

U.S. policymakers increasingly have turned to the social sciences, hoping for some clarity on the terrorism phenomenon.[1] Social-science research often confirms what subject-matter experts suspect to be true but other times contradicts widespread beliefs.[2] Social science also identifies second- and third-order effects. It is interesting to note that although some social-science disciplines (such as political science and economics) have been used regularly to inform national security, others (such as sociology, psychology, and anthropology) have fit less comfortably into the realm of national security research. That situation has been changing since September 2001.[3]

For example, methodologies such as social network analysis, associated with anthropology, have been used at the tactical level in Iraq and Afghanistan. Conferences sponsored by the U.S. government increasingly have included participants from academia and, particularly, social scientists. Similarly, well-known social scientists, such as Quintan Wiktorowitz and David Kilcullen, have provided strategic advice to the U.S. national security community.[4] Because social scientists have made significant contributions at both tactical and strategic levels, it is logical that policymakers would look to them for answers to even more complex questions.

Unfortunately, social scientists do not always speak the same language and, even when they do, disagreements abound. For example, "empirical research" tends to mean quantitative research to an economist but not necessarily to an anthropologist. Nor are quantitative methods necessarily considered by all scholars to be more academically rigorous. These methodological disagreements also affect how much credence social scientists place on other types of studies. Moreover, some questions cannot be answered sufficiently by academia and are better suited to the intelligence community. These realities present U.S. policymakers with the challenge of determining when to turn to social science and when to turn away, separating social-science knowledge into the useful and the not useful, and weighing findings from one study against another. It is not an easy challenge.

Previous chapters have attempted to alleviate this challenge somewhat by providing background discussions on what social science has brought to the understanding of terrorism over the past 30 years. Clearly, this task is daunting and although the chapter authors have been as thorough as possible in their analyses—given the time and resources available—they have likely overlooked some studies. Given the nature of the research, it also can be challenging to relate the chapters to each other. As discussed in the Introduction and in Chapter Eleven by Paul Davis, we asked the authors to construct simple graphical models at the end of each chapter to summarize important factors and their relationships. We also asked authors to identify gaps in the social-science base for their chapters. It was our hope that these discussions would provide a basis for comparison across the various chapters but still provide enough specificity to be useful to policymakers interested in one specific question or another.

This chapter on policy implications attempts to be cross-cutting. Rather than attempting to summarize everything or to synchronize the various factors, therefore, it highlights what we found to be the greatest points of tension and points of agreement across the literature. It also attempts to delineate the most imperative policy implications following from these tensions and agreements.

Points of Tension

Our review of the social-science literature revealed a number of tensions in the terrorism-related research base. Although some might be interesting from an academic perspective, they are not necessarily relevant to U.S. national security policy. For example, as mentioned in the Introduction to this monograph, the U.S. government is likely going to attempt to improve the U.S. image overseas, regardless of whether individuals are motivated to become terrorists as a reaction against U.S. foreign policy. Some tensions are important, however:

- debates on whether the supply of recruits far exceeds the demand from terrorist cells, organizations, or networks
- debates on whether it is more useful to think in terms of terrorist organizations or terrorist networks
- debates on the significance of religion as an influence on individual motivations or permissive environments.

The next sections explore these points of tension in greater detail. We address these specific tensions because our analysis suggests that taking one side of the debate to its extreme without accounting for alternative findings could negatively affect U.S. counterterrorism policy. Therefore, subsequent paragraphs function primarily as cautions to U.S. policymakers who might be aware of the arguments on one side of each debate but not the other.

Supply Versus Demand

The Conflicting Views. One of the most noteworthy tensions is whether the supply of terrorist recruits far outstrips demand. This debate is important because it affects the likely prioritization of counterterrorism resources. How much of its resources should the United States focus on reducing the number of recruits?

Claude Berrebi, in his chapter, argues that many terrorist groups will always be able to find more potential fighters. Indeed, both he and Eli Berman have argued that it is common for societies to have a large supply of the kinds of individuals who are potential fodder (that is, young men interested in action and perhaps with various grievances).[5] However, the terrorist organization may need far fewer recruits than those available, allowing it to filter selectively. Further, it has been argued that individuals choosing to become terrorists are doing so consistent with a rational-choice model that would probably apply to many other individuals of a similar nature. If one accepts these arguments, it would be foolish to expend resources in a doomed effort to minimize the number of recruits. In the extreme, one would cease related efforts.

In contrast, Todd Helmus's findings on terrorist motivations reveal that individuals follow a radicalization trajectory as they progress from sympathy toward a willingness to become a fighter. His research suggests, however, that not every terrorist recruit completes this trajectory. Similarly, Brian Jackson suggests in his chapter that the quality of the recruit is perhaps more important to terrorist groups than the quantity (a point that arises in Berrebi's chapter as well). Gaga Gvineria, in his chapter on the decline of terrorism, observes that part of decline may be the drying up of societal support and the flow of recruits that this provides. Thus, these other discussions could be interpreted to suggest that efforts to affect supply are important—in the extreme, critical to the success of counterterrorism policies.

Toward Resolving the Apparent Tension. This supply versus demand debate is a "point of tension," not total disagreement. Most social scientists would agree that their findings should not be pushed to extremes, because shortcomings exist. These tensions also are often the result of limited or poor-quality data. Terrorists are difficult to collect data on because they function as clandestine organizations. As a result, most terrorism datasets are event- or incident-based, focusing on observable attacks, which might not be useful in answering complex questions. It is also common for analysts to rely heavily on data from the Palestinian-Israeli conflict or the fighting in Northern Ireland. These conflicts have evidenced high levels of violence over a long period of time and have been monitored extensively, providing considerable data. It is questionable, however, how far such findings can be generalized to other areas of the world.

Similarly, studies on radicalization trajectories tend to rely heavily on interviews and interrogation reports, which often include strong testimony about what the terrorists saw as lofty motivations. As with quantitative datasets, however, these qualitative resources also can be unreliable. That is, the methods focus exclusively on individuals who have already chosen to become terrorists and ask them to retroactively explain the experiences and feelings that led to such a choice. Over time, social scientists have learned that retroactive accounts tend to be less than reliable whether they are oral or written. It would be problematic to rely exclusively on autobiographies, former jihadis' testimony, detainee interviews, and interrogation reports to design barriers to terrorist recruitment or even as a decision point to emphasize countering terrorist motivations and the associated flow of volunteers as the core component of counterterrorism policy. Thus, arguments for focusing on motivations and related supply issues are at least as problematic as arguments suggesting that supply is not the limiting factor.

So how can we resolve this tension? First, consider that it has become apparent that U.S. security forces simply cannot capture or kill enough al-Qaeda fighters worldwide to defeat the movement. Recent efforts by the United States and its allies, in locations such as Saudi Arabia, Singapore, Yemen, Egypt, and even the United Kingdom have nominally been attempting to halt the flow of new recruits into al-

Qaeda and affiliated groups. Most such efforts attempt to halt the flow of new recruits by undermining terrorist motivations.[6] This approach may be quite sensible, but is it posed properly given the possibility that the supply of new recruits may not be a limiting factor?

One important distinction is between potential recruits and the flow of new terrorists who have been not only recruited but indoctrinated, trained, and organized. The latter is the *output* of recruiting, and it is much more relevant ultimately than the former.

If so, one approach would be to think in terms of focusing resources on the rest of the process: on limiting the number and effectiveness of recruiters, recruiting, indoctrination, training, and organizing.[7] One aspect of doing so might still involve attacking motivations for becoming a terrorist (for example, so-called inoculation efforts), but other aspects might well have more leverage. These include making the recruiters "hunker down," attacking training facilities and deterring or precluding the opening of new ones, and constant disruptions to interfere with all aspects of the process (to include arrest or direct attacks on recruiters and trainers). People who are running and hiding have much less opportunity to seek out, filter, train, and organize recruits.

Such an alternative perspective might affect the orientation of numerous U.S. efforts, including those of Special Operations Forces. It might increase the relative attention given to the fringes of the al-Qaeda organization, say in Maghreb, rather than the hot spots of Iraq and Afghanistan. It might change the value seen in disruptive activities, which many operators believe in but have difficulty measuring. It might also suggest changes of geographic priority for building partner capacity to counter al-Qaeda-affiliated groups, intelligence collection, and even diplomatic efforts by the Ambassador for Counterterrorism in the State Department. And it might suggest additional metrics, as discussed in the appendix by Ben Bahney.[8]

The imagery of this approach is suggested in Figure 10.1—even with a large supply of raw recruits, the approach would seek to minimize the flow through the process. To some, such as those who have long argued for the power of disruption, our observations may be obvious. To others, they will not be.

Figure 10.1
Reducing Flow to a Trickle?

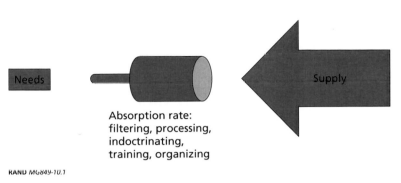

Absorption rate:
filtering, processing,
indoctrinating,
training, organizing

RAND MG849-10.1

Organizations Versus Networks

Another point of tension revealed in the previous chapters is whether it is more useful to think in terms of terrorist organizations or networks. This debate affects how counterterrorism policymakers and other experts weigh the success or failure of counterterrorism policies, as well as the overarching terrorist threat facing the United States.

The subject is also topical. For example, some of the top experts on terrorism and al-Qaeda in the U.S. academic community have been engaged in a debate as to the status of al-Qaeda in 2008. Marc Sageman (2004, 2008) has argued that the influence of al-Qaeda senior leaders has eroded and that the main threat to U.S. interests comes primarily from networked cells present in the Western world. He bases this argument in part on the denunciation of al-Qaeda by top ideological thinkers and partly on the limited number of successful spectacular attacks against U.S. interests in recent years, which he attributes in part to a reduction in the quality of al-Qaeda recruits primarily in the West. He has referred to them as "a bunch of guys" very different from the recruits of the 1990s. In contrast, Bruce Hoffman has argued that al-Qaeda still resembles a structured organization and that some of the seemingly disaggregated cells actually take direction from al-Qaeda central leadership. According to Hoffman (and intelligence-community conclusions in 2007), this structure means that al-Qaeda presents as great a threat as ever to the United States. He similarly points to the numerous failed attacks by al-Qaeda, including some near-misses, as evidence of its continued threat.[9]

A different slant on the network versus organization perspective comes in connection with "direct action" discussion. If one thinks in terms of a centralized organization with traditional command and control, then decapitation by attack of leaders seems likely to be fruitful. However, someone focused on the network image may stress the organic, adaptive, self-healing nature of networks and may disparage the plausibility of decapitation.

Brian Jackson tackles this tension somewhat abstractly in his chapter on terrorist decisionmaking. He focuses on decisionmaking cells, which could be as small as a handful or as large as a few thousand and which might or might not be physically collocated. Size and structure do not matter as much for Jackson's analysis as that the entity can functionally make and implement decisions. Arguably, his approach presents a way to account for both the network and the central-organization views of al-Qaeda. That is, al-Qaeda senior leaders might make decisions based on their beliefs that cells and affiliated groups will or will not implement their guidance. At the same time, those same cells might make their own calculations, taking input from al-Qaeda senior leaders, but also making their own assessment of various factors. Jackson's findings suggest that neither "organization" nor "network" alone represents an adequate construct for thinking about terrorist entities.

The tension also is implicit in Darcy Noricks's chapter on how and why terrorism emerges. In her chapter, Noricks attempts to account for broader social movements that give rise to political violence. But her findings also highlight the importance of organizational structures that emerge to *harness* that violence and give it greater direction. By pushing to the extreme, one could interpret her findings to mean that disaggregated networks alone cannot keep up a substantial momentum of attacks and, therefore, threat. However, she also suggests that even well-organized terrorist structures must have some wider movement behind them as they emerge and sustain themselves.

Is lack of data fundamentally the problem with this second tension as well? The paucity of data presents a challenge to past studies reviewed in both Jackson's and Noricks's chapters, but other issues exacerbate the tension. Fundamentally, subject-matter experts and social scientists have long struggled with the definitions of terrorism versus insurgency. In this study, we did not want to impose artificial definitions on the researchers, so we defined terrorism more broadly as the use or threatened use of violence to achieve terror. This allowed researchers to draw on a wider variety of literatures, including those about insurgencies that employed terrorism. Nonetheless, the organization versus network tension seems to have affected the degree to which authors relied on information about subnational or nonstate terrorist groups specifically or political violence in general. Therefore, this tension represents another caution to those in the analytical community. Subject-matter experts with different backgrounds, such as terrorism versus insurgency work, will have different perspectives and assumptions.

As with our discussion of supply versus demand, the tension between network and organizational views can be useful in broadening our perspectives and seeing different facets of the problem.

More generally, we believe that it is best to view the terrorism phenomenon in "system" terms. RAND and the Institute for Defense Analyses were early champions of this perspective in 2002, noting that it was wrong-headed to see al-Qaeda as a monolith. It was better to recognize that the al-Qaeda system has many critical components, each with its own vulnerabilities (Davis and Jenkins, 2002). The components include, for example, leaders, lieutenants, foot soldiers, financiers, logisticians, sources of moral support, state supporters, and the supporting elements of the relevant population. David Kilcullen took influential next steps in his work on global insurgency in 2004 and subsequently.[10] In fact, a system approach often requires numerous perspectives on the system, leading to different insights and tacks. More work of this character is needed (see also National Academy of Sciences, 2002).

Possible Implications of the Systems Approach. In the world of al-Qaeda, terrorist organizations clearly have an influence in day-to-day operations as well as in strategic guidance. Al-Qaeda senior leaders, for example, even in the expert community's most austere interpretation of their relevance, still issue statements on strategic direction. Moreover, al-Qaeda collaborators, such as al-Qaeda in the Maghreb, claim to be conducting attacks in the name of senior leaders, even if their day-to-day operational decisions are independent. Both of these components of the system are separate entities but organizations nonetheless. Additionally, networks of individuals also clearly exist within this system, such as for bringing foreign fighters into Iraq and Afghanistan. They function separately from al-Qaeda senior leaders, but they contribute to the whole. Finally, within this system, disparate cells might also operate either with direction from al-Qaeda leaders or on their own.

One implication of a systems approach is that policy should be designed to specifically target components of the system as appropriate. For example, rather than focusing on efforts to counter al-Qaeda in Iraq specifically, in isolation from efforts to counter al-Qaeda in the Maghreb, policies might focus on thematic threats, such as financial and recruitment networks, without geographic constraints.[11] Thus, policymakers might identify financial networks, for example, as the most important priority. Then, law-enforcement instruments available to Interpol would be used against the fraud and other petty crimes that fund disparate cells in Europe. At the same time, diplomatic efforts could be used with countries in the Gulf to tighten restrictions on monetary flows. Alternatively, policymakers might choose to prioritize the movement of people across borders to areas of conflict outside their immediate country of residence. If this path were chosen, then the Department of Homeland Security might identify countries across which these individuals flow and work with host nations to improve their border security. At the same time, Special Operations Forces would work through "train and equip" programs to improve the counterinsurgency capabilities of security forces in remote and hard-to-monitor border areas. Of course, these organizations, networks, and cells likely interact to a certain extent within the system. Thus, second- and third-order effects would need to be accounted for and adjusted to within the system. This system approach also allows policymakers to identify one component of the system—for example, recruitment or financial networks—and understand more fully its relationship to the whole.

Significance of Religion. The final point of tension that emerged from our analysis is the significance of religion as a factor of influence on individual motivations, terrorist behaviors, or even permissive environments. Of course, questions on the significance of ideology in general are not new to the study of al-Qaeda in the counterterrorism world. Subject-matter experts on Latin American militancy have long pondered the extent to which the Shining Path in Peru, for example, or the Revolutionary Armed Forces of Colombia (FARC) were motivated by communism.[12] But the issue of radical, or revivalist, Islam has received particular emphasis in studies of violence since September 2001.[13]

The tension that religion exhibits within social science is perhaps both the least and most obvious in our chapters. It probably reflects some discomfort within the scholarly community. This tension may not at first be apparent in the earlier chapters. After all, every chapter addresses the topic to some degree. But *how* they do so differs and provides insight into tensions within the academic community. The reader should recall that the authors were asked to approach their subjects in a way that remained relatively true to the particular literatures they addressed, even when the authors were aware of the various tensions as a result of project-level discussions and general knowledge.

Perhaps the first observation is that, despite covering numerous literatures thought to be relevant, none of the chapters explicitly addresses recent studies on revivalist Islam from the field of religious studies. Todd Helmus, in his chapter, does cite Quintan Wiktorowicz and both Darcy Noricks and Chris Paul refer to Mark Juergensmeyer in their chapters. However, others, such as Charles Kurzman and Ibrahim Abu Rabi, experts in Islamic studies, have not been included.[14] In retrospect, this reflects the fact that the issue of religion does not currently integrate comfortably into the wider disciplines and, as a result, was not addressed as well as it might have been. Also, there remains a marked difference in how ideology and religion are treated from chapter to chapter.

For example, in his chapter, Chris Paul treats ideology as a *tool* of validation for political organizations as they attempt to harness popular support. Paul argues that "normative acceptability of violence" contributes to popular support for terrorism, but his chapter—reflecting the literature he consulted—does not treat revivalist Islam or even ideology more generally as contributing to that acceptability. This absence is interesting, because the effect is often asserted by subject-matter experts and policymakers. Claude Berrebi similarly argues in his chapter that neither religion nor ideology provides an explanation for the emergence of terrorism. In this sense, Berrebi and Paul both minimize the importance of religion, with Paul emphasizing that it is accompanied by or simply reinforces feelings of humiliation or a broader sense of being under attack by outsiders. If these arguments were taken to the extreme, it might suggest that efforts to counter ideological extremism are unlikely to have an effect on popular support for terrorism or, indeed, affect the success or failure of any particular terrorist group.[15]

Darcy Noricks conveys a different story in her chapter. It emphasizes disagreement in the social sciences on the importance of religion, but she gives this factor more weight than either Paul or Berrebi, noting also that religion appears to be more important when external threats take on sacred meaning to members of a community. Noricks draws heavily on work by Mark Juergensmeyer and Jessica Stern in her chapter. Most interestingly, she places "ideology" in her model directly under "facilitative norms on the use of violence." Todd Helmus, in his chapter, breaks "religion" into separate pieces when addressing how it relates to individual motivations for violence. For example, religion is treated as a factor contributing to bottom-up peer group radicalization. But religion also affects perceived rewards and "desire for change." Thus, he places even greater emphasis on the role of religion when it comes to individual radicalization than does Noricks, especially when it comes to providing a permissive environment for the emergence of terrorism. Taken to their extremes, Helmus's and Noricks's findings would suggest that efforts to counter radical ideology could act as preventative measures to halt the emergence of terrorism at either a societal or individual level and, thus, should be given priority in counterterrorism policy.

Of the three points of tension discussed in this chapter, the role of religion is perhaps the most troubling. Significant resources have been placed toward building moderate Muslim networks and countering propaganda. Subject-matter experts, such as Peter Bergen, have lauded the recent criticisms heaped on al-Qaeda by well-respected ideologues such as Sayyid Imam al-Sharif (Dr. Fadl) and Sheikh Salman al-Oudah.[16]

However, the chapters in this monograph suggest that the effect of such activities is somewhat debatable. Given the state of the literature, it is difficult to provide policymakers with strong conclusions about the benefits of or approaches to countering ideology, including religion.

However, our study does suggest some different analytical approaches to the question of religion. One would be to distinguish between leaders and followers in studies of individual motivations, especially ideological motivations. Logic suggests that patterns of recruitment may differ between leaders and followers within the al-Qaeda organization. That is, level of analysis matters here. Making these distinctions in research would probably bring more insight into the role of religion, as would distinguishing between, for example, fundamentalist and more tolerant strands of the various religions, and, as a separate factor, distinguishing between strands that have or do not have patterns of violence in enforcing their perspectives on others. Similarly, research efforts to distinguish between the importance of religion in communities with minority Muslim populations and majority Muslim populations would be worthwhile. Because religion may contribute to a broader sense of being under attack or threat from outsiders, distinguishing between minority and majority populations in this regard could also be enlightening, especially in thorough qualitative comparative analyses.

Summary. In sum, this chapter has identified three thematic tensions in our monograph. We did not attempt to resolve these tensions, as many of them are the result of the different perspectives across disciplines that are worth recognizing: limited data, definitional struggles, and even methodological constraints within social science. Nonetheless, our probes into these tensions suggested some alternative policy approaches (and also research approaches):

- To address the flow of new members into al-Qaeda or other affiliated terrorist groups, focus resources on the process of recruiting, training, and organization, rather than more narrowly on original motivations of those being recruited.
- Recognize that *disruption* of al-Qaeda and its affiliates is probably valuable in this regard and use associated metrics.

- Disregard debates about whether al-Qaeda is an organization or a network in preference to a systems approach in which both facets appear naturally.
- To better explore ideology and prioritize countering ideological support for terrorism, distinguish between leaders and followers in terrorism studies.
- To better explore ideology, distinguish between minority and majority Muslim populations.

Having explored points of tension, the next section addresses points of agreement or consensus across our chapters.

Points of Agreement

Although our review of the literature exposed disagreement across the social-science community, it also identified important points of consensus. The following sections explore four of these in greater detail.

- In identifying factors that contribute to terrorism, context matters and no specific "one size fits all" recipe exists.
- In exploring terrorism life cycles, factors that contribute to its emergence do not always contribute to its sustenance.
- Popular support is often very important to terrorist organizations, and loss or popular support can be very important in their decline. However, there is no silver bullet here: Terrorist groups weigh popular support against many other factors when making decisions and taking actions.
- Descent does not necessarily mirror ascent.

There are many other points of consensus, of course, as discussed in the other chapters. However, we have highlighted these four points of agreement because they have important policy implications.

Context Matters

In identifying the myriad of factors that contribute to the emergence or persistence of terrorism, it appears clear from these papers that "historically specific circumstances" matter. Although this may seem banal, it is not: At any given time, there tend to be strong inclinations to overgeneralize and apply solutions broadly. This is often counterproductive, as noted consistently by people coming back from theaters of operation such as Iraq and Afghanistan.

Citing Frances Fox Piven (1979), Darcy Noricks first raises the importance of context in her chapter on how and why terrorism rises as a caution against globally systemic explanations. That is, it is too simplistic to understand the rise of al-Qaeda as part of a global wave of religious violence. Al-Qaeda in Iraq has little to do with Hamas and Hizballah. And so on.

Chris Paul, in his chapter on how terrorists generate and maintain support, reaches a similar conclusion. He argues that there is no "one size fits all" formula to popular support for terrorism: Specific circumstances matter. For example, it is simplistic to explain the nature and extent of support for al-Qaeda in the Muslim world as a response to U.S. foreign policy—a typical explanation with posited policies ranging from the Gulf War in the early 1990s to the Palestinian-Israeli conflict. Paul regards context so important as to list it as a recommendation: The first order of business should be to understand a particular context rather than attempting to apply a generic template.

Finally, and perhaps most important, Gaga Gvineria notes in his chapter on how terrorism ends that the success or failure of state counterterrorism measures often depends on context as much or more as the policy itself.

A related dynamic is playing itself out in the global war on terrorism. For example, Operation Enduring Freedom—Philippines (OEF-P) is cited frequently as a success story. But U.S. security officials in Pacific Command and the Philippines caution against using OEF-P as a model for other regions, because its success is very much context-specific. Before OEF-P, the U.S. government had a good relationship with Manila. The Abu Sayyaf Group already had marginalized most of the residents of the southern Philippines by turning to crime. And, perhaps most important, the Moro Islamic Liberation Front (MILF) already had entered into peace negotiations with Manila, creating a relatively permissive environment for U.S. intervention and activities related to the global war on terrorism. Therefore, it is difficult to ascertain how much the success of OEF-P had to do with U.S. counterterrorism policy instruments used and how much to a historical convergence of events.

Unfortunately, this finding—"context matters"—is not particularly useful for policymakers. Key questions then become "To what extent?" and "In what circumstances?" Although more research of this nature could be done, our review and ongoing RAND research on other projects suggest a number of instances in which context is likely to be especially important.

- The utility of democratic reforms may be more useful for majority Muslim populations than for minority populations.
- The utility of economic measures may be more useful to counter terrorism than to counter radicalization.
- The tolerance for terrorism may be higher in societies with a history of violence than for those without.
- Deradicalization efforts using religious arguments may be more effective in rural areas, whereas arguments to encourage disengagement may be more persuasive in urban areas.

Of course, context likely matters across a wide variety of factors and issues, but these four conditions emerged repeatedly in our chapters and thus are worthy of special attention. Additionally, it seems clear that the actual manifestation of terrorism is also correlated to the specific motivational cause for which it is undertaken. Put differently, the evolving dynamic of a terrorist campaign will depend not only on the environmental context in which it takes place but also on the political agenda against which it is justified. This needs to be recognized and taken into account when framing a counterterrorism plan of action.

Finally, the fact that no single root cause exists for terrorism strongly suggests that there is similarly no single "silver bullet" for eliminating outbreaks of such violence. In addition, this finding would seem to lend support to the notion that trying to eradicate the existence of terrorism is an untenable goal. Rather than trying to "defeat" terrorism, the goal should be to reduce outbreaks of extremist violence to manageable levels.

Root Causes Do Not Always Sustain Terrorism

It also appears fairly certain, looking across the chapters, that significant factors in the emergence of terrorism are not necessarily important to how terrorist groups sustain themselves.

For example, Table 10.1 lists factors from Darcy Noricks's chapter on the root causes of terrorism and compares them with those of Christopher Paul's chapter on how terrorist groups generate and maintain support. Those chapters list a great many more factors than those below. We have chosen to replicate only those factors for which substantive agreement exists in the social sciences that they are, indeed, relevant to the emergence and continuation of terrorism.

Table 10.1
Contrasting Lists of Factors

Root Cause	Maintain Support
Perceived illegitimacy of state	Perceived illegitimacy of state
State repression	State repression
Lack of opportunity	Lack of opportunity
Curtailed civil liberties	Humiliation and alienation
Elite disenfranchisement	Resistance as a public good
Ethnic fractionalization	Defense of self or community
	Identification with group
	Kinship ties with group members
	Intimidation by group
	Group provision of services
	Perceived legitimacy of group

In theory, one might expect that factors allowing terrorism to arise in the first place would also be significant in sustaining popular support for terrorism over time, because both in essence relate to how people feel about the application of violence against civilians in opposition to the state. And, indeed, some commonality exists between the lists in the table: perceived illegitimacy of the state, state repression, and lack of opportunity for political freedom. It is interesting to note that all three common factors could be interpreted as a disjuncture between the nation-state and its citizens. That is, these factors have less to do with the circumstances in which people find themselves and more to do with their relationship with the nation-state itself.

These preliminary finding suggest that, no matter the cause of a particular campaign of terrorism, state countermeasures need to conform to and uphold accepted constitutional parameters of the rule of law if the situation is not to be unduly exacerbated. Excessive repression, it would seem, encourages more extreme reactions on the part of the militants and, even if it does not directly encourage terrorist recruitment, it at least serves to delegitimize the state in the eyes of the wider populace. It is essential therefore that all government responses be directed only against the terrorists themselves (rather than extended to elements that are merely supportive or sympathetic to militant designs) and are appropriately calibrated to be proportionate to the situation at hand.

At first glance, it is difficult to see how U.S. policy interventions can affect these factors. Would it not be better for the host nation of terrorist groups to address its own perceived illegitimacy? Stop repressive measures? Provide greater political freedom to its citizens? Nevertheless, the U.S. global war on terrorism has included some measures to affect these three factors. For example, foreign internal defense (FID) programs not only teach basic military tactics but also attempt to steer recipient militaries and nations away from collective punishment and other harsh repressive measures. Even in Iraq, efforts are made by military commanders to "put the Iraqis out front" in an effort to bolster the legitimacy of the Iraqi government in the mind of the population. Of course, it is unclear how much effect these programs can have on the long-term behavior of recipient nations and their military commanders, especially after U.S. forces depart the country. That question alone merits further study. But, until we know the answer to that question, as these programs also have other purposes, it is also worthwhile considering them in the context of how they can be used to reduce state repression and bolster legitimacy in the minds of terrorist supporters.

Beyond FID, the U.S. government has other policy options to affect perceived illegitimacy, state repression, and lack of opportunity for political freedom. Fundamentally, the U.S. government can pressure nation-states to negotiate with terrorists to address these three factors. This approach is somewhat controversial, of course, but it has been taken in the past with the Provisional Irish Republican Army in Northern Ireland and the Palestine Liberation Organization in the West Bank and the Gaza Strip. So it is worth mentioning in this context.

More important, Table 10.1 also illustrates that divergences exist between factors that contribute to the emergence of terrorism and those that sustain popular support. Although factors affecting the relationship between the nation-state and its citizens emerged prominently in the "root causes" of terrorism, support for terrorism appears to hinge at least as much on the relationship between citizens and the terrorist group. That is, six factors listed under "maintain support"—defense of self or community, identification with group, kinship ties, intimidation, provision of services, and perceived legitimacy—all reflect ties between the terrorist group and the population. This finding suggests that popular support for terrorism has as much to do with how individuals view the terrorist group as how they view the nation-state. In this context, the traditional emphasis on counterinsurgency on the population dovetails nicely with our findings.

Popular Support: Very Important But Not a Silver Bullet

Although there is consensus that popular support is often a very important factor in either generating or sustaining terrorism, or in bringing about its decline (seethe chapters by Chris Paul and Gaga Gvineria, for example), there is also consensus that the factor is not a silver bullet. First, not all terrorist groups depend significantly on a support base. Second, terrorist groups balance many factors in deciding on their actions. Brian Jackson lists a variety of such factors that terrorist leaders consider as they choose targets or tactics. These include

- scope of positive reaction the terrorist activity will produce in a relevant population (even if not the broad public)
- amount the terrorist activity will advance group strategy or interests
- scope of positive reaction the terrorist activity will generate inside the group
- acceptability of risks associated with the terrorist activity
- level of resources the group is willing to commit to the terrorist activity
- perceived sufficiency of information to make decision.

One conclusion is that, if terrorists weigh their decisions across these six top-level factors, then counterterrorism activities should be undertaken with as thorough an understanding as possible of how a particular terrorist group values these factors. Another conclusion from Jackson is that there is value in analyzing terrorist group decisionmaking at the operational level, rather than focusing exclusively at the strategic level. One reason is that terrorist leaders at the operational and tactical levels must meet more regularly—virtually or physically—to weigh these factors than do those terrorist leaders who provide generalized, broad guidance to their followers. Thus, there may be more opportunities to observe and to infer how the groups value these factors. Changes in those values are more likely to take place first at the operational level and then filter upward to leaders separated from the day-to-day survival of the group.

Of course, it is difficult to determine the relative weight of one factor over the other for any given terrorist group. That is, just because a terrorist group weighs one factor against others does not mean that the factor is unimportant. For example, some evidence suggests that al-Qaeda leaders have sometimes viewed popular support as the preeminent consideration (at least temporarily): This preeminence is clear in the letters written by al-Qaeda in the mid-1990s and provided by the Combating Terrorism Center at West Point.[17] Yet, at other times, al-Qaeda leaders have forsaken popular support to accomplish their immediate operational objectives. The most well-known examples are the attacks against fellow Muslims by al-Qaeda in Iraq under Abu Mus'ab al-Zarqawi's leadership, including hotels in Amman, Jordan. So, clearly, it would be useful to explore the relative weight of al-Qaeda's priorities further, either through academic social-science research or within the intelligence community. In the meantime, Jackson's and Berrebi's chapters do provide initial insight.

It appears from these chapters to be generally accepted that isolated leaders within terrorist groups tend to be driven more by internal dynamics. In contrast, broadly dispersed movements tend to be driven by strategies or initiatives posted on the Internet by members and through popular support. Similarly, terrorist groups heavily involved in criminal activities tend to be less reliant on populations for their survival. In contrast, terrorist groups who draw on resources from their supporters are similarly more likely to weight popular support heavily. These findings, for the most part, confirm what experts have generally believed about the role that popular support can or cannot play in terrorism. Nevertheless, it is worth underscoring that neither Jackson's nor Berrebi's study diminishes the importance of popular support. They rather should be understood as a plea for policymakers to consider other factors as well.

Descent Does Not Necessarily Mirror Ascent

If root causes do not always sustain terrorism, then it is equally evident from our research that terrorism does not always end the way it begins. Table 10.2 lists factors derived from Noricks's chapter with modes of decline identified by Gaga Gvineria. For example, it is possible to argue that attempts by the nation-state to increase legitimacy, reduce repressive measures, and address the lack of civil liberties could all be construed as success or preliminary success for the terrorist group, depending on the strategic objectives of the terrorists. Similarly, the elite disenfranchisement often articulated in root causes literature tends to have these elites taking a leadership position in the terrorist group. Thus, burnout, poor succession, and the loss of leaders could relate to elite disenfranchisement. Still, even with these parallels, the loss of state support and

Table 10.2
Comparison of Root-Cause Factors and Modes of Decline

Root Cause Factor	Mode of Decline
Curtailed civil liberties	Success or preliminary success
Elite disenfranchisement	Burnout, poor succession, loss of leaders
Ethnic fractionalization	N/A
Illegitimacy of state	Success or preliminary success
State repression	Success or preliminary success
Lack of opportunity	Success or preliminary success
N/A	Loss of popular support
N/A	Counterterrorism activities
N/A	Loss of state support

NOTE: N/A means not applicable.

popular support in particular seems disconnected from the root-cause factors derived from the social-science literatures.

This disconnect makes sense, since root causes do not account for terrorist decisionmaking or the relationship that emerges between the terrorist group and support populations. It can be argued, setting aside the cases of terrorist success, that the most likely situations of decline relate to terrorist group decisionmaking and their relationships to society—more so than relationships to root-cause factors.

Our conclusions, thus far, have related primarily to the terrorist group itself as well as to terrorist leaders. But it is equally evident from our examination of how and why people become terrorists that the descent from terrorism does not necessarily parallel the path to becoming a terrorist actor. Todd Helmus, in his chapter on how and why individuals become terrorists, concludes that socialization processes—for example, group affirmation that the path chosen is correct—are a necessary precondition for individuals to become terrorists. He further argues that two additional factors increase the likelihood that an individual will become a terrorist: an individual's desire for change and perception of duty. One or the other (or both) must be present.

Darcy Noricks similarly addresses patterns of disengagement and deradicalization. It appears (judging significantly by the work of John Horgan) that individuals can disengage from terrorist groups while still maintaining a desire for change or even a perceived duty. Thus, the key to disengagement and deradicalization is in the descent or the process itself, rather than the belief system. And yet, according to Noricks, the descent from terrorism does not necessarily parallel the ascent. Noricks's findings suggest that, in contrast to radicalization processes, individuals who choose to reject terrorism do so in isolation of their "community" of other terrorist group members. Thus, disengagement and deradicalization can be an isolating process rather than a socialization process.

Summary

This chapter has identified four overarching points of agreement that can be derived from our review of the social-science literature on terrorism: (1) context matters; (2) root causes do not necessarily sustain terrorism; (3) popular support is very important, but terrorists also weigh other important factors in decisionmaking; and (4) descent does not necessarily mirror ascent. These points of agreement should not be a surprise to most experts, as the social-science literature for the most part simply confirms what subject-matter experts suspect to be true. Nonetheless, the points have implications for the policymaking and analytic communities.

For example, they suggest that the emphasis on determining root causes might be less interesting, provided that one expects policymakers to be interested in these factors only as a means to counter terrorism. Some of the factors are worth resolving on their own, of course, but it appears that terrorists rely on other factors beyond root causes to sustain themselves.

Many of the factors that distinguish the emergence of terrorism from the sustainment of terrorism are about the relationship between terrorist groups and their support population. Although the U.S. policymaking community has begun to focus on undermining popular support in Iraq, it is worth underscoring the importance of monitoring the relationships between terrorists and these communities, because these relationships can ebb and flow and so opportunities exist to further drive a wedge between them.

Similarly, although we would not want readers to take away from this study that the emphasis on undermining popular support is unnecessary, the findings suggest that other factors are also important for policymakers to address, specifically as they attempt to affect terrorist decisionmaking and deter terrorist attacks. Perhaps the most important of these factors relates to how senior and operational leaders assess priorities and risks, especially if senior and operational leaders weight options differently.

Finally, it is noteworthy that a number of the factors that appear to influence the end of terrorism are beyond the control of policymakers. This finding is discouraging to say the least, but it suggests that the success or failure of counterterrorism might be a matter of being aware and taking advantage of opportunities as they arise within the terrorist organization itself and the wider movement. Although we have attempted to derive some overarching implications from our review of the social-science literature, we also found some specific implications for policy instruments: These more-detailed and specific instruments are discussed in the next section.

Specific Policy Instruments

This section addresses implications for various counterterrorism policy instruments, such as diplomatic, intelligence, military, political, and economic measures. The discussion is by no means exhaustive but rather is illustrative.

Military Instruments

Counterterrorism activities frequently include actions that target terrorists. For example, activities to capture terrorist leaders and operators are often considered direct action. All other activities tend to be referred to generically as "indirect actions," although the line often blurs between direct and indirect. Similarly, although social-science research is thought of as contributing to indirect policies, it has some relevance to direct action as well.

In his chapter on how terrorism ends, Gaga Gvineria addresses the arrest or assassination of key terrorist leaders as a factor in groups' decline. As he suggests, some debate exists as to whether hierarchical or networked groups are more vulnerable to the removal of key leaders. Although he highlights some logical inconsistencies, most experts believe that networked groups are less vulnerable to the removal of key leaders. If one were to consider the world of al-Qaeda as a system, then this naturally would suggest that some elements of the system—the more structured elements—are more vulnerable to direct action than others. In this context, Todd Helmus also provides insight about direct action but from a more cautionary perspective. That is, in his chapter on why some people become terrorists and other do not, Helmus cites revenge and a desire for retribution for personal attacks directed at self or loved ones as key factors for individual radicalization. These findings suggest that a recruitment backlash is possible in response to direct counterterrorism actions. This issue has emerged since September 2001. U.S. counterterrorism activities have emphasized capturing or killing key al-Qaeda leaders in such places as Pakistan, Yemen, Saudi Arabia, Indonesia, and Iraq. In some places, these attacks appear to have degraded al-Qaeda capabilities but less so in others. For example, the capture of Jemaah Islamiyyah leaders in Indonesia apparently has diminished that group's capabilities, but al-Qaeda in the Islamic Maghreb and the Islamic Army of Aden-Abyan have been able to successfully continue their campaigns. In other instances, there has been painful collateral damage and, quite possibly, backlash.

Given recent experiences, it is less useful to debate the utility of direct action unless social science can provide some insight into when it is useful and when it is best avoided. Although more room for study exists, our research suggests that direct action against terrorists is most useful when the benefits (operational degradation) outweigh the backlash measured in future recruits and societal support. Thus, direct action should be considered under the following conditions:

- if removing key leaders or operators will diminish the effectiveness of the group or a specific operational cell
- if key leaders do not have a wide support base and they can be captured or killed without collateral damage
- if operators do not have celebrity-like status in the community and they can be captured without high collateral damage.

Diplomacy, Political and Economic Reform

The degree to which political reform and economic programs can stave off the emergence of terrorism or counter its influence is still very much in debate. Our analysis suggests that the effect of these activities might be different in populations with a majority Muslim population rather than a minority population. Similarly, a number of chapters address the issue of elites in the community and the important contribution that they make to the terrorist organization themselves. Thus, reforms that address the concerns of these elites might similarly collapse the leadership structure of a terrorist organization. But neither of these points should be taken to the point of policy prescriptions, because the social sciences tend to be cautionary rather than confirmatory on these specific types of instruments. However, this caution is in part a response to the shifting nature of terrorism and the paucity of data on the subject. Nonetheless, the following could be useful in the consideration of political or economic reform to counter terrorism:

- if reforms bring into the system disenfranchised elites, who similarly form the core of terrorist decisionmakers
- if reforms are directed toward countries with majority Muslim populations, or if minority communities receive the benefits of reform
- if the promise of reform "rewards" outweighs the rewards of terrorism and does not raise popular expectations higher than should reasonably be expected.

Additionally, the role and efficacy of diplomatic instruments appear to be the least explored in the social sciences. Michael Egner has taken a step toward meeting this gap with his chapter on strategic communications. But other diplomatic instruments, such as visiting scholars programs run by the State Department or the funding of media outlets, or even diplomatic pressure on active or passive state sponsors, might prove worthy of further exploration. In this context, although disengagement and counterradicalization are separate components in the wider "war of ideas," they can (and should) usefully complement one another. As Noricks discusses in her chapter, the experience of countries in Southeast Asia and the Middle East, for example, shows that former terrorists who have been successfully rehabilitated act as more-effective agents for counterradicalization (than religious moderates or government-sponsored spokesmen), especially charismatic individuals who enjoyed widespread influence in their former movements. For example, Indonesia has managed to rehabilitate 30 Afghan-trained members of Jemaah Islamiyyah, who are now at the forefront of Jakarta's counterradicalization program and who have enjoyed considerable success in dampening militant propaganda in their home villages and mosques. That is, we acknowledge that these programs are ongoing as part of U.S. policy currently, but the social science literature does not address the likely utility of such programs or inform how they might best be used. This lack could be simply another data problem; nonetheless, it seems another field of potential utility for a greater contribution of the social sciences.

Intelligence Activities

Finally, intelligence collection in the global war on terrorism informs a wide range of tactical, operational, and strategic analysis. It is easy to state that intelligence collection across this spectrum needs to be improved, but identifying how it can be improved and what specifically needs to change is more problematic. Our findings provide some insight into the application of intelligence to help better inform our understanding of al-Qaeda and other terrorist threats.

For example, Brian Jackson, in his chapter on organizational decisionmaking by terrorist groups, addresses decision points for terrorist leaders as they consider the adoption of violence and implementation of terrorist measures. He argues that if we better understand how terrorists make these decisions, we can better anticipate how they might respond to counterterrorism measures. Similarly, in his appendix on measures of effectiveness, Ben Bahney argues that it is important to continue to monitor characteristics of terrorist groups and support populations so that U.S. counterterrorism activities can adjust as necessary.

Indeed, terrorist actions do not take place in a vacuum—they are invariably influenced by the reactions they generate in target audiences. This "feedback loop" is critical to a comprehensive understanding of terrorist decisionmaking (especially in an era of mass and largely instantaneous communications) and is one that the state absolutely needs to take account of when formulating its counterterrorism responses. A failure to do so risks the danger of allowing terrorists to dictate the nature, scope, and extent of a government's mitigation strategies (which, by definition, eliminates the possibility for instituting concerted, proactive approaches for ameliorating terrorist violence).

The dynamics of al-Qaeda make identifying and monitoring these characteristics even more important. Over the past several years, we have witnessed fractures in the relationship between al-Qaeda senior leaders and local insurgent leaders, as well as between well-known Salafist ideologues and al-Qaeda senior leaders. Terrorist experts continue to argue about whether or not these fractures will affect the effectiveness and direction of al-Qaeda, either from a strategic point of view or even at the level of grassroots recruitment.

In this context, key intelligence questions can be derived from our findings, as follows:

- How do al-Qaeda senior leaders perceive the threat from both the U.S. military and local leaders to themselves?
- How do al-Qaeda senior leaders consider the provision of support to local leaders and operational cells?
- What is the distribution of that support and how has this changed since 2001, 2003, and 2007?
- What risks do al-Qaeda leaders perceive in their support for al-Qaeda in Iraq and the Taliban?
- Who is al-Qaeda's primary audience and to what extent do senior leaders care about secondary audiences?

- What are al-Qaeda's priorities and how do the priorities in public documents differ from clandestine priorities?

In many ways, the intelligence community likely has this information available. Thus, our implications have less to do with the collection of specific intelligence as its assessment and asking the right questions to reach that assessment.

Finally, a couple cautions for the intelligence community also emerge from our analysis. First, inaction should not automatically be seen as a sign of terrorist weakness, as it may merely reflect a cost/benefit calculation on the part of a terrorist organization. That is, under certain circumstances, desisting from an attack could imaginably be more instrumental to an organization's aims than continuing to engage in an active campaign of terrorist violence. And, second, governments should be wary of personalizing terrorist groups in the guise of single leaders or a narrow cohort of operational commanders. One negative result of this personalization is that it could cause planners and publics to place undue confidence in the utility of decapitation as an effective instrument of counterterrorism.

Conclusion

In many ways, this study confirms what subject-matter experts often have emphasized as important in counterterrorism. In many respects, terrorists are rational actors. Undermining popular support is an effective tool to affect terrorist decisionmaking. Context has a significant influence both within a terrorist organization and among terrorist recruits and supporters. Therefore, some of the most significant levers discussed throughout this chapter should not be a surprise to experts, as they are hardly revolutionary. The value comes not so much in the repetition but rather in the way that the chapters have attempted to pull together existing knowledge and demonstrate how the levers likely relate to each other.

Nevertheless, this study also poses some provoking new ideas that have yet to gain significant attention in the policymaking community.

Recruiters themselves might be a significant vulnerability in terrorist organizations. Countering root causes might not be the best way to undermine popular support. Intelligence assessments that examine internal decisionmaking among al-Qaeda senior leaders and operational leaders might provide insight into future priorities as well as new vulnerabilities. Some of these statements could be hypotheses for further studies. But, throughout this chapter, we have also have attempted to provide ways for policymakers to contextualize what we suspect and what social science knows about the terrorism phenomenon.

Finally, a number of questions are still left unanswered. Some of the most important questions relate to al-Qaeda decisionmaking, individual motivations, and even how policymakers might truly implement the concept of a "systems" approach. Although U.S. policymakers clearly must make decisions in the absence of social science research, a role exists for social scientists to help improve strategy, to help operationalize that strategy, and even provide a framework to think through second- and third-order effects. This requires that U.S. policymakers ask the questions and it also requires willingness on the part of social scientists to work with U.S. government sponsors, using classified information and providing concrete recommendations.

Bibliography

Abu-Rabi, Ibrahim M., *Intellectual Origins of Islamic Resurgence in the Modern Arab World,* Albany, N.Y.: State University of New York, 1996.

Abuza, Zachary, *Crucible of Terror: Militant Islam in Southeast Asia,* Boulder, Colo.: Lynne Rienner, 2003.

Arnson, Cynthia, ed., *Comparative Peace Processes in Latin America,* Washington, D.C.: Woodrow Wilson Center Press, 1999.

Bergen, Peter, and Paul Cruickshank, "Special Report: Is Al-Qa'ida in Pieces?" *The Independent,* June 22, 2008.

Berman, Eli, and David D. Laitin, "Religion, Terrorism and Public Goods: Testing the Club Model," NBER Working Paper 13725, 2008.

Cragin, Kim, "The Early History of al-Qa'ida," *The Historical Journal,* December 2008.

Davis, Paul K., and Brian Michael Jenkins, *Deterrence & Influence in Counterterrorism: A Component in the War on Al Qaeda,* Santa Monica, Calif.: RAND Corporation, 2002. As of January 8, 2009:
http://www.rand.org/pubs/monograph_reports/MR1619/

Felter, Joseph, et al., *Harmony and Disharmony: Exploiting al-Qaeda's Organizational Vulnerabilities,* New York: West Point Counter Terrorism Center, 2006.

Felter, Joseph, and Brian Fishman, *Al-Qa'ida's Foreign Fighters: A First Look at the Sinjar Records,* New York: West Point Counterterrorism Center, 2007.

Gerges, Fawaz A., *Journey of the Jihadist: Inside Muslim Militancy,* Orlando, Fla.: Harvest Books, 2007.

Hafez, Mohammed M., *Why Muslims Rebel: Repression and Resistance in the Islamic World,* Boulder, Colo.: Lynne Rienner, 2003.

Hoffman, Bruce, "The Myth of Grass-Roots Terrorism: Why Osama bin Laden Still Matters," *Foreign Affairs,* Vol. 87, No. 3, May/June 2008.

Kilcullen, David, "Countering Global Insurgency," *Small Wars Journal,* November 2004. As of January 8, 2009:
http://www.smallwarsjournal.com/documents/kilcullen.pdf

———, "Countering Global Insurgency," *The Journal of Strategic Studies,* Vol. 28, No. 4, August 2005, pp. 597–617.

Kurzman, Charles, ed., *Liberal Islam: A Sourcebook,* Oxford: Oxford University Press, 1998.

McClintock, Cynthia, *Revolutionary Movements in Latin America: El Salvador's FMLN and Peru's Shining Path,* Washington, D.C.: USIP Press, 1998.

National Academy of Sciences, *Making the Nation Safer: The Role of Science and Technology in Countering Terrorism,* Washington, D.C.: National Academies Press, 2002.

Rashid, Ahmed, *Jihad: The Rise of Militant Islam in Central Asia,* New York: Yale University Press, 2002.

Sageman, Marc, *Understanding Terror Networks,* Philadelphia, Pa.: University of Pennsylvania Press, 2004.

———, *Leaderless Jihad: Terror Networks in the Twenty-First Century,* Philadelphia, Pa.: University of Pennsylvania Press, 2008.

Sciolino, Elaine, and Eric Schmidt, "A Not Very Private Feud over Terrorism," *The New York Times,* June 8, 2008.

Strong, Simon, *Shining Path: Terror and Revolution in Peru,* New York: Random House, 1992.

U.S. Army and Special Operations Research Office, "The US Army's Limited-War Mission and Social Science Research," Washington, D.C.: American University.

Wiktorowitz, Quintan, ed., *Islamic Activism: A Social Movement Theory Approach*, Bloomington, Ind.: Indiana University Press, 2003.

Endnotes

[1] Perhaps the most striking example is the development of Human Terrain Teams and their use in Iraq and Afghanistan. The Minerva project, which provides funding to universities for academic research on violent phenomena, similarly, is indicative of this trend.

[2] For example, Todd Helmus's paper on individual-level radicalization refers to social-science research that has demonstrated that terrorists are not psychopaths, as believed by most terrorist experts. In contrast, other social-science research, discussed in Claude Berrebi's paper, suggests that poverty might not have quite the effect always believed by terrorism experts.

[3] Notably, in 1962, the U.S. Army Special Operations Research Office commissioned a symposium entitled, "The US Army's Limited-War Mission and Social Science Research." So the potential role that social science can play in counterterrorism and insurgency has always existed, even if it was forgotten for a time.

[4] Wiktorowitz (2003) and Kilcullen (2004).

[5] For more research by Eli Berman, see Berman and Laitin (2008).

[6] For more information on these programs, see the paper on disengagement and deradicalization by Darcy Noricks.

[7] For additional information on the recruitment of al-Qaeda foreign fighters, see Felter and Fishman (2007).

[8] This suggestion amounts to saying that the portfolio of efforts should be more balanced rather than being tilted almost exclusively to the top-priority matters. Such imbalance is a notorious problem in government and other organizations.

[9] For more information on this debate, see Hoffman (2008) and Sciolino and Schmidt (2008).

[10] Kilcullen (2004).

[11] Admittedly, al-Qaeda networks have subcomponents that are geographically linked, but the point is to address these higher-level components globally because of

the reality of networking and the feasibility of relatively easy substitution effects if only some nodes are attacked.

[12] See, for example, Strong (1992), McClintock (1998), and Arnson (1999).

[13] See, for example, Rashid (2002), Gerges (2007), Hafez (2003), Abuza (2003), and Wiktorowitz (2003).

[14] Abu-Rab (1996) and Kurzman (1998).

[15] The matter is, however, complex. For example, Berrebi notes for the Palestinian context that attacks by religiously motivated groups are more effective in terms of lethality and damage.

[16] Bergen and Cruickshank (2008).

[17] Felter et al. (2006).

CHAPTER ELEVEN

Representing Social-Science Knowledge Analytically

Paul K. Davis

Introduction

This paper asks how we can represent social-science knowledge about terrorism analytically, and then communicate the results effectively across interdisciplinary boundaries.[1] Doing so is nontrivial because of the diversity of backgrounds and styles among those who need to communicate.

I present ideas for taking the first steps to do so, with an emphasis on causal system models. The intent here is to express approximate knowledge in understandable terms independent of any particular programming language, mathematical formalism, or disciplinary background.[2] After introducing the ideas and pointing to relevant multidisciplinary literature, I illustrate the methods' integrative value by drawing on companion papers that employed some of the suggested methods. I end by suggesting next steps in using the methods to sharpen the analytical base of social science for counterterrorism.[3]

Contrasting Approaches to Analytic Knowledge Representation

To better understand the tack taken in the paper, it is useful first to contrast two very different approaches to scientific inquiry generally; both are "analytic" in using structured reasoning, models, equations, and charts. These approaches are the "data-driven" (sometimes called "atheoretical" or "empirical") and the "theory-driven" approaches, as summarized in Table 11.1. The titles are misleading in that both depend on empirical information, but they use it differently.[4]

Table 11.1
Contrasting Approaches

Data-Driven	Theory-Driven
Specialization on one or a few factors	System approach
Focus on empirical data and theory based on readily measurable factors	Focus on factors underlying the phenomena, whether or not easily measured
Statistical modeling	Causal modeling
Discussion about correlations	Causal explanations
Data-driven inquiry ("let the data speak")	Theory-driven inquiry, with data used to test and calibrate theories

The Data-Driven Approach. Much of the quantitative counter-terrorism social-science literature is of the data-driven variety (left column). Practitioners commonly specialize, focusing on establishing empirically the significance of one or a few factors (that is, variables). Different papers may address, democratization, religion, and nationalism, for example. As discussed by Barbara Geddes (Geddes, 2003), a result is that people view issues through different paradigm-related lenses, without seeing the whole.

Data-driven practitioners analyze available data statistically. The data may not be ideal. They may be highly aggregated, for example, as with counting the number of incidents of all kinds of terrorism by country in a year. A given type of data may be a mere proxy for the factor of ultimate interest. As another example, survey questions may suggest the prevalence of religious extremism, but—for obvious reasons—the questions will seldom be as concrete as "Are you a member of a religiously motivated terrorist group?" The data sample may also be biased, as with interviews of captured terrorists, who may be relatively incompetent. And, of course, the data may be invalid: People may lie when responding to surveys; and some governments lie about many things, including the incidence rate of terrorism.

Such challenges are described in texts and good social scientists are skilled at ferreting out information and adjusting cleverly for data imperfections. The data are then analyzed statistically to evaluate hypotheses about the factors of interest in a given study. An illustrative result might describe the degree to which the incidence rate of terrorism correlates with those factors.

Philosophically, the data-driven approach is one of "letting the data speak." Practitioners are often skeptical about "theory," even using the term derisively to mean "mere speculation." This attitude is understandable, since so many widely held CT beliefs have proven empirically to be wrong (see Sageman, 2008, Chapter 3; Berrebi, 2009), such as the belief that jihadis must be poor, uneducated, unfamiliar with democracy, or mentally imbalanced.

Much data-driven work says little about causality* and the results of a given study may not apply outside the realm for which data are available, such as to a different country, subgroup, or historical context. Data-driven researchers may refer to "explanation" when describing results, but the "explanation" often is to be understood in a purely statistical sense, as in "70 percent of the variance in data is predicted by this regression equation." That would not constitute "explanation" to a layman.

The Theory-Driven Approach. The theory-driven approach is common in mature areas of physical sciences and engineering (right side of Table 11.1).[5] Here the term "theory" means a set of principles that comprehensively explains much or all of the phenomena at issue. If several factors matter, then theory will include all of them as a matter of course. Whether the term is used explicitly or not, a "system perspective" is often taken so as to address the system or phenomenon holistically.

It is also usual in the physical sciences to focus on the "real" variables, even if proxy variables are sometimes used. For example, a physician may measure a patient's temperature (a proxy) but will then reason in terms of underlying causes, such as infection and a weakened immune system.

---

* An exception is that economists have methods for inferring causality in data-rich circumstances. Some involve exploiting "natural experiments" (Angrist and Kreuger, 1999). Others are more complex (Angrist and Pischke, 2009). Note, however, that economists use the term "causal variable" somewhat differently than do researchers in other disciplines.

In the theory-driven approach, causal explanation guides reasoning about the likely effects of interventions, as in "We understand this system. Its behavior is determined under all circumstances of interest by three variables A, B, and C. We can reduce the system's output D by making changes in A, B, and/or C in any of several combinations. The question is how to do so conveniently or economically." This is different from saying "Well, based on past data, it seems that if all else is unchanged (that is, ceteris paribus), then increasing A will decrease D."

Whenever possible, the theory-driven approach is based on empirical data, but data are used to test a theory, to calibrate its parameters if the theory seems to work, and to suggest enhancements if the theory is falsified. Where the theory comes from is one of the mysteries of creative science. It comes from some combination of observation, reasoning, intuition, leaps of imagination, and guesswork. As one philosopher of science noted with only some exaggeration, "Anything goes" (Feyerabend, 1975, pp. 27–28). However, such theory is not just ad hoc speculation. It reflects an effort to think about all the factors that matter, although often simplifying with idealizations. Theory becomes increasingly complex realism and a wider range of circumstances are considered.

A great deal of Department of Defense (DoD) analysis is theory-driven. The North Atlantic Treaty Organization (NATO) and the Warsaw Pact never went to war, but they amassed and operated sophisticated militaries based on past experience, war gaming, modeling, and analysis. Interestingly, given that so much is made currently about "nonkinetic" considerations, much of military science has always been "soft" and people-related (for example, the quality of an army and its generals).

The Need for a Mix of Approaches

Many researchers viscerally prefer one or the other of the approaches described above, but both are needed because of their different virtues (Table 11.2). The data-driven approach (left column) provides hard information and cautions. It can disconfirm false theories or claims. It often demonstrates that effects are dominated by only one or a few

Table 11.2
Relative Strengths of Theory-Informed and Atheoretical Empirical Work

Issue	Data-Driven (Atheoretical) Work	Theory and Theory-Informed Empirical Work
Cautions	••••	
Empirical disconfirmation of theory or hypotheses	••••	
Explaining phenomena more simply than with existing theory	••••	
Making predictions where no respectable theory exists	••••	
Finding evidence that a previously unappreciated factor matters	••••	••
Disconfirmation of theory or hypotheses by drawing on deeper, settled theory		••••
Predicting possible results for new situations	••	••••
Tightening and calibrating a model to make it more useful		••••
Clarifying underlying principles and mechanisms		••••
Laying the basis for causal reasoning in policy making		••••

NOTE: Strength increases with the number of bullets.

factors even though existing theory suggests great complexity. The best examples of data-driven work permit predictions within the domain for which data exist. These predictions may be the best available because theory does not exist or the inputs to theory are unknown. Empirical work can also flag the importance of a previously unappreciated factor, motivating improvements in theory.

The theory-driven approach is best for the lower rows of Table 11.2. It can disconfirm by using logic and established principles. It may be able to make predictions well beyond the realm for which data exist (a crucial consideration when relevant data are lacking), to sharpen explanatory models, to clarify principles and mechanisms, and to support causal, explanatory reasoning of the type needed by decisionmakers.

The general conclusion is that a *combination* is needed of (1) causal system theory where it is feasible, (2) empirical work informed by that theory, and (3) more purely empirical work to inform, temper, and motivate. More specifically, however, my conclusion, based on RAND's experience reviewing the social-science CT literature, is that

- On the one hand, much more empirical work is needed, especially if the United States will be involved for years in CT/COIN [counterinsurgency] activities. Further, existing data should be made available to those who can analyze them well.*
- On the other hand, current scholarly literature in quantitative social science (as distinct from history, anthropology, and so on) is imbalanced toward the data-driven approach. *More theory-driven work is badly needed to improve communications and increase the quality and coherence of related social science.*

Structure of the Remainder of the Paper

Against this background, the next section of this paper sketches ideas on how to represent social-science knowledge in an approach aspiring to good theory. As suggested in Figure 11.1, the themes include (1) causal system modeling with qualitative variables, multiresolution "factor trees," graphical and tabular depiction of interactions, and treatment of "random" effects; (2) interactive and iterative modeling for knowledge discovery and validation; and (3) a style of work called exploratory analysis. The goal of analysis should be to identify likelihoods and trends, and combinations of factors to be influenced, but without the expectation of reliable predictions. After describing the primary concepts of these themes, I sketch their use and suggest a way ahead for CT research of this character.

---

* Relevant data are often restricted but could be sanitized and released. The value of doing so is illustrated by what has been learned from extensive data available about the long-running Israeli-Palestinian conflict (Boaz, 2005; Berrebi, 2009; Berman and Laitin, forthcoming).

Figure 11.1
Key Elements of an Approach to Representing Knowledge

Causal system modeling with

   – qualitative variables

   – multiresolution "factor trees" and influence
     diagrams

   – graphical/tabular depiction of interactions
     and processes

   – random processes resulting from hidden variables

Using approximations
to see forests from
trees

Interactive exploratory modeling for knowledge
discovery and validation

Exploratory analysis amidst uncertainty

Identifying likelihoods
and trends, not
certainties

RAND *MG849-11.1*

Representing Knowledge with Causal System Modeling

What Is Feasible?

Before proceeding, it should be noted that we do not know how far causal system modeling can be taken in the CT realm. The sheer complexity of the phenomena will limit what can be accomplished. Some might see an analogy to economics, where—after years of pursuing computer-modeling approaches based on causal theory—leading practitioners have largely embraced the empirical methods of econometrics (see Angrist and Pischke, 2009, for example). The prevailing view is that these have proven more fruitful, given sufficient data.

    As discussed above, my own view is that empirical approaches are valuable and even essential but that CT research suffers from a lack of sufficient theory-driven analytic work. This paper is largely about ways to do better in that regard. As a minimum, good causal system modeling should prove valuable for structuring knowledge and improving communication. I expect it to be useful as well for qualitative reasoning and option assessment. High-confidence quantitative prediction is another matter altogether and may be beyond the pale, as discussed below. I note further that econometricians benefit enormously from a rich shared background in economic theory that allows them to communicate easily about, for example, the equilibrium theory of supply and demand (David Ricardo), input-output relationships (Wassily Leontief), and the role of product differentiation in global trade (Paul Krugman). This assists their empirical research. CT research does not yet have such a foundation of core theory. We should also recognize how frequently imperfect theory is very useful. Decisionmakers in all walks of life use simple devices such as doctrine manuals to remind them of the many things to which they need to attend. Those reflect a de facto theory. The recent counterinsurgerncy field manual (Nagl et al., 2007) is a good example.

Principles and Aspirations

A first principle of the suggested theory-oriented approach is to reject the search for simplistic conclusions such as "It's all about X" (for example, it's all about poverty, repression, or radical Islam). In fact, *numerous* factors contribute to CT phenomena. A second principle is to strive for understandability. Decisionmakers need analysis based on causal models allowing them to understand and reason about the phenomena in question, including the effects of multiple factors, potential interventions, changes of circumstance, or changes in the system itself. Such analysis should lend itself to encapsulation in a "story." It should integrate separate streams of knowledge. This is recognized in the pleas of national governments for what is variously referred to as a "comprehensive approach," a "DIME/PMESII approach," or a "whole-of-government approach."[6]

Relationships to Past Work

How does one go about describing complicated systems? Methods certainly exist. System engineers have well-developed methods for dealing with exceedingly complicated projects (Sage and Cuppan, 2001; Haimes, 1998). Systems dynamics has been used in a wide range of policy applications (Forrester, 1963, 1969; Sterman, 2000). My own work on strategic planning and analysis describes ways to identify the critical components of a system and to then ensure portfolio-style investments in all of them, rather than in just those currently most fashionable (Davis, 2002a; Davis, Shaver, and Beck, 2008).

Some crucial distinctions should be noted, however. In domains such as classic system engineering, the components can be comprehensively and precisely defined; their interactions can be specified so that work can proceed in parallel on the components. That is, integration can be accomplished by modular decomposition and careful specification of interfaces (Baldwin, 2000). In problem domains such as counterterrorism, and even in engineering when dealing with "systems of systems" that include humans, matters are not so straightforward (Sage and Cuppan, 2001).[7] The natural modules may not all be recognized, may change with context, and may have subtle interactions. Model composition is much more difficult than normal software engineering (Davis and Anderson, 2003). The system may be dynamic, even "organic." Technically, there is a need for "variable-structure modeling" with different structures and decompositions at different times.[8] This may be unsettling to those with a desire for neatness and stability, but it comes more naturally to those familiar with the realities of human behavior, networking, and complex adaptive systems generally.[9]

It follows that there are many lessons to be drawn from past work but that describing social-science knowledge poses special challenges.

Features of an Approach to Knowledge Representation

Some key features of the approach I will sketch are (1) qualitative modeling, (2) relating variables (factors) to each other, (3) depicting the combining logic of multifactor interactions, including feedbacks and nonmonotonicties, (4) dealing with uncertainty (including random effects), (5) and dealing with dynamics, such as learning and adaptation. Taken together, these items represent a significant, albeit approximate, first step in representing knowledge. Let us address them in turn.

Qualitative Modeling. The best way to express social-science knowledge is often with qualitative modeling—not as a poor second choice tolerated by necessity but because qualitative factors are often natural. This means accepting soft and squishy variables; to ignore them would be as foolish as for a military commander to ignore the morale of his troops—acting as though it had no effect.[10] Much of the CT subject area is about comparably soft factors, as is evident in the social science coming from historians, anthropologists, and psychologists.

Qualitative variables may be given a degree of rigor—for example, by assigning them discrete values such as in the set Low, Marginal, High and by then describing the circumstances in which the different values apply. To avoid circularity, the distinctions drawn must be observable in principle, even if observations are rare (as when intelligence uncovers secret documents). Over time, the values of such qualitative variables can be more precisely defined.

In the spirit of causal system modeling, we should focus on the purest elements of the phenomena in question, rather than thinking in terms of dubious surrogate factors (that is, proxies) that may be more easily measured. For example, a region's level of democratization is not well captured by data on whether elections occur. We cannot avoid using surrogate measures if we are to test our knowledge empirically, but we can defer doing so as long as possible so as to focus on the deeper concepts.[11]

Relating Factors with Factor Trees and Influence Diagrams. Many factors, mostly qualitative, affect CT phenomena. How can they and their relationships be represented comprehensibly? One mechanism is what may be called "factor trees." Figure 11.2 illustrates the idea. If a subject's experts identify an alphabet soup of relevant factors, say A, B, . . . Z, then the hope is to identify relationships among those factors so that the overall causal structure can be represented as shown: with only a few independent high-level factors mattering, but with those factors dependent on lower-level factors. In the example of Figure 11.2, A, K, and P are independent from a structural perspective (the values of these factors may still be correlated). In contrast, R has some effect on P as well as on K (the dashed line indicates a weaker effect). Similarly, N has some effect on both D and R. The result is that the structure is a "nearly" hierarchical decomposition (weak interactions exist among branches indicated with dashed lines). Such simple depictions can sometimes make relative order out of chaos.

The factor-tree method draws on past work. Multiresolution modeling (Davis and Bigelow, 1998) is a kind of systematic approximate

Figure 11.2
Rearranging Factors in a Nearly Hierarchical Decomposition

Relevant factors   A, B, C, D, E, F, G, H, I, J, K , L, M, N, O, P, Q, R, S, T, U, V, W, X, Y, Z

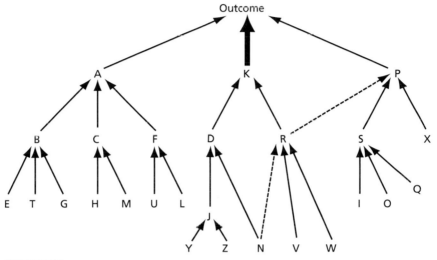

abstraction useful in both hard and soft applications. In the social-science domain, it has been successfully used to build behavioral models of adversaries, for example (Davis, 2002b; National Research Council, 1996). The ubiquity of "nearly hierarchical decomposition" is described in a classic essay by the late Nobelist Herbert Simon (Simon, 1978). Graphical depictions such as Figure 11.2 are variants of the *influence diagrams* of Jay Forrester's system dynamics in which if two variables are connected by an arrow, it means that an increase in the first variable (at the arrow's tail) will tend to increase the second variable (at the arrow's head). A negative sign on top of an arrow, as between B and A, indicates that the effect is reversed—that an increase in the first variable tends to decrease the second. The variables, or factors, are usually thought of as having "levels" (for example, the degree of a population's discontent or the degree of an individual's religious ardor).[12] As I use diagrams in this paper, they may include dashed lines to indicate a weak effect, as in Figure 11.2, or thicker lines to indicate a stronger effect (e.g., factor K's effects in this figure). This modest extension of influence-diagram notation has been proven in research, collaboration, and discussion with policymakers.[13]

Representing Combining Logic of Multifactor Interactions. Diagrams such as Figure 11.1 describe what factors operate together and affect others, but they do not say *how*. Do the factors have independent effects or do they interact? Are the effects direct or indirect? Is there some order in which they must arise?

One simple device for describing interactions at an elementary level is a combining logic diagram (CLD) illustrated in Figure 11.3. A CLD adds some "and/or" notation to an influence diagram, implicitly assuming binary values such as yes or no (or true and false). The figure indicates that A and B are substitutable for each other but that factor C has independent importance. According to Figure 11.3, a positive outcome (yes) occurs if either A or B is yes (that is, true), *and* C is also yes. That is, C is a necessary condition, whereas A and B are alternative conditions.

The assumption of binary values is crude but useful in conveying approximate knowledge.[14] Fine-tuning can be deferred to model builders, who need more precision. Such a cavalier attitude would be

Figure 11.3
A Combining Logic Diagram

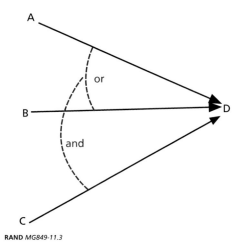

RAND *MG849-11.3*

inappropriate in a more exact science, but the baseline of CT/COIN social-science theory is arguably confusion calling out for sense-making, even if approximate.

Nonmonotonicities. A second challenge is to describe simply how an effect changes as a causal variable continues to increase. As in Figure 11.3, assume that an effect, D, depends on A, B, and C, but hold B and C constant. Figure 11.4 contrasts three ways in which an effect, D, might then vary with A. Case 1 illustrates monotonicity; cases 2 and 3 illustrate nonmonotonicity. Mathematically, a monotonic function's derivative never changes sign.

In principle, the functional form could be arbitrarily complex, but much social-science knowledge can be captured by merely relating to one of these types of relationship. Something like Case 3 appears in the CT literature when researchers observe an inverted "U" relationship, as when the incident rate of terrorism apparently increases with democratization but eventually decreases again.[15]

Figure 11.4
Monotonic and Nonmonotonic Effects (for constant values of B and C)

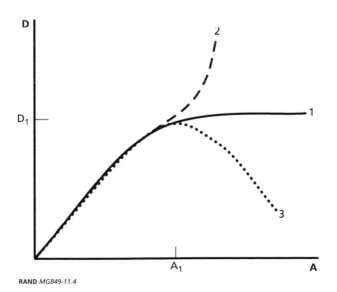

A great deal of effort could go into finding "best" functions to represent effects observed empirically. Social-science knowledge is often quite rough, however, especially when based on statistical analysis of data averaging over a range of contexts. It may be necessary to allow for the reversal of sign (that is, in an influence diagram, the arrow from A to C would have to bear a +/- because the sign of the effect is not constant). That may happen as a result of random effects discussed below.

Feedbacks. A special case of great interest involves "feedback," where increasing a factor causes an effect, which then affects the process giving rise to the effect. If increased fervor increases the likelihood of terrorist attacks, and if increased terrorist attacks in turn arouse even more fervor, then a "positive feedback effect" exists. However, at some point, the attacks may go too far, kill the wrong people, or cause severe retribution, in which case fervor for attacks will decrease, as will the attacks themselves. This would be an instance of negative feedback.

Feedback effects are easily denoted in an influence diagram as in Figure 11.5. Feedbacks have long been a core element of engineering's control theory and system dynamics (Forrester, 1963, 1969; Sterman, 2000). Again, the notation of Figure 11.5 means that although increasing A and B leads to an increase in C, there is a feedback effect: As C increases, A is reduced (indicated by the negative sign), thereby decreasing the subsequent effect on C.

Feedbacks are ubiquitous in natural systems. If all feedback loops were shown in social-science influence diagrams, they might clutter the diagram hopelessly and render the very concept of "causality" troublesome. I suggest that two simplifications save the day:

- Many feedback effects are relatively small over the time scale of interest.
- Many feedback effects need not be addressed explicitly because they occur on very short, even "instantaneous," time scales until the combined effects of the several independent variables "settle down."

**Figure 11.5
A Simple Feedback Diagram
with Feedback**

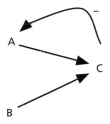

RAND *MG849-11.5*

The second item addresses the common perception that social phenomena are too complicated to deal with analytically. Imagine an influence diagram describing the social outcome of an encounter between two people. Such factors as physical appearance, scent, motions, verbal comments, facial expressions, and handshake strength are among the many that could interact. There could also be feedbacks: What appeared to be a grimace could cause the other person to back off, stutter, or weaken the handshake. Or, conversely, an expression of joy might lead to a rapidly warmer encounter. However, suppose that these occurred in combination: a grimace (due to being interrupted), followed by recognition, followed by joy, followed by embarrassment at being found excessively dressed, followed by. . . . The instantaneous dynamics might be quite complex, but, after a period of seconds, the outcome would be determined. Figure 11.6, then, suggests that such complex feedbacks occurring on a fine time scale may be ignorable on a coarser time scale.

Thresholds and Ceilings. Representing threshold and ceiling effects is important. Human beings may slough off risks, for example, until they reach some level of apparent significance (Slovic, 2000; Shoemaker, 1980). This may help explain historical incidents of "unreasonable" risk-taking, such as that of Saddam Hussein (Davis and Arquilla, 1991; National Research Council, 1996). At the same time, many effects have a saturation point. Significantly:

Figure 11.6
Feedbacks and Time Scale

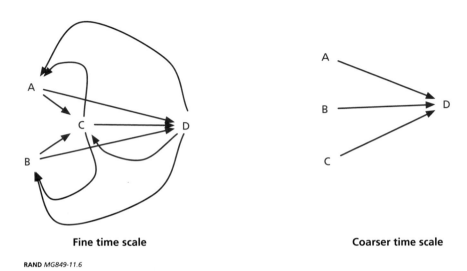

- There is special value in treating thresholds and ceilings in the modeling of counterterrorism because they may play a role in *theories of victory.*

To illustrate, it is probably unnecessary to reduce the materiel and human-capability components of a terrorist group to zero before effectiveness drops to negligible proportions: We may reasonably hope to see critical-mass effects. Unless such matters are represented analytically, we could underestimate the potential value of such tactics as disrupting an organization by targeting its leaders or forcing it to change operational locations and processes frequently.

It is straightforward to represent the nonlinear phenomena of thresholds and ceilings. The concepts can be conveyed with simple diagrams, such as Figure 11.7. The first curve (solid curve) has a strict threshold, followed by a linear increase of D with A, followed by a constant maximum value (ceiling). The alternative (dashed curve) is generated by using a standard mathematical function called an S-curve (or sigmoid). The shape can be tuned by adjusting parameters in the function if precision is needed.

Figure 11.7
Thresholds and Ceilings

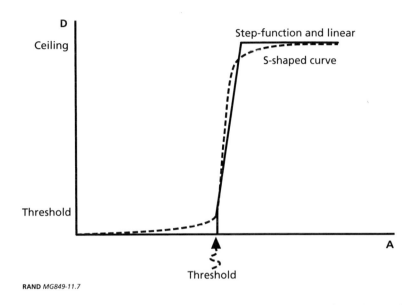

RAND *MG849-11.7*

Randomness and the Need for Humility. Even if we have done a good job identifying factors and combining relationships, social phenomena will often yield surprises. This may be the consequence of "hidden variables" (the problem of omitted-variable bias in econometrics), which might or might not be knowable in advance. Such variables include the health and mood of protagonists, perceptions about exogenous events in the world, and the order of events. Some events are truly random, however, and humans display some inherent inconsistency—as studied by social psychologists. There is something amusing in the way critics of the intelligence community expect a higher degree of omniscience about events in faraway lands than exists in our own nation when we try to project the winner of an upcoming election months in advance. History should also be sobering. Probably no crisis has been better studied than the Cuban Missile Crisis of 1963, but research is still uncovering material that demonstrates how a number of microscopic events could easily have changed the outcome, perhaps unleashing a war that no one wanted (Dobbs, 2008).

How should randomness be handled in CT research? The first principle is humility: We should aspire to estimating the *odds of* being correct rather than getting predictions "right." Analytically, we can add explicit random variables just to remind us constantly of uncertainty, which can work either positively or negatively (Davis, Bankes, and Egner, 2007). Alternative important approaches, not discussed in this paper, involve Bayesian nets or influence nets—methods for which have evolved substantially over the last decade or so.[16]

Figure 11.8 illustrates representation of the randomness issue. In this, factors A, B, and C affect outcome D, but so also do factors that have not been explicitly identified, called "other." The presence of those factors might either increase or decrease the likelihood of the outcome D (hence the +/-). The heaviness of the arrows indicates that A is especially important, that B and C are less so, and hidden variables even less so. Much can be conveyed with diagrams with only this level of complexity.

There are limits to what can be expressed diagrammatically without excessive complication. The next step, arguably, is to use simple outcome tables. Table 11.3 is intended to reflect more precisely the same thinking as in Figure 11.8. Factor A is especially important; if it is Low, then the claim is that outcome D will be Low. If factor A is High, then outcome D will *probably be* High if at least one of B and C are High, and *very likely* be High if both are High. A modeler

Figure 11.8
A Combining Logic Diagram with Hidden Factors

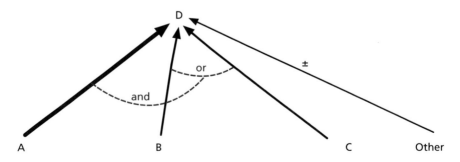

Table 11.3
A Simple Outcome Table, Consistent with Figure 11.8
But More Precise

A	B	C	D	Confidence
High	High	High	High	High
High	High	Low	High	Moderate
High	Low	High	High	Moderate
High	Low	Low	Low	High
Low	High	High	Low	High
Low	High	Low	Low	High
Low	Low	High	Low	High
Low	Low	Low	Low	High

familiar with expert judgments could give these a bit more precision, such as associating "moderate" with ~60 percent and "high" with ~80 percent in probability terms, or odds of, say, 3:2 and 4:1. Subject-matter experts could ponder about whether those would be "about right."

The continuing point is that simple representations of knowledge may have a degree of imprecision consistent with the knowledge itself. Even if the factors must be allowed more discrete values (say three, as in Low, Marginal, High), experience in developing artificial-intelligence models using highly structured rules demonstrates that sophisticated but comprehensible models can be built using these techniques (see also the appendix). With straightforward mathematical techniques and appropriate spot-checking by human experts, much can also be done to verify and even validate—relative to subjective expert knowledge.

The logic table (Table 11.3) is equivalent to the mathematics, described in pseudo code as:

    If A is High and (B is High and C is High)
    Then: D is High; Confidence is High
    Else
      If A is High and (B is High or C is High)

Then: D is High; Confidence is Moderate
Else: D is Low; Confidence is High.

In this case, the pseudo-code is as good as or better than the table, but communication is usually better with table structures—especially as dimensionality increases.

The Dynamics of Competition, Learning, and Adaptation. Competition, learning, and adaptation are crucial in social phenomena, but representing them is not always easy. Competition can in some cases be represented by game-theoretic methods. In the simplest form, these do not purport to describe the actual dynamics of interaction but rather to show what outcomes would be like if competitors act most effectively in their own interest. This can be done sequentially.[17] Modern game theory includes cooperation and competition and can include agent-based modeling (for example, Axelrod, 1997; Epstein, 2007), as discussed below.

Agent-based modeling (ABM) is closely associated with the study of complex adaptive systems (CAS) generally (for example, Holland and Minmaugh, 1998; Epstein and Axtell, 1996; Epstein, 2006; Bar-Yam, 2003; Gilbert, 2007). ABM can be included in any of a number of CT simulation environments, such as REPAST (North and Macal, 2007), SEAS (Chatuverdi et al., 2000), and COMPOEX.[18]

As described elsewhere (Davis, 2005), entity-level ABM research should be very helpful in developing lower-resolution models useful in policy analysis. It may not be fruitful in policy work to follow the dynamics of complex adaptive systems to see the details of precisely how "emergent phenomena" such as insurgencies arise. That may be left for separate research, with the fruits of the research being reflected in simpler models identifying when situations should be expected to be unstable. It will probably prove necessary to have much more sophisticated agents and representations of the environment to develop a good microscopic understanding of phenomena, including emergence.[19] In any case, making the connection between the worlds of micromodels and macromodels is both exciting and challenging.[20]

Some researchers believe that detailed agent-based simulations can be used predicatively. They sometimes disparage the feasibility of social-science modeling that does not include agent-based modeling. Such researchers have far more faith in the validity of the current ABMs than I do and far less confidence that the consequences of the various "emergent phenomena" can be represented macroscopically. It is an interesting theoretical debate that will be resolved over time with experience.

Other classes of model may also be important, such as models describing the consequences of a conservation law or of aspects of a system that are constant after a steady state has been reached. Some of these types are familiar in the physical sciences, economics, and other social-science disciplines. I mention them because the common use of computer simulations sometimes crowds out simpler depictions.

A Vision of Analysis Amidst Uncertainty

The preceding sections have emphasized the special difficulties associated with uncertainty and soft, qualitative knowledge as occurs in social science. It follows from an appreciation of these that

- The objective of analysis in social science should often not be reliable "prediction," but rather an understanding of possibilities and perhaps of rough probabilities, or odds.

This admonition applies to much policy analysis generally. The Department of Defense has come increasingly to recognize that massive uncertainty exists about such fundamental issues as who will be the future adversaries of the United States, where and under what circumstances future wars will occur, and what strategies and tactics will be employed. The result has been an emphasis on capabilities-based planning (Rumsfeld, 2001).

Much has been done to develop uncertainty-sensitive methods of analysis (Davis, 2002a; Davis, Shaver, and Beck, 2008) with applications to defense. Similar methods have been brought to bear on policy debates about climate change and regional natural-resource planning (Lempert, Popper, and Bankes, 2003; Lempert, Groves, Popper, and

Bankes, 2006). The philosophy reflected in these efforts is consistent with what has become a key element in strategic planning: preparing for adaptiveness (Davis, 2002a; Light, 2004; Alberts and Hayes, 2003).[21] All of these applications relate to people and organizations, that is, to social phenomena.[22]

A key element of these approaches is *exploratory analysis,* as illustrated in Figure 11.9. Consider first the flow along the top of the figure. Given some alternative packages of strategies, tactics, and investments, which packages we can call Options 1, 2, 3, and 4, we seek to assess them despite extraordinary uncertainty about "everything" (see the assumption classes at the bottom of the figure). To do this, we develop an experimental plan that systematically varies the assumptions—even assumptions about the functional form of the model being used for evaluation! The experimental plan then drives computational experiments. In each experiment, the model (an "engine" for generating cases) has a set of inputs and produces outputs (which may be stochastic). The plan may call for huge numbers of runs, but actually conducting the

Figure 11.9
The Vision of Analysis

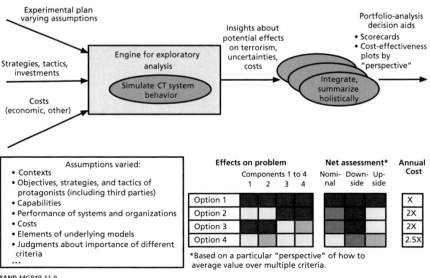

runs is a "mechanical" matter behind the scenes. The results, then, must be analyzed like "data" to see if patterns emerge. The results can be integrated and summarized for comprehensible displays such as the colored scorecards used in RAND's approach to portfolio analysis (bottom right). Analysts examine the results in a myriad of ways using such methods as standard statistical regressions, motivated meta-models, and data mining (Davis, Bankes, and Egner, 2007; Lempert, Groves, Popper, and Bankes, 2006).

This "exploratory analysis" approach represents a very different paradigm than starting with a baseline scenario, simulating the consequences for a few alternative strategies, and then conducing a handful of excursions with different assumptions.

- The fundamental concept in such work is to find strategies that are likely do well across much of the uncertainty space rather than performing well only for best-estimate assumptions.

The approach seeks "FAR strategies," that is, strategies that are Flexible, Adaptive, and Robust (see also National Research Council, 2006).

Some traditional analysts view such an image with horror because they are used to spending months working out details of a baseline model and database. How could they be asked to work on an entire scenario space? Fortunately, in exploratory analysis the premium is on achieving a synoptic view rather than precision. Simpler models suffice for initial work, so that running huge numbers of cases may be relatively straightforward, occurring behind the scenes. The fruits of exploratory analysis can be shown in displays that identify the circumstances in which outcomes are, for example, favorable or unfavorable, and where the boundary lines lie, that is, defining different *regions*. The intellectual content has to do with learning how many importantly different regions exist and where they lie in the n-dimensional assumptions space (also called factor space, scenario space, and parameter space).

In some cases, the regions can be identified by clever analysts without much computation, in which case it is even easier to identify a small "spanning set" of analytical cases, one or two for each important region. In assessing alternative courses of action, it is then necessary to test only against the spanning-set cases because they "stress" the alternative in all of the most important ways (Davis, Shaver, and Beck, 2008; National Research Council, 2008; Davis, Johnson, Long, and Gompert, 2008). Under budget pressures, policymakers could decide to deemphasize some of the case, but they would do so with recognition of the risks.

Once the synoptic view has been accomplished with low-resolution exploratory analysis, detail can be added selectively to better understand implications. This process is most compelling and rigorous when combined with multiresolution modeling.

This vision is ambitious, but 15–20 years of experience now exists with exploratory analysis, which has proven its viability and usefulness—assuming sufficient relevant knowledge (even if uncertain).

Illustrative Application to Integrating Social-Science Knowledge

The previous sections have described generic methods for communicating social-science knowledge. What follows illustrates how they can be used.

Companion papers in the larger study of which this paper is part (Davis and Cragin, 2009) reviewed the social-science literature on terrorism and counterterrorism with regard to root causes (Noricks, 2009), individual radicalization (Helmus, 2009), achieving and maintaining public support (Paul, 2009), and how terrorist organizations make decisions (Jackson, 2009). Each included a factor tree relating the factors identified in the respective reviews. These are reproduced here as Figures 11.10 through 11.13.[23]

Noricks's root-causes tree (Figure 11.10) has three approximately necessary conditions for overall root-cause pressures to be significant: norms that tolerate violence, perceived grievances (for example, foreign occupation), and the mobilizing structures for terrorism. These top-

**Figure 11.10
Factor Tree for Root Causes**

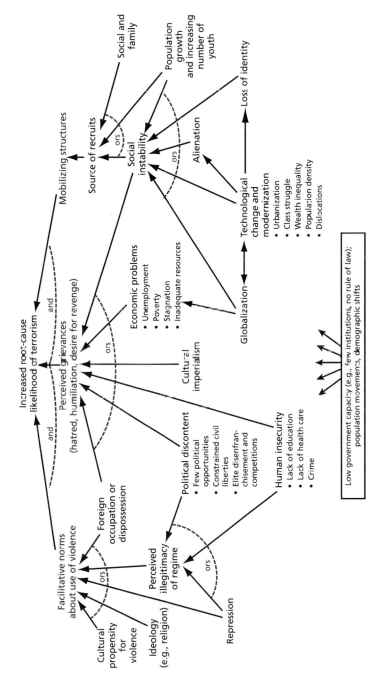

level factors are determined by a myriad of complex and subtle factors indicated lower in the tree.

In comparing this with Helmus's tree (Figure 11.11) on individual-level radicalization, we see considerable overlap despite the differences in the perspectives. Helmus sees radicalizing groups as necessary but also perceived rewards in one form or another (not included in Noricks's discussion of root causes), grievances, and a desire for change. Once again, the richness of discussion depends on factors farther down the tree, but we are interested in the high-level abstractions.

Paul's tree (Figure 11.12) relates to public support. Despite having begun by thinking about the group level of analysis, Paul found himself recasting issues in terms of individuals' propensity to support terrorism—but with individuals affected by social pressures and incentives. His factor tree again has much in common with the earlier ones. However, he puts more emphasis on "identity" and social pressures. He also highlights the "negative" kind of social pressure caused by intimidation. If the terrorist organization is able to credibly threaten a population or individuals within it, then that will increase social pressures to support the terrorist activity—perhaps only in a passive manner, perhaps by providing materiel support, or perhaps by becoming a terrorist (more the subject of Helmus's paper). Paul also includes future-benefit considerations (which, implicitly could be negative, reducing public support). Individuals trade off "benefits" and "costs," but how they do so varies a great deal. They may value such intangibles as prestige; they may merely be swept along in a fervor. The result may be less rational-analytic than emotional behavior.

Figure 11.13 is my adaptation from Jackson (2009), which examines how terrorist organizations decide whether to take particular actions. I have translated that into a factor tree analogous to the ones above. Here, again, we see factors for benefits and risks but also reference to costs and risks; we also see the importance of resources and information. Is the potential action going to provide benefits in terms of advancing group interests or strategy? If so, is it presumably consistent with interests and ideology and also (an "and" condition) in alignment with constraints of external influences, such as the interests of state sponsors, cooperating groups, or supportive social movements?

Figure 11.11
Radicalization Factor Tree

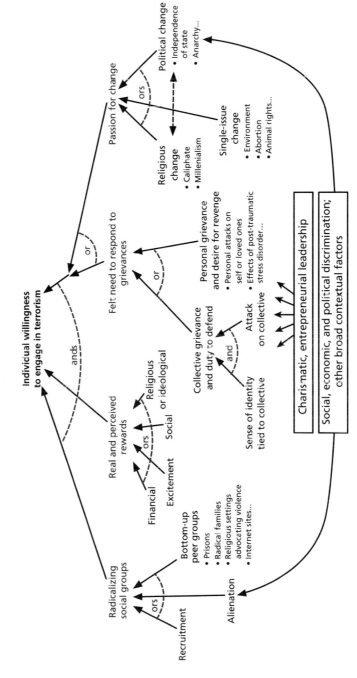

SOURCE: Adapted from Helmus (2009).

**Figure 11.12
Public Support Tree**

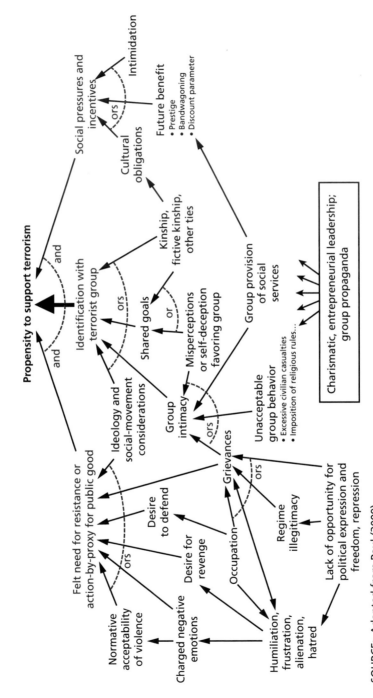

SOURCE: Adapted from Paul (2009).
RAND MG849-11.12

Figure 11.13
Decisionmaking

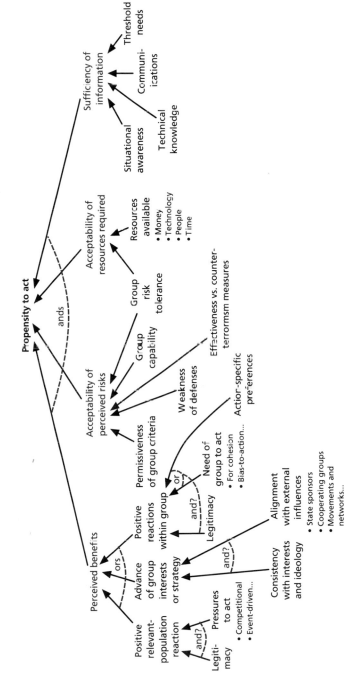

SOURCE: Adapted and simplified from Jackson (2009).

RAND MG849-11.13

Claude Berrebi's paper (Berrebi, 2009), written from an economist's perspective, also emphasizes the rational-choice model and its explanatory power. As one would expect in an economist's discussion, it considers benefits, costs, and risks.

Considering the observed overlap, it is natural to consider whether a high-level synthesis exists for these various "views of the elephant." Such a synthesis is not straightforward because the appropriate abstraction depends on what insights one is looking to highlight or which of several stories one wants to tell. I will merely illustrate one synthesis. The story is expressed in terms of influence on individuals and groups.

For this, it is useful to begin with a system-level influence diagram (Figure 11.14, adapted from Gvineria [2009] and Davis [2006]). In this diagram, the terrorist organization already exists, but its operational capabilities (central oval) may increase or decrease as a function of the resources and organizational structures available to it, which in turn depend on support obtained from states (for example, Iranian support

Figure 11.14
A System Diagram Relating to Terrorism

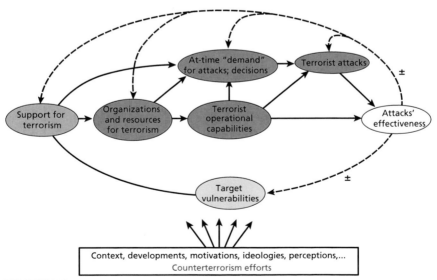

for Hizballah), general populations (for example, broad popular sentiment support for al-Qaeda), or more specific popular support (support of expatriate communities in western Europe for al-Qaeda or local affiliates).[24]

Given a degree of operational capability, the terrorist organization has the potential to conduct attacks, but the potential effects depend also on the targets' vulnerabilities. If support for action is strong enough, and if operational capability is adequate, then attacks will ensue. Those will have effects, which in turn will affect subsequent support. Another spectacular event akin to the attacks on the U.S. World Trade Center and the Pentagon might increase support for what would be seen as a revitalized al-Qaeda. Or it might spark back-reaction because of the loss of human life and retaliation. Or both. The consequences, then, might have positive or negative feedback effects (hence the ± symbology).

The primary function of Figure 11.14 is to illustrate how support for terrorism matters. Support, however, comes in many different forms. Suppose that we put aside state support, which is a subject unto itself, and consider only public support. That also varies markedly. Support may be so great that individuals will actually become terrorists (see Helmus's discussion); or it may come in the form of active or passive public support without direct participation in terrorism attacks (see Paul's discussion). Such public support is widely regarded in social science as a key to the success or decline of terrorism (see the classic, Galula [1963], for example; and discussions in Paul [2009], Gvineria [2009], and Stout, Huckabey, and Schindler [2008, p. 52], which is based on perspectives from within al-Qaeda).

On reflection and on comparing the discussions, it seems that the high-level factors contributing to either radicalization or active public support are very similar, although sometimes with different names depending on the author.

This suggests a composite view as shown in Figure 11.15, which shows the propensity for participating in terrorism or public support of the terrorist effort (an aggregation for simplicity) to be a function of four primary factors:

Figure 11.15
A High-Level Factor Tree Relating to a Population's Support for Terrorism

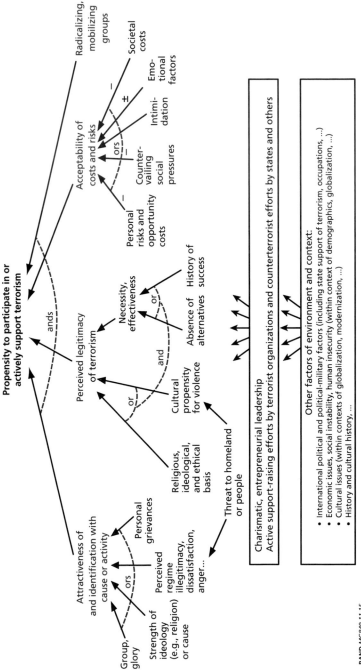

- attractiveness of and identification with a cause or other action
- perceived legitimacy of terrorism
- acceptability of costs and risks
- presence of radicalizing or mobilizing groups.

The first of these is my renaming in "positive terms" the factor related to motivation. As has been repeatedly noted by terrorism scholars (for example, Sageman, 2008; Jenkins, 2006), terrorists do not see themselves as "terrorists." They often see themselves as warrior heroes supporting either a cause (religious or otherwise) or, at least, an activity that they find exciting. The second factor uses the term "legitimacy." As we know from accounts of terrorists' internal debates, such matters are important—even if they conveniently discover rationale for doing what they are motivated to do anyway (Sageman, 2008; Stout, Huckabey, and Schindler, 2008, pp. 232 ff.). The third factor, acceptability of costs and risks, is implicit in some of the companion papers and explicit in others (for example, Jackson's, as well as Berrebi's). The fourth factor appears explicitly in Noricks's and Helmus's work and implicitly in the others. Note also, at the bottom, that charismatic, entrepreneurial leaders can be very important (Gupta, 2008).

At the second and third levels of detail, Figure 11.15 shows more than a dozen constituent factors. All of these are discussed in one form or another in the papers by Noricks, Helmus, Paul, Jackson, and Berrebi. The papers offer different perspectives as to how they come into play, but the differences are arguably not of first-order importance.

As discussed above, "ands" and "ors" are important in Figure 11.15. To first order, the research base suggests that all of the top-level factors must surpass some threshold or terrorism will decline. However, there are different ways that a cause may be seen as attractive and that terrorism can be seen to be legitimate. Similarly, there are a number of factors affecting the "negatives," that is, determining the perception of costs and risks. Although only one of several possible high-level perspectives,* the figure highlights overarching factors that appear repeatedly in the research literature in one form or another. Further, it does so holistically rather than asserting, for example, that participation in or support of terrorism is just a consequence of a cost-benefit calculation or that the current wave of terrorism is supported by popular sympathy driven only by Salifism or only by political grievances. To put matters otherwise, the intent of the diagram is to cover all of the available respectable explanations, not just the one deemed currently by some particular experts to be dominant in a particular time and place.

At first glance, it may appear that the factors of Figure 11.15 are assumed to combine via rational choice: Is there value to the terrorism, is it legitimate, are the costs and risks acceptable, and is there a mechanism? That might, in the instance of some individuals and groups, be correct and, as discussed in a companion paper (Berrebi, 2009), much can be understood with the rational-choice model. Social science tells us, however, that that model is often not descriptive. Even if we consider an individual or group contemplating terrorism, so that the concept of "decision" is perhaps apt, the more general concept is arguably *limited rationality*. People attempt to be rational, that is, to take actions consistent with their objectives, but they are affected by many influences, which include†

- the constraints of *bounded rationality*, which include erroneous perceptions, inadequate information, and the inability to make the complex calculations under uncertainty demanded by strict "rational choice"; the result is often heuristic decisionmaking, which employs simplified reasoning and may even accept the first solution that appears satisfactory

---

* For example, if one wises to emphasize the differences between root-cause factors and factors affecting public support or causing the decline of terrorism (see Cragin, 2009), then the "super aggregation" represented by Figure 11.16 is inappropriate.

† See the Nobel Prize lectures Simon (1978) and Kahneman (2002) for discussions of bounded rationality and cognitive biases. Davis, Kulick, and Egner (2005) review modern decision science with extensive citations to the rational-analytic and "naturalistic" decision-making literatures.

- the consequences of *cognitive biases*, such as the tendency to demonize opponents, to select information that bolsters what one wants to believe, to ignore risks below a threshold of apparent likelihood, and to make use of information that is most readily "available" cognitively (for example, the most recent report)
- the consequence of *naturalistic decisionmaking*, which is more intuitive and dependent on situation-dependent heuristics than evaluation of alternatives.

In still other cases, behavior can scarcely be called rational; it is driven by emotions at the time (whether fervor for action or vengeance on the one hand or unreasonable fear on the other) and is strongly affected by events and social context (as when an unhappy crowd turns into a rioting mob). Figure 11.15 is agnostic about such matters. The acceptability of costs and risks, in particular, could be determined by a rational calculation, heuristics, cognitive biases, or emotions at the time.

Figure 11.15 is simplified in other ways. First, it glosses over level-of-analysis issues; second, it treats many important issues as features of the surrounding context (see the boxes at the bottom, which refer to topics discussed in more detail above). Third, it is intended as a first approximation, recognizing that any such depiction will have some counterfactuals, which is why the individual papers cited include numerous cautions. Finally, a different top-level perspective and decomposition would be appropriate if the question of interest were different. There is no way to evade the complexity suggested by the larger review of relevant social science (that is, the complexity addressed in the papers by Noricks, Helmus, Paul, Jackson, Berrebi, and Gvineria). Nonetheless, Figure 11.15 conveys one broadly correct story.

Notional Results of Analysis

If Figure 11.15 were correct, and if it could be used as the basis for a more extensive exploratory analysis, one result might be the kind of diagram shown in Figure 11.16. This "region chart" shows the expected propensity to participate in or actively support terrorism (represented

Figure 11.16
A Region Plot for Participation in or Active Support of Terrorism (Notional)

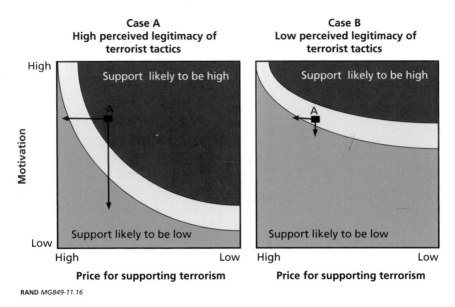

by color) as a function of the attractiveness of and identification with cause (abbreviated as "motivation" along the vertical axis), the price (acceptability of cost and risk along the horizontal axis), and the perceived legitimacy (left versus right panel). Radicalizing groups are assumed to exist in this example.

The notional plot asserts that if motivation and a sense of legitimacy are high enough, support for terrorism is likely to be high (red) (for example, the top right in either panel). However, if the sense of legitimacy is reduced (such as by the terrorists killing too many of the wrong people or by continuing to kill despite political and social progress within the relevant community), then the level of support will be much less, given the same motivation and sense of price (right panel). The notional plot suggests that the perceived-legitimacy factor has high leverage. For point A, for example, moving into the desirable regime of low support would require much less in terms of raising perceived price or reducing motivation if legitimacy were deemed low (right panel versus left).

Counterterrorism, then, should seek to reduce motivations, to increase the sense of *illegitmacy,* and to impose increasing costs on those who participate or support. Disrupting the organizations for radicalization and mobilization would also have great value (not shown).

Conclusions and Suggestions About Future Research

Ultimately, this paper is an integrative, theory-oriented think piece, one suggesting major features of an approach to representation of knowledge and also an approach to analysis. It has illustrated how factor-tree methods of decomposition can be used to modularize problems so that they can be addressed separately but seen as part of a whole. At the same time, it has illustrated how the vision of the "whole" depends on the perspective taken. That is, which representation one uses depends on the challenge being addressed, such as "understanding the terrorist phenomenon" versus laying out a counterterrorism campaign and allocating resources wisely. Different decompositions would be suitable for different questions. For example, Cragin (2009) draws on the several companion papers to argue that sustainment of terrorism is quite different from terrorism's rise or decline and that focusing on the sustainment phase is valuable in identifying the components of a counterterrorism campaign.

Table 11.4 summarizes the paper's themes in a procedural framework that anticipates a number of contributing individuals pooling knowledge, followed by movement toward actual modeling, exploratory work, refinement, and testing. This paper has focused on concepts and tools for Tier One efforts—representation of knowledge before rigorous model development and programming (if appropriate).

An important element of the suggested procedure is the development of a few system perspectives suitable for different aspects of counterterrorism. Each would require its own decomposition into modules, which could then be addressed in multiple independent or cooperative research efforts, the fruits of which could be discussed and debated using simplified methods of knowledge representation such as discussed in this paper. Small models can be encoded to generate results

Table 11.4
Procedural Elements of Methodology

Tier One
Collect factors (variables): Focus on concepts, not just convenient measurables
Define factor levels meaningfully, either with measurable criteria or meaningful descriptions
Organize variables in abstraction trees, allowing as necessary for cross-branch interactions (hopefully weak)
Consider alternative trees for different representations
Translate trees into influence diagrams, preferring variables giving positive effects; indicate feedback loops; use dashed lines for small effects
Amend diagrams to indicate first-order combing logic and criticality
Review, debate, iterate; refine

Tier Two
Module by module, characterize combining function with diagrams, logic tables, "operator math," and pseudo code—whatever works best; indicate potential need for thresholding and S-shaping
Implement simple module-level models; exercise; refine
Conduct validation exercises using simplified table methods and graphs to elicit key assessments
Do as above, but for a system-level model, with a focus on integration and top-down exploratory analysis (with selective zoom) rather than pure bottom-up modeling with results excessively dependent on uncertain low-level details

as a function of assumptions, to assist in iteration and review. My colleagues and I have done prototype work of this nature using *Analyica* models (Davis, 2006; Davis et al., 2008).

Further, it would be possible to see how the work on different modules relates to the others. Indeed, it is possible to construct system models connecting small computer models representing knowledge module by module and at differing levels of detail. A number of simulation environments make this possible,[25] although the validity of model composition is much more problematic than that of the more familiar composition of software components (Davis and Anderson, 2003). Such an integrated depiction could be quite interesting, although—as discussed throughout this paper—it would be unwise to take "predictions" seriously because of uncertainties.

An alternative approach would be for the relevant community to agree on a common set of decompositions akin to those illustrated above and to then go about building independent models of the various conceptual models using a variety of programming languages and environments. Using representation techniques such as those described in this paper, it should be possible to have discussion, debate, and convergence on first-order matters of substance, but without requiring commonality of programs. Most vigorous science has proceeded in that type of approach, rather than one calling for a high degree of centralization or rigid standards. Although the DoD has come to focus much of its combat analysis on a suite of well-controlled models managed and operated by the Office of the Secretary of Defense (Program Analysis and Evaluation), it has been argued by independent panels that future developments even in that relatively mature domain would be better served by more decentralization, competition, and smaller, specialized models.[26]

Appendix: Specifying Qualitative Knowledge with Logic Tables

The size of logic tables grows rapidly with the number of variables and the number of their values. Table A.1 illustrates this with the same assumption as in Table 11.3 except that A, B, and C can each have three values: Low, Marginal, and High. Instead of eight rows, the table must now have, 3^3 or 27 (not counting the header). Combinatorial explosion can be a problem.

Fortunately, the apparent complexity can be reduced with mathematical logic. After all, the underlying rationale for Table A.2 is not 27 different assertions but rather a shorter form of approximate reasoning. This is illustrated by Table A.2, which uses some special notation to permit truncation.

The notation in Table A.2 is that a dash means "any value," and an operator such as > means "a value greater than." Further, the table is to

Table A.1
A Possible Logic-Table Summary of Knowledge
for Three Factors with Three Values Each

A	B	C	D
Low	Low	Low	Low
Low	Low	Marginal	Low
Low	Low	High	Low
Low	Marginal	Low	Low
Low	Marginal	Marginal	Low
Low	Marginal	High	Marginal
Low	High	Low	Low
Low	High	Marginal	Marginal
Low	High	High	Marginal
Marginal	Low	Low	Low
Marginal	Low	Marginal	Low
Marginal	Low	High	Low
Marginal	Marginal	Low	Low
Marginal	Marginal	Marginal	Low
Marginal	Marginal	High	Low
Marginal	High	Low	Low
Marginal	High	Marginal	Low
Marginal	High	High	Marginal
High	Low	Low	Low
High	Low	Marginal	Low
High	Low	High	High
High	Marginal	Low	Low
High	Marginal	Marginal	Marginal
High	Marginal	High	High
High	High	Low	High
High	High	Marginal	Marginal
High	High	High	High

Table A.2
Truncated Logic Table Using Shorthand

A	B	C	D
Low	≥ Marginal	High	Marginal
Low	High	≥ Marginal	Marginal
Low	—	—	Low
Marginal	> Low	> Low	Marginal
Marginal	—	—	Low
High	High	—	High
High	—	High	High

be read like If-Then-Else logic, starting with the first row and working downward. The postulated underlying logic is actually clearer in Table A.2. The claim is being made that because A is so important, if A is Low, D will at most be Marginal. That will occur only when at least one of B and C is High and the other is at least Marginal. If A is Marginal, then D will at most be Marginal, and that will be achieved only if neither B nor C is Low. However, if A is High, then D will be High if *either* B or C, or both, is High.[27]

Note that logic tables such as this may be intentionally non-linear, in which case the results (right column) will not quite match the results of any linear-sum calculation. However, such comparisons (which must include rounding rules) can be quite useful in pointing out particular cases (rows) where results are "unusual" in that sense and thereby worthy of being double-checked by subject-matter experts. It is common for heuristic rules used initially by experts to have flaws that would be highlighted by such checking. Thus, checking might correct the original heuristics or confirm that nonlinearity was intended.

This said, it can be very useful to express a relationship such as D = F(A, B, C) in linear-sum terms and compare results to logic tables generated as described above. In my experience with qualitative modeling (for example, Davis and Arquilla, 1991), this will frequently identify unintended inconsistencies and sharpen the model. In other cases, what appear to be inconsistencies are valid in representing the subjective knowledge or, in some cases, experimentally verifiable knowledge. Many of the phenomena observed in behavioral-psychology experiments exhibit effects of thresholds and path-dependence that are indeed inconsistent with simple linear-sum algebra.

Bibliography

Alberts, David S., *Planning Complex Endeavors*, Washington, D.C.: Command and Control Research Program, Department of Defense, 2007.

Alberts, David S., and Richard E. Hayes, *Power to the Edge: Command and Control in the Information Age*, Washington, D.C.:, Department of Defense, 2003.

Allen, Thomas, et al., *Foundation for an Analysis Modeling and Simulation Business Planning*, Arlington, Va.: Institute for Defense Analyses, P-4178, 2007.

Angrist, Joseph D. and Alan B. Krueger, "Empirical Strategies in Labor Economics," O. Ashenfelter and D. Card, eds., in *Handbook of Labor Economics*, Vol. 3, Amsterdam: Elsevier Science, 1999.

Angrist, Joshua, and Jorn-Steffen Pischke, *Mostly Harmless Econometrics: An Empiricist's Companion*, Princeton, N.J.: Princeton University Press, 2009.

Axelrod, Robert, *The Complexity of Cooperation: Agent-Based Models of Competition*, Princeton N.J.: Princeton University Press, 1997.

Baldwin, Carliss Y., *Design Rules: The Power of Modularity*, Cambridge, Mass.: MIT Press, 2000.

Bar-Yam, Yaneer, *Dynamics of Complex Systems*, Boulder, Colo.: Westview Press, 2003.

―――, *Making Things Work: Solving Complex Problems in a Complex World,* Brookline, Mass.: Knowledge Press, 2005.

Berman, Eli, and D. David Laitin, "Rational Martyrs vs. Hard Targets: Evidence on the Tactical Use of Suicide Attacks," in Eva Meyersson Milgrom, ed., *Suicide Bombing from an Interdisciplinary Perspective*, Princeton, N.J.: Princeton University Press, forthcoming.

Berrebi, Claude, "The Economics of Terrorism and Counterterrorism: What Matters and Is Rational-Choice Theory Helpful?" in Paul K. Davis and Kim Cragin, *Social Science for Counterterrorism: Putting the Pieces Together,* Santa Monica, Calif.: RAND Corporation, 2009. As of January 28, 2009: http://www.rand.org/pubs/monographs/MG849/

Bigelow, James H., and Paul K. Davis, *Implications for Model Validation of Multiresolution, Multiperspective Modeling (MRMPM) and Exploratory Analysis,* Santa Monica, Calif.: RAND Corporation, 2003. As of January 17, 2009: http://www.rand.org/pubs/monograph_reports/MR1750/

Blaha, Michael, and James Rumbaugh, *Object Oriented Modeling and Design with UML,* Upper Saddle River, N.J.: Pearson Prentice Hall, 2005.

Boaz, Ganor, *The Counter-Terrorism Puzzle: A Guide for Decision Makers,* Edison, N.J.: Transaction Publishers, 2005.

Chaturvedi, Alok R., Mukul Guptak, Shailendra Raj Mehta, Karl E. Weick, and Wei T. Yu, "Agent-Based Simulation Approach to Information Warfare in the Seas Environment," *Proceedings of 33d Hawaii International Conference on System Sciences,* Wailea, Maui, Hawaii, 2000.

Clancy, Tom, Anthony Zinni, and Tony Klotz, *Battle Ready (Commanders Series),* N.Y.: G.P. Putnam's Sons, 2004.

Cragin, Kim, "Cross-Cutting Observations and Some Implications for Policymakers," in Paul K. Davis and Kim Cragin, eds., *Social Science for Counterterrorism: Putting the Pieces Together,* Santa Monica, Calif.: RAND Corporation, 2009. As of January 17, 2009: http://www.rand.org/pubs/monographs/MG849/

Davis, Paul K., *An Analyst's Primer for the RAND/ABEL Programming Language,* Santa Monica, Calif.: RAND Corporation, N-3042-NA, 1990. As of January 18, 2009: http://www.rand.org/pubs/notes/N3042/

———, *Analytic Architecture for Capabilities-Based Planning, Mission-System Analysis, and Transformation,* Santa Monica, Calif.: RAND Corporation, 2002a. As of January 17, 2009: http://www.rand.org/pubs/monograph_reports/MR1513/

———, "Synthetic Cognitive Modeling of Adversaries for Effects-Based Planning," *Proceedings of the SPIE,* Vol. 4716, No. 27, 2002b, pp. 236–250.

———, "New Paradigms and New Challenges," keynote presentation, *Proceedings of the 37th Winter Simulation Conference,* Orlando, Florida, 2005.

———, "A Qualitative Multiresolution Model for Counterterrorism," in Dawn Trevisani, ed., *Proceedings of Enabling Technologies for Modeling and Simulation Science X,* SPIE, Paper 6227-15, 2006.

———, *Exploratory Analysis for Strategy Problems Amidst Uncertainty,* Santa Monica, Calif.: RAND Corporation, unpublished, 2008.

Davis, Paul K., and John Arquilla, *Deterring or Coercing Opponents in Crisis: Lessons from the War with Saddam Hussein,* Santa Monica, Calif.: RAND Corporation, 1991. As of January 17, 2009:
http://www.rand.org/pubs/reports/R4111/

Davis, Paul K., and James H. Bigelow, *Experiments in Multiresolution Modeling (MRM),* Santa Monica, Calif.: RAND Corporation, 1998. As of January 17, 2009:
http://www.rand.org/pubs/monograph_reports/MR1004/

Davis, Paul K., and Robert H. Anderson, *Improving the Composability of Department of Defense Models and Simulations,* Santa Monica, Calif.: RAND Corporation, 2003. As of January 18, 2008:
http://www.rand.org/pubs/monographs/MG101/
The summary appears also in *J. Defense Modeling and Simulation,* Vol. 1, No. 1, 2004, pp. 24–36.

Davis, Paul K., and Amy Henninger, *Analysis, Analysis Practices, and Implications for Modeling and Simulation,* Santa Monica, Calif.: RAND Corporation, OP-176-OSD, 2007. As of April 20, 2009:
http://www.rand.org/pubs/occasional_papers/OP176/

Davis, Paul K., and Kim Cragin, eds., *Social Science for Counterterrorism: Putting the Pieces Together,* Santa Monica, Calif.: RAND Corporation, 2009. As of January 17, 2009:
http://www.rand.org/pubs/monographs/MG849/

Davis, Paul K., Jonathan Kulick, and Michael Egner, *Implications of Modern Decision Science for Military Decision Support Systems,* Santa Monica, Calif.: RAND Corporation, 2005. As of March 12, 2009:
http://www.rand.org/pubs/monographs/MG360

Davis, Paul K., Steven C. Bankes, and Michael Egner, *Enhancing Strategic Planning with Massive Scenario Generation: Theory and Experiments,* Santa Monica, Calif.: RAND Corporation, TR-392-OSD, 2007. As of January 17, 2009:
http://www.rand.org/pubs/technical_reports/TR392/

Davis, Paul K., Russell D. Shaver, and Justin Beck, *Portfolio-Analysis Methods for Assessing Capability Options,* Santa Monica, Calif.: RAND Corporation, 2008. As of January 17, 2009:
http://www.rand.org/pubs/monographs/MG662/

Davis, Paul K., Richard Hillestad, Horacio Trujillo, and Sam Neill, "Modeling Terrorism and Counterterrorism for Policy Analysis," unpublished briefing, 2005.

Davis, Paul K., Stuart E. Johnson, Duncan Long, and David Gompert, *Developing Resource-Informed Strategic Assessments and Recommendations,* Santa Monica, Calif.: RAND Corporation, 2008. As of January 17, 2009:
http://www.rand.org/pubs/monographs/MG703/

Dobbs, Michael, *One Minute to Midnight: Kennedy, Khrushchev, and Castro on the Brink of Nuclear War,* New York: Knopf, 2008.

Dobbins, James, John G. McGinn, Keith Crane, Seth G. Jones, Rollie Lal, Andrew Rathmell, Rachel M. Swanger, and Anga R. Timilsina, *America's Role in Nation-Building: From Germany to Iraq,* Santa Monica, Calif.: RAND Corporation, 2003. As of January 18, 2008:
http://www.rand.org/pubs/monograph_reports/MR1753/

Epstein, Joshua, *Generative Social Science: Studies in Agent-Based Computational Modeling,* Princeton, N.J.: Princeton University Press, 2007.

Epstein, Joshua, and Robert Axtell, *Growing Artificial Societies: Social Science from the Bottom Up,* Cambridge, Mass.: MIT Press, 1996.

Feyarabend, Paul, *Against Method,* London: Verso, 1975.

Forrester, Jay W., *Industrial Dynamics,* Cambridge, Mass.: The MIT Press, 1963.

———, *Urban Dynamics,* Cambridge, Mass.: Wright Allen Press, 1969.

Galula, David, *Pacification in Algeria, 1956–1958,* Santa Monica, Calif.: RAND Corporation, 1963. As of January 17, 2009:
http://www.rand.org/pubs/monographs/MG478-1/

Geddes, Barbara, *Paradigms and Sand Castles: Theory Building and Research Design in Comparative Politics,* Ann Arbor, Mich.: University of Michigan, 2003.

Gilbert, Nigel, *Agent-Based Models (Quantitative Applications in the Social Sciences),* Thousand Oaks, Calif.: Sage Publications, 2007.

Gupta, Dipak, *Understanding Terrorism and Political Violence,* New York: Routledge, 2008.

Gvineria, Gaga, "How Does Terrorism End?" in Paul K. Davis and Kim Cragin, eds., *Social Science for Counterterrorism: Putting the Pieces Together,* Santa Monica, Calif.: RAND Corporation, 2009. As of January 17, 2009:
http://www.rand.org/pubs/monographs/MG849/

Haimes, Yacov, *Risk Modeling, Assessment, and Management,* New York: John Wiley & Sons, 1998.

Hall, H. Edward, Mark LaCasse, Robert H. Anderson, and Norman Z. Shapiro, *The RAND-ABEL Programming Language: History, Rationale, and Design,* Santa Monica, Calif.: RAND Corporation, 1985. As of January 19, 2009:
http://www.rand.org/pubs/reports/R3274/

Hausman, Daniel M., "Philosophy of Economics," in Edward N. Zalta, ed., *The Stanford Encyclopedia of Philosophy,* Fall 2008. As of January 17, 2009:
http://plato.stanford.edu/archives/fall2008/entries/economics/

Helmus, Todd C., "Why and How Some People Become Terrorists," in Paul K. Davis and Kim Cragin, eds., *Social Science for Counterterrorism: Putting the Pieces Together,* Santa Monica, Calif.: RAND Corporation, 2009. As of January 17, 2009:
http://www.rand.org/pubs/monographs/MG849/

Hillestad, Richard, and Paul K. Davis, "A Prototype Game-Theoretic Model for Counterterrorism," RAND Corporation, unpublished, 2007.

Holland, John H., and Heather Mimnaugh, *Hidden Order: How Adaptation Builds Complexity,* New York: Perseus Publishing, 1996.

Jackson, Brian A., "Organizational Decisionmaking by Terrorist Groups," in Paul K. Davis and Kim Cragin, eds., *Social Science for Counterterrorism: Putting the Pieces Together,* Santa Monica, Calif.: RAND Corporation, 2009. As of January 17, 2009:
http://www.rand.org/pubs/monographs/MG849/

Jenkins, Brian Michael, *Unconquerable Nation; Knowing Our Enemy, Strengthening Ourselves,* Santa Monica, Calif.: RAND, 2006. As of January 19, 2009:
http://www.rand.org/pubs/monographs/MG454/

Kahneman, Daniel, *Maps of Bounded Rationality: A Perspective on Intuitive Judgment and Choice* (Nobel Prize Lecture), Stockholm, 2002. As of March 12, 2009:
http://nobelprize.org/nobel_prizes/economics/laureates/2002/kahneman-lecture.html

Keane, Michael P., "Structural versus Atheoretical Approaches to Econometrics," keynote address, Conference on Structural Models in Labor, Aging and Health Duke University on September 17-19, 2005. As of January 17, 2009:
http://gemini.econ.umd.edu/jrust/research/JE_Keynote_7.pdf

Kulick, Jonathan, and Paul K. Davis, *Modeling Adversaries and Related Cognitive Biases,* Santa Monica, Calif.: RAND Corporation, 2003. As of March 12, 2009:
http://www.rand.org/pubs/reprints/RP1084/

Lempert, Robert J., David G. Groves, Steven W. Popper, and Steve C. Bankes, "A General Analytic Method for Generating Robust Strategies and Narrative Scenarios," *Management Science,* Vol. 17, No. 1, April 2006, pp. 73–85.

Lempert, Robert J., Steven W. Popper, and Steven C. Bankes, *Shaping the Next One Hundred Years: New Methods for Quantitative, Long-Term Policy Analysis,* Santa Monica, Calif.: RAND Corporation, 2003. As of January 17, 2009:
http://www.rand.org/pubs/monograph_reports/MR1626/

Lempert, Robert J., Michael E. Schlesinger, and Stephen C. Bankes, "When We Don't Know the Costs or the Benefits: Adaptive Strategies for Abating Climate Change," *Climatic Change,* Vol. 33, No. 1996, pp. 235–744. As of January 18, 2009:
http://www.rand.org/pubs/reprint/RP557/

Light, John, *The Four Pillars of High Performance,* New York: McGraw Hill, 2004.

Military Operations Research Society, *Human Behavior and Performance as Essential Ingredients of Realistic Modeling of Combat, MORIMOC II, Vols. 1–2,* Alexandria, Va., 1989.

Ministry of Defence, *The Comprehensive Approach,* United Kingdom, Joint Discussion Note 4/05, 2005.

Nagl, John A., David Petraeus, David Amos, and Sarah Seall, *U.S. Army/Marine Counterinsurgency Field Manual,* Chicago: University of Chicago Press, 2007.

National Research Council, *Post–Cold War Deterrence,* Washington, D.C.: National Academies Press, 1996.

———, *Defense Modeling, Simulation, and Analysis: Meeting the Challenge,* Washington, D.C.: National Academies Press, 2006.

———, *U.S. Conventional Prompt Global Strike: Issues for 2008 and Beyond,* Washington, D.C.: National Academies Press, 2008.

Noricks, Darcy M.E., "The Root Causes of Terrorism," in Paul K. Davis and Kim Cragin, eds., *Social Science for Counterterrorism: Putting the Pieces Together,* Santa Monica, Calif.: RAND Corporation, 2009. As of January 17, 2009:
http://www.rand.org/pubs/monographs/MG849/

North, M. J., T. R. Howe, N. T. Collier, and J. R., Vos, "The REPAST Simphony Development Environment" 2007. As of January 19, 2009:
http://agent2007.anl.gov/2005procpdf/Agent_2005_North_Development.pdf

North, Michael J., and Charles M. Macal, *Managing Business Complexity: Discovering Strategic Solutions with Agent-Based Modeling and Simulation,* Oxford, UK: Oxford University Press, 2007.

Pate-Cornell, M., and Robin Dillion, "The Respective Roles of Risk and Decision Analysis in Decision Support," *Decision Analysis,* Vol. 3, No. 4, 2006.

Paul, Christopher, "How Do Terrorists Generate and Maintain Support?" in Paul K. Davis and Kim Cragin, eds., *Social Science for Counterterrorism: Putting the Pieces Together,* Santa Monica, Calif.: RAND Corporation, 2009. As of January 17, 2009:
http://www.rand.org/pubs/monographs/MG849/

Pearl, Judea, *Causality: Models, Reasoning, and Inference,* New York: Cambridge University Press, 2000.

———, "Causal Inference in Statistics: A Gentle Introduction," *Computing Science and Statistics Proceedings of Interface '01,* Vol. 33, 2001.

Pourret, Olivier, Patrick Naim, and Bruce Marcot, *Bayesian Networks: A Practical Guide to Applications,* Hoboken, N.J.: John Wiley, 2008.

Quinlivan, James T., "Force Requirements in Stability Operations," *Parameters,* Winter, No. 1995, pp. 59–69.

Ragin, Charles C., *The Comparative Method: Moving Beyond Qualitative and Quantitative Strategies,* University of California Press, Berkeley, Calif., 1989.

————, *Fuzzy-Set Social Science,* Chicago: University of Chicago Press, 2000.

Rumsfeld, Donald, *Report of the Quadrennial Defense Review,* Washington, D.C.: Department of Defense, 2001.

————, *Report of the Quadrennial Defense Review,* Washington, D.C.: Department of Defense, 2006.

Sage, Andrew P., and C. D. Cuppan, "On the Systems Engineering and Management of Systems of Systems and Federations of Systems," *Information, Knowledge, and Systems Management,* Vol. 2, No. 4, 2001, pp. 325–345.

Sageman, Marc, *Leaderless Jihad,* Philadelphia, Pa.: University of Pennsylvania Press, 2008.

Shapiro, Norman Z., H. Edward Hall, Robert H. Anderson, Mark LaCasse, M. S. Gillogly, and Robert Weissler, *The RAND-ABEL Programming Language: Reference Manual,* Santa Monica, Calif.: RAND Corporation, N-2367-1-NA, 1988. As of January 19, 2009:
http://www.rand.org/pubs/notes/N2367-1/

Shoemaker, P.J.H., *Experiments on Decisions Under Risk: the Expected Utility Hypothesis,* Boston, Mass.: Marinus-Nijhoff, 1980.

Silverman, Barry, "Curriculum Vitae," Web page, no date. As of February 13, 2009:
http://www.seas.upenn.edu/%7Ebarryg/cv-short.html

Silverman, Barry, G. Bharathy, and B. Nye, "Gaming and Simulating EthnoPolitical Conflicts," *Proceedings of the Descartes Conference on Mathematical Modeling for Counter-Terrorism (DCMMC),* New York: Springer, 2007.

Simon, Herbert, *Rational Decision-Making in Business Organizations,* Nobel Prize Lecture, Singapore: World Scientific Publishing, 1978.

————, *Sciences of the Artificial,* 2nd ed., Cambridge, Mass.: MIT Press, 1981.

Slovic, Paul, *The Perception of Risk,* London: Earthscan, 2000.

Sterman, John D., *Business Dynamics: Systems Thinking and Modeling for a Complex World,* Boston, Mass.: McGraw-Hill/Irwin, 2000.

Stout, Mark E., Jessica M. Huckabey, and John R. Schindler, *The Terrorist Perspectives Project: Strategic and Operational Views,* Naval Institute Press, 2008.

Symposium on Complex Systems Engineering, Santa Monica, Calif.: RAND Corporation, 2008. As of February 13, 2009:
http://cs.calstatela.edu/wiki/index.php/Symposium_on_Complex_Systems_Engineering

Uhrmacher, Adelinde, "Reasoning about Changing Structure: a Concept for Ecological Systems, *Applied Artificial Intelligene*, Vol. 9(2), 1005, 157-180.

Wagenhals, Lee, Insub Shin, and Alexander Levis, "Exccutable Models of Influence Nets Using Design/CPN," Fairfax, Va.: Systems Architectures Laboratory, George Mason University, 2001.

Zeigler, Bernard, Herbert Praenhofer, and Tag Gon Kim, *Theory of Modeling and Simuation*, 2nd ed., *Integrating Discrete Event and Continuous Complex Systems*, San Diego, Calif.: John Wiley, 2000.

Endnotes

[1]   Most of the larger volume of which this paper is part are scholarly literature reviews. This paper is different in kind and style; it is more of a think piece, one stimulated by the challenges of trying to pull together disparate segments of the social-science literature.

[2]   This is a return to the past, when models were designed before being implemented. Much of the paper is an attempt to use simple graphical and tabular techniques as a first-order, approximate, "specification language."

[3]   I thank Ben Wise and Claude Berrebi for comments on an earlier draft. The paper also benefited from an internal research and development project (Davis, 2006; Davis, Hillestad, Trujillo, and Neill, unpublished, 2005).

[4]   Confusion about this dates back centuries, as discussed in the philosophy-of-economics literature (Hausmann, 2008). The 18th century work of John Stuart Mill, for example, has sometimes been contrasted with empiricism, but Mill believed that the principles in a domain should be empirically established. (Hausmann, 2008, section 3.2.)

[5]   I refer to "mature" areas, because the data-driven approach is quite important in areas of physical science where theory is less well developed..

[6]   The British refer to the "Comprehensive Approach" (Ministry of Defence, 2005); the DoD refers to PMESII factors (i.e., political, military, economic, social, infrastructure and information systems) and DIME instruments (i.e., diplomatic, information, military, and economic instruments of power).

[7] See, for example, Holland and Mimnaugh (1996), Alberts (2007), and Alberts and Hayes (2003).

[8] See, for example, Zeigler, Praenhofer, and Kim (2000) and Uhrmacher (2006).

[9] The discipline of systems engineering is struggling with how to deal with systems of systems and complex adaptive systems. See, for example, papers presented at the Symposium on Complex Systems Engineering.

[10] This point was made in the early work of system dynamics (Forrester, 1963). Military "soft factors" are discussed in Military Operations Research Society (1989).

[11] This section contrasts with econometrics, where practitioners insist on measurable variables and seek to "explain" data with as few variables as possible.

[12] In some fields, such as in Bayesian belief nets, the term "influence diagram" has a somewhat different meaning from the one used here (see Pearl, 2000, for example).

[13] See Davis and Arquilla (1991), National Research Council (1996), and Kulick and Davis (2003).

[14] For exploitation of binary-logic modeling, see Ragin (1989). See also his extension (Ragin, 2000).

[15] My own suspicion is that this empirical phenomenon is an artifact of omitting such important variables as the effectiveness of the state's internal security apparatus, of averaging over very heterogeneous data for different countries or time periods, or both.

[16] See, for example, Pourret, Naim, and Marcot (2008), Wagenhals, Shin, and Levis (2001), and Pate-Cornell and Dillion (2006).

[17] My colleague Richard Hillestad has illustrated how game-theoretic concepts can be represented in CT modeling (Hillestad and Davis, 2007).

[18] COMPOEX is an environment, and suite of tools, generated in a major effort sponsored by the Defense Advanced Research Projects Agency (DARPA) (program manager: Sean O'Brien). Not much about it has been published so far in the scholarly literature, although some information is available on DARPA's Web site.

[19] CAS research has demonstrated that simple rules can predict emergent phenomena that appear markedly similar to real-world behaviors (for example, the flocking of geese or the start of a riot). However, it does not follow that only a small set of simple rules is sufficient to do justice to the actual phenomena. For example, human beings are not solely driven by simple cost-benefit calculations and, even when they are, they may act as though they use complicated, idiosyncratic, and context-dependent utility functions.

[20] Statistical physics relates the molecular and thermodynamic domains for equilibrium and some nonequilibrium systems. My expectation is that exploratory analysis using CAS models will lead to macroscopic "laws" in social-science behavior. CAS research relevant to CT is ongoing under sponsorship of U.S. Joint Forces Command (Chaturvedi et al., 2000) and other DoD organizations. A great deal of published research on related matters can be found on the Web site of Penn Professor Barry Silverman. Recent work has also been done as part of DARPA's COMPOEX activity.

[21] Colleague Ben Wise mentions a distinction between the ability to "learn" (and therefore behave differently) without changing structures and the ability to adapt even though it means changing structures (for example, changing from a centralized to decentralized operation or working with a faction that had previously been an adversary).

[22] Various terms have been used for related methods. These include "adaptive planning," a shorthand for planning for flexible, adaptive, and robust capabilities (Davis, 2002a; National Research Council, 2006); "robust adaptive planning" (Lempert, Groves, Popper, and Bankes, 2006), and "planning for agility" (Alberts, 2007).

[23] See also a recent book taking a life-cycle approach to describing terrorism (Gupta, 2008).

[24] The nodes in Figure 11.10 have subcomponents. For example, "Support for terrorism" might apply to each of a number of groups. Aggregating upward is nontrivial. For example, a given terrorist organization might need only a modest amount of support to succeed. Thus, "Support for terrorism" might have a high value (indicating adequacy) if even a relatively small subpopulation provided it.

[25] Four such environments are DARPA's COMPOEX (see the DARPA Web site), SEAS (associated with Alok Catuverdi of Simulex), an environment developed by Barry Silverman, and REPAST, an extensively exercised environment developed at Argonne National Laboratory and the University of Chicago (North et al., 2007).

[26] Related suggestions have been made by Dell Lunceford, previously of DARPA and the Army's Modeling and Simulation Office, and by members of a working group contributing to a white paper for DoD (Davis and Henninger, 2006; Allen et al., 2007).

[27] This refers to experience with the RAND-ABEL language (Hall, Lacasse, Anderson, and Shapiro, 1985; Shapiro, Hall, Edward, and Anderson, 1988; Davis, 1990).

CHAPTER TWELVE

Conclusions

Paul K. Davis and Kim Cragin

This review of the base in social science relating to terrorism and counterterrorism began with a set of chapters taking different perspectives: root causes of terrorism (Noricks, Chapter Two), individual radicalization (Helmus, Chapter Three), the sustainment of public support for terrorism (Paul, Chapter Four), the economics of terrorism (Berrebi, Chapter Five), how terrorist organizations make decisions (Jackson, Chapter Six), how terrorism ends (Gvineria, Chapter Seven), what we know about deradcialization and disengagement (Noricks, Chapter Eight), insights about strategic communications (Egner, Chapter Nine), cross-cutting observations (Cragin, Chapter Ten), and an analytical representation of social-science knowledge about terrorism and counterterrorism (Davis, Chapter Eleven). We have also included an appendix taking a first cut at identifying measures of effectiveness applicable to the various topics studied throughout the volume (Bahney, Appendix B).

Because our volume includes an extensive executive summary, we have chosen to keep the conclusions section short. However, our overarching conclusions are the following:

- Social science does well in identifying the key factors affecting terrorism and counterterrorism. However, special analytical techniques are needed to bring some order out of the resulting chaos. We have developed and illustrated a first set of such techniques, with an emphasis on being able to communicate and debate ideas and assumptions across disciplinary boundaries.

- A key element in doing so is taking a causal system perspective—at least for the purposes of establishing a framework of communication.
- Consistent with this system perspective, it is essential to distinguish sharply among different *contexts*: Failure to do so has probably been the biggest single problem impeding coherent scientific discussion of terrorism and counterterrorism. Terrorist organizations differ enormously, even "affiliates" of al-Qaeda differ, and many key issues are local.
- Because multiple factors affect terrorism and counterterrorism, and because their relative importance depends so much on context, social scientists must often answer simple questions with "It depends."
- Fortunately, it is possible to go well beyond "It depends" and to identify different "types" of situation, that is, different "types" of context. This volume has identified the way ahead for doing so in future work.
- Even where social-science knowledge is quite good, however, the aspiration should be one of anticipating *possibilities* and improving the *odds* of correct predictions, as distinct from seeking reliable prediction. Reliable prediction is not only infeasible in most cases, it is a counterproductive goal to the extent that it discourages an emphasis on monitoring, feedback, and adaptation.
- The style of analysis should be determined accordingly. In particular, it should inform finding strategies that are flexible, adaptive, and robust rather than finely tuned to dubious assumptions and fragile to random developments.

About the Authors

Paul K. Davis is a principal researcher at RAND and a professor of policy analysis in the Pardee RAND Graduate School. He has published widely on strategic planning (especially defense planning), strategy, advanced methods of modeling and analysis, deterrence theory, decisionmaking theory, cognitive modeling of adversaries, and analytical support of high-level decisionmakers. He was the lead author of the 2002 monograph *Deterrence and Influence in Counterterrorism: A Component in the War on Al Qaeda*. Before joining RAND, Paul was a senior executive in the Office of the Secretary of Defense (Program Analysis and Evaluation). He has a B.S. from the University of Michigan and a Ph.D. in chemical physics from the Massachusetts Institute of Technology. He has served on numerous panels of the National Research Council and Defense Science Board and is a past member of the Naval Studies Board.

Kim Cragin is a senior policy analyst at RAND. As a cultural historian, her research focuses on terrorism-related issues, including arms trafficking by the FARC (a left-wing rebel group in Colombia), suicide bombings, anti-U.S. extremism, the relationship between terrorism and development, terrorist groups' operational requirements, and border security. She was the lead author of a recent study, *Curbing Militant Recruitment in Southeast Asia*. Before coming to RAND in 2000, Kim received a master's degree in public policy from Duke University, where she studied Hamas and Israeli right-wing extremism and wrote her thesis on U.S. counternarcotics policy with regards to Plan Colombia and the FARC. She has also studied at Hebrew University, the Moscow Economic Institute, and Xinjiang University. Kim received her Ph.D. from the Faculty of History, Clare College, Cambridge University, writing on the transformation of Hamas over three decades.

Darcy M.E. Noricks is an associate political scientist at RAND, where she works on terrorism and counterterrorism, homeland security, defense strategy and planning, and international politics. Before joining RAND, she was project manager for the Global Transnational Terrorism Project and DFI Government Practice. She directed research and collection efforts on jihadi cells in Indonesia and Australia, as well as on al-Qaeda, and led a number of homeland security projects beginning in 1999. She received a B.A. in political science from the University of California, Berkeley, and an M.S. in foreign policy from Georgetown University; she is completing her Ph.D. in political science at the University of Washington.

Claude Berrebi is an associate economist at RAND, with expertise in terrorism and counterterrorism, labor and household behavior, workforce and labor markets, and international social and economic issues. He authored the recent papers "Human Capital and the Productivity of Suicide Bombers" (*Journal of Economic Perspectives*) and "Are Voters Sensitive to Terrorism?" (*American Political Science Review*), among others. Claude received his B.A. in economics and M.B.A. in finance from Hebrew University, and his M.A. and Ph.D. in economics from Princeton University (his dissertation was entitled "Terrorism: Causes and Consequences").

Todd C. Helmus is a behavioral and social scientist at RAND and was the lead author of the 2007 RAND monograph *Enlisting Madison Avenue: The Marketing Approach to Earning Popular Support in Theaters of Operation*. He has also written monographs on combat stress reduction, the challenges of operations in an urban environment, and curbing militant recruitment. Todd received a B.A. in psychology from Hope College and a Ph.D. in clinical psychology from Wayne State University.

Christopher Paul is a behavioral and social scientist at RAND. He is the author of *Information Operations—Doctrine and Practice: A Reference Handbook* and has written on the history of U.S. military interventions, press-military relations, urban warfare, simulation training, and acquisition. Chris received his Ph.D. in sociology from the University of California, Los Angeles.

Brian A. Jackson is a senior physical scientist at RAND, where he has written widely on organizational learning in terrorist groups, including terrorists' use of technology. His research focuses on homeland security and terrorism preparedness. He is also associate director of RAND's homeland security research program. Brian received a Ph.D. in bioinorganic chemistry from the California Institute of Technology and an M.A. in science, technology, and public policy from George Washington University.

Gaga Gvineria is currently a doctoral candidate in the Pardee RAND Graduate School. Before coming to RAND, he was a consultant in national security affairs at the Center for Strategic Research in Georgia (former USSR). He has also been a consultant with the World Bank and with the Post-Privatization Restructuring Project. He has received degrees from Tblisi State University, the Georgian Institute of Public Administration, and the Tbilisi State Institute of Economic Relations.

Michael Egner is currently a doctoral candidate in the Pardee RAND Graduate School, where he has conducted research on counterterrorism, modern decision science, and strategic communications. Before coming to RAND, Michael performed research for the Urban Institute, the White House Council of Economic Advisers, and Goldman Sachs. He received his B.A. in economics at Princeton, where he won the President's Award for Academic Achievement.

Ben Bahney is a project associate at RAND. He holds a B.A. in history from the University of Pennsylvania and an M.A. in international economics from the University of California, San Diego. His primary interests are counterinsurgency, international politics, and homeland security.

Analytic Measures for Counterterrorism and Counterinsurgency

Benjamin Bahney

Introduction

Objectives

This appendix provides initial suggestions about ways to measure the factors discussed in the main body of the monograph. Well-conceived metrics should be an important part of a counterterror or counterinsurgency (CT/COIN) campaign. The appendix suggests a number of possible metrics but provides little detail, because the intention is to provide only a first-cut look. Just a few of the metrics discussed here will be relevant to any particular campaign plan, and their specifics will likely depend on both global conditions and conditions in the area of operations (AOR). The following pages deal primarily with measures for factors that are amenable to influence.

The Measures Literature

The relevant literature on analytical measures for CT/COIN is surprisingly modest with some exceptions, such as a School of Advanced Military Studies monograph by then Major Douglas Jones (Jones, 2006), which in turn drew on work by Bard O'Neill (O'Neill, 1990). In this discussion, I adopt the terminology and framework of a recent dissertation on effects-based operations (EBO) (Bullock, 2006). This effort stands out for discussing measures of effectiveness and providing both a mathematical perspective and a typology relevant to military applications. Much of the other existing work on measures has been from a security-studies perspective, which tends to highlight the pitfalls of using either well-defined or poorly chosen measures (Murray, 2001) and determining what characteristics can be attributed to an adequate measure (Jones, 2006; Murray, 2001; Eisenstadt, 2005).

I will also incorporate some lessons learned in the measurement of CT/COIN that have not been incorporated in the exiting literature. The Hamlet Evaluation System (HES) from the Vietnam War provides one interesting (if antiquated) test case in measuring COIN; it was designed by the Central Intelligence Agency and conducted by the U.S. Military Assistance Command Vietnam to measure rural development and pacification (Hunt, 1995; Long, 2006; Thayer, 1985; Corson, 1968; and Sweetland, 1968). This system of measures was designed to assess 12,000 individual South Vietnamese hamlets by grading them on a six-point scale across 18 indicators relating to security or development, then complementing groups of these indicators with confidence scores. The HES was a pioneering effort in developing analytical measures to apply to CT/COIN, but its misuse and contentiousness in both military and civilian circles caused it to be in low repute by the end of the campaign. However, a recently declassified study on the HES showed that it was a statistically valid, modest system for measuring rural pacification (Sweetland, 1968), which may be rich in lessons-learned that could potentially be applied to measuring the global war on terror.

Distinctions and Definitions

It is useful in what follows to distinguish among what will be referred to in acronyms as MOPs, MOEs, and MOOs (Figure B.1). A measure of process (MOP) relates how inputs of a system are transformed into outputs, corresponding with tactical levels of operation.[1] A measure of effectiveness (MOE) measures system changes resulting directly from creating certain outputs, which correspond with operational tasks that are determined to be necessary to realize strategic effects.[2] A measure of outcome (MOO) characterizes higher-order conditions created by numerous system effects and thus captures only strategic-level outcomes.

As I use the term, MOEs relate cause and effect and describe how well actions are achieving their objectives. MOEs fit into a vertical framework of linking measures to fundamental system objectives, where the objective is an overall desired end-state (Keeney, 1992). In

Figure B.1
A Taxonomy of MOEs

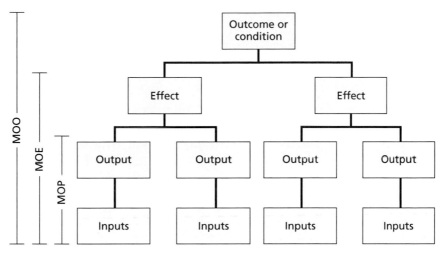

SOURCE: Adapted from Bullock (2006).
RAND MG849-B.1

this sense, MOEs can be distinct from MOPs, which measure such attributes as efficiency or technical qualities of system components contributing to intermediate objectives (Artley, 2001; Army, 2006; and Keeney, 1992).[3] MOOs gauge the indirect conditions created by system effects (DSMC, 1994), and thus may be strongly affected by factors that are not subject to influence. Thus, an intervention or action may have high performance in the sense of using good technology and efficient processes, various outcomes may be achieved, and yet the action may be ineffective overall.

The quality of different proposed measures may depend on their accessibility or on the ease of creating and operationalizing the measure from the concept being measured. Any single measure also fits into an accessibility type; certain measures are easily quantified and linked to a system end-state; others may need to be constructed and may be correlated only with the end-state. In this context, measures can be categorized by the nature of their construction and the degree to which they measure the objective end-state. Measures can be natural (that is, commonly understood) or constructed (that is, developed for a specific purpose), and direct (that is, linked to an objective) or proxy (that is, correlated with an objective). This typology allows us to classify measures by how accessible they are in monitoring the system's end-state, with natural-direct measures being most accessible and constructed-proxy measures being the least (Kirkwood, 1997).

Particular attributes of a measure must also be considered to gauge its usefulness as a tool to measure the system in question. A measure must reflect the system attributes it is supposed to represent; if it does, the measure is considered valid. A measure must also be precise (that is, sharply defined) to be meaningful and reliable. Finally, an analytical measure should be sensitive to changes in policy or action; if it is, it is considered to be *responsive*.[4] With a developed theory laid out for understanding the nature and use of measures, I can now turn to applying specific measures to the most important factors delineated in each of the preceding author's chapters. Of course, I am primarily concerned with understanding MOEs that are sensitive to policy and action over both the short and the long term.

Although this appendix approaches the search for metrics rather directly, other RAND work has emphasized that metrics are best developed as spinoffs of operational analysis to understand and build models of relevant processes, such as combat operations—including the consequences of adaptations by the adversary. Metrics developed in that way have a better-defined relationship to what is actually of interest and are more likely to be defined carefully than are metrics arrived at through intuitive leaps. Their shortcomings are also more likely to be understood, which is especially important in dynamic contexts where adversaries adapt, those whose work is being assessed learn to "play the game," and circumstances change.[5] With this caveat expressed, let us now proceed to discuss possible metrics for many of the subjects discussed in the main text.

Root-Cause Factors

The elements that constitute root causes of terrorism are fairly well understood, if not their relative strength and relationships (see Chapter

Two, by Noricks). It is useful, in terms of warning, to monitor trends in these elements or factors contributing to these elements. For root causes, the desired end-state is the prevention or avoidance of terrorism and insurgency. The elements that I consider in subsequent paragraphs are regime legitimacy, repression, demography, grievances, and mobilizing structures. (See Table B.1.)

Regime Legitimacy

I can divide the measures of legitimacy between those that are predictive (or "leading") and those that lag shifts in popular perception. Upward movement or an acceleration in inflation (consumer price index or CPI), unemployment, and income inequality (measured by the Gini index or by concentration of land ownership) or downward shifts in the prices of exported goods all may have a degree of predictive power in measuring shifts in popular opinion of a regime (Long, 2006). The activity of small- to medium-sized enterprises (SMEs) may also be a leading indicator of insurgent or terrorist action (Army, 2006), and changes in the availability of goods may also be of interest if these groups oppose the presence of particular goods, such as alcohol or tobacco.[6] Dramatic increases in the prices of goods that are uniquely from a specific area may be indicative of insurgent taxation (direct or indirect) and thus may indicate the weakness of the regime in that locale (Fall, 1966). Further, changes in flows of foreign investment or domestic private investment into an area may be a leading indicator of a change in political or social stability. Economic indicators may be reliable indicators of shifts in the perceived legitimacy of a regime but are not valid measures in this context, because they reflect system processes that are not directly connected to the end-state of perceived legitimacy and are thus mostly MOPs.

Because the desired end-state in this instance is a regime that is legitimate in the eyes of the populace, indicators that capture government control over administrative processes may reflect changes in popular perceptions of the regime—once government action is controlled for. Bernard Fall found that government appointment of schoolteachers and collection of taxes reflected the weakness of the government of Vietnam in the late 1950s when faced with the Vietminh (Fall, 1956). Because these measures are responsive and capture both

Table B.1
Root-Cause Measures

Concept	Measure Group	Measure Type	Comments
Regime legitimacy	Leading economic indicators (e.g., consumer price index, unemployment, foreign direct investment flows, consumer price indices)	MOP	Moderate data-collection difficulty; reliable and responsive; no direct connection to end-state
Regime legitimacy	Lagging economic indicators (e.g., gross domestic product, wealth inequality [i.e., Gini index], stocks of foreign direct investment)	MOP	Moderate data-collection difficulty; reliable and responsive; no direct connection to end-state
Regime legitimacy	Measures of regime control (e.g., taxes collected per capita, school teacher assignments)	MOE	Moderate data-collection difficulty; reliable and responsive; direct connection to end-state
Regime legitimacy and grievances	Polled support for government and support for government among key opinion leaders	MOP	Difficult data-collection; reliability is questionable; no direct connection with end-state
Regime legitimacy	Social ferment (e.g., numbers and intensity of demonstrations, desertion rate from military, rise of subversive groups)	MOE	Moderate data collection difficulty; reliable and responsive; well connected to end-state
Repression	Disproportionate flows of government funds	MOP	Difficult data collection; validity and reliability are questionable; no direct connection with end-state
Repression	Disenfranchisement of at-risk social groups	MOP	Very difficult data collection; validity is questionable; no direct connection with end-state
Repression	Limitation of access to key resources for social groups	MOP	Difficult data collection; validity is questionable; no direct connection with end-state
Demographics	Increase in number of young males in the population	MOP	Easy data collection; not responsive; no direct connection with end-state but should be used in conjunction with other measures

Table B.1 (continued)

Concept	Measure Group	Measure Type	Comments
Grievances	Antigovernment graffiti	MOE	Moderate data collection difficulty; reliable and responsive; connection with end-state
Mobilizing structures	Hazardous social group membership	MOE/ MOP	Very difficult data collection; reliability is questionable; possible direct connection with end-state; must be considered alongside other social ferment indicators
Mobilizing structures	Population's willingness to talk about insurgency	MOE	Moderate data-collection difficulty; reliable and responsive; direct connection with end-state

government action and population response, they may be valid MOEs for regime legitimacy.

Measures related to social ferment may also be directly connected to the end-state, because social movements may indicate that substate actors seek to undermine the regime. The frequency or intensity of public demonstrations, riots, and desertions from the military may all be measures of shifts in the social ferment, depending on local conditions (Tanham, 1988). These measures are reliable and valid for gauging the end-state and constitute MOEs, in that they indicate a population response. Further, polls indicating that perceptions of the legitimacy of the regime are moderating may also be valid and reliable as MOPs, although they do not indicate that the populace is necessarily willing to take action. Measures of the perceived illegitimacy of the regime should be considered in tandem with measures of repression, because direct action by the regime may affect the population's willingness or ability to act.

Repression

Repression can have many manifestations, which suggests the need for multiple measures. Measures may include incidents or intensity of indiscriminate military or police actions against particular religious or political groups, disproportionate flows of government economic support to unpopular or minority institutions, estimated changes in the numbers of disenfranchised voters, or changes in access to key resources for certain subpopulations. Repression measures should be considered as MOPs because of their lack of a direct connection to the end-state and lack of responsiveness. These MOPs feed into perceptions of regime illegitimacy, and thus the population responses mentioned above should be closely monitored as MOEs that react to repressive processes.

Demography
Demographics interacts with disruptive aspects of modernization and resultant changes in the social ferment. Census or survey data showing increases in the number of youth in the population may track the potential for terrorism or insurgency when combined with other factors. Detailed surveys of at-risk subpopulations may be of interest in monitoring reactions to certain policies or actions. Although demographics cannot be influenced, demographic trends should be monitored in conjunction with other metrics, such as selection measures of the social ferment.

Grievances
Grievances provide a trigger for violent action. In situations such as Iraq, where the policy of the counterinsurgent is considered to be a communal grievance, changes in attitudes and activities of the populace should be considered as important indicators of CT/COIN effectiveness. Polling, monitoring local or regional media sources, and analysis of information gathered in the process of conducting information operations from power brokers or elites may be informative about shifts in beliefs or attitudes about the counterinsurgent force. However, these indicators may provide only MOPs, because they are not directly linked to the end-state despite having high responsiveness. Other indicative measures, such as the incidence of anticounterinsurgent graffiti, may be more closely tied to the end-state, because it combines opinion with some form of antigovernment action.

Mobilizing Structures

Mobilizing structures are clearly an important conduit for recruitment, and monitoring the nature of these structures will be of interest to CT/COIN officials. The relative levels of personal ties that individuals have with known terrorists are clearly proxy measures for terrorism, because there is a high correlation between the ratio of number of personal ties with group members to other personal ties and the individual's propensity to join the group. In this sense, measuring personal ties is of interest to the counterinsurgent/counterterrorist; monitoring areas or groups where the aforementioned ratio is high may afford some ability to warn, and decreasing this ratio in at-risk populations may be an important MOE. Numerous experts on COIN have noted the warning power of the general populace's willingness to talk about possible insurgent activity with counterinsurgent forces (Hosmer and Crane, 1962; Sweetland, 1968; and Tanham, 1985), and this measure may also capture the strength of insurgent mobilization once it has become powerful enough to coerce the populace.

Radicalization Factors

Because radicalization is a multicausal process and a multiple-fields approach is necessary to understand the phenomenon, the measures used to monitor and assess radicalization must be drawn from the individual, group, and environmental levels that are all critical to the process of radicalization (see the chapter by Helmus). Developing measures for radicalization is less an exercise in trying to predict or warn of terrorist events or an incipient insurgency than an effort to understand at-risk subpopulations and the factors that influence the radicalization process of individuals within a given subpopulation. The measures for radicalization may be most useful for monitoring terrorist groups in stronger states where full-scale insurgency seems unlikely, although they will also have utility for possible insurgent scenarios as well. For radicalization, the desired end-state is the prevention or avoidance of the radicalization of at-risk subpopulations. The elements that I consider in subsequent paragraphs are social groups, a desire for systemic change, and duty. (See Table B.2.)

Social Groups

Taking into account that social groups are a critical factor in the radicalization process, identifying and tracking them is important to successfully predict and disrupt threats. Existing research supports the efficacy of monitoring those social and familial groups with known connections to terrorists, as well as groups of foreigners who may feel culturally alienated. The monitoring of networks is inherently difficult and controversial in liberal democracies, because it requires extensive

Table B.2
Radicalization Measures

Concept Being Measured	Measure Specification	Measure Type	Comments
Social groups	Number of social groups or individuals with ties to known terrorists	MOE	Very difficult data collection; reliability is questionable; direct connection with end-state
Desire for systemic change	Polled dissatisfaction on systemic issues	MOP	Difficult data collection; reliability is questionable; direct connection with end-state
Desire for systemic change	Polled desire for Sharia law	MOE	Difficult data collection; reliability is questionable; direct connection with end-state
Desire for systemic change	Number of protests or riots	MOE/ MOO	Easy data collection; reliability is questionable; connection with end-state may depend on political system
Duty	Number of killed or wounded civilians	MOP	Moderate data-collection difficulty; validity is questionable; direct connection with end-state
Duty	Internet traffic from specific geographic areas on extremist Web sites	MOE/ MOP	Difficult data collection; validity is questionable; possible connection with end-state

personal-level data and may tempt authorities to push the limits of personal privacy laws. It may be useful to aggregate group-level social network data to find the number of groups with known contacts to terrorists or places where terrorist influences are significant, or the number of at-risk groups that seek connections with known terrorist groups. Such a metric would be valid for gauging the effectiveness of efforts to socialize potentially radical groups and for gauging the basic risk of the formation of groups with intentions to commit terrorism in a given area; there is agreement in the literature that socialization processes are a necessary precondition for radicalization. Measures of at-risk socialization may also be directly linked to perceived duty and perceived or real rewards in the radicalization process. These concepts can be measured only by personal interviews or intelligence and should be understood in light of the fact that rewards may differ even within a single group. But increasing levels of perceived or real rewards combined with increasing numbers of at-risk individuals or groups may indicate that radicalization is on the rise.

Desire for Systemic Change

Factors related to widespread desire for systemic change are widely recognized as being important to driving radicalization and terrorism, but it is difficult to directly measure any form of personal desire. Polls of at-risk populations, structured interviews with group opinion leaders, or interrogation reports taken from suspected terrorists may illuminate common points of desire for change among radicals. Monitoring the relevant issues and the strength of these desires may be valid MOEs for shifts in government policy aimed at the radicalization process or CT/COIN psychological operations (PSYOPs). Polls that show a desire for Sharia law and reveal the strength of Muslim identity may also be integral measures linked to the radicalization process when combined with measures of group processes and rewards. Qualitative indicators of a desire for Sharia law may also be relevant; notable absence of cigarettes or alcohol in stores may actually be a superior measure of radicalization, although only where open insurgency is not occurring, because it could also measure the ability of insurgents to coerce the population. Also, the numbers of riots, rallies, and protests may be powerful measures of a desire for political change (Tanham, 1988), although they may constitute only MOPs, because they do not directly link to the end-state of terrorism and available avenues of open protest may in fact decrease the risk of terrorism. It has been shown that the desire for systemic change is a potent factor leading to terrorism when combined with group processes and rewards, so measures of the desire for change should be considered in tandem with measures of these two associated concepts. But group processes and rewards may also be analytically powerful when combined with perceived duty, which is discussed below.

Duty

Duty in this context is conceptually both a desire to exact revenge and a perceived need for a collective defense of Islam, so measures of these concepts are likely to be of interest to counterterrorism policymakers. Those who have experienced personal attacks against loved ones, those who are likely to have post-traumatic stress disorder (PTSD), and those who have been convinced of a duty to defend Islam against threats are clearly more likely to engage in terrorism when other critical factors are present. It is difficult to conceive of operable measures to capture changes in these factors, even if specific populations are targeted. However, ideal proxy measures that would be relevant to these concepts would include the number of civilians killed or wounded during combat and changes in Internet traffic to extremist Islamic Web sites.

Although it is difficult to develop measures to monitor radicalization processes, these factors are critical to pinpoint changes in an individual's propensity to become a terrorist within certain populations and thus are most critical in domestic campaigns against homegrown terrorism or for foreign governments whose populations emigrate to join terrorist groups elsewhere. Policies to counter radicalization will be inherently difficult to assess in this manner, but the metrics listed above may shed some limited light on how well government actions are affecting movements toward extremism and terrorism.

Decisionmaking Factors

Factors that influence group decisionmaking in this context should be considered with the group's decisionmaking unit as the unit of analysis, because this level of command is relevant to the proximal decisions of whether, how, and when to conduct an attack. Because a large number of factors have been determined to be substantial contributors to group decisionmaking (see the chapter by Jackson), I will concentrate here on the most crucial factors that both lead a group to certain decisions and may be measurable longitudinally. Further, the topic of terrorist decisionmaking is fundamentally different from the topics discussed above, because the outcome of decision processes—behaviors—are observable, whereas the factors that contribute to decisions are more difficult to measure. Thus, I will first consider measures of relevant outcomes that may follow decisionmaking processes and determine their utility in understanding how terrorist or insurgent groups operate. Then I will examine critical and observable factors that lead to decisionmaking and evaluate the potential use of these measures. For decisionmaking, the desired end-state is being able to influence the decisions of existing terrorist or insurgent groups. The elements that I consider in subsequent paragraphs are outcomes, lessons learned with outcome measures, decisions, organizational structure, and resource stocks. (See Table B.3.)

Outcomes

For our purposes, outcomes of interest at the operational and tactical levels primarily involve attack data; these data include attack types (numbers of fighters, tactics, and weapons), attack locations or targets, and numbers of attacks across time. These simple MOOs are tracked in quarterly reports about the war in Iraq made by the DoD to the U.S. Congress in the forms of weekly attack trends broken down by attack type (infrastructural attacks, improvised explosive devices, small arms fire, indirect fire) as well as by attack levels for a given period broken

Table B.3
Decisionmaking Measures

Concept Being Measured	Measure Specification	Measure Type	Comments
Decision outcomes	Attack data (e.g., types, size, targets, numbers of attacks)	MOO	Easy or moderate data collection; should be analyzed alongside calendar events, CT/COIN action, and qualitative indicators of second- or third-order effects
Decisions	Financial activity (e.g., revenue sources, expenditure types)	MOO	Very difficult data collection; reliability is questionable; other indicators of group processes are also necessary
Organizational structure	Organizational structure type	MOP, MOE, or MOO	Very difficult data collection; reliability is questionable; can be MOP, MOE or MOO depending on how it is used
Resource stocks	Order of battle (e.g., organization structure, weapons, training, leadership)	MOP or MOE	Difficult or very difficult data collection; sensitivity is questionable; connection with end-state may be mediated by many other factors

down by Iraqi province (U.S. Congress 2005, 2006, 2007, 2008). David Galula once noted that the size of guerilla engagements and the guerillas' ability to conduct complex attacks against counterinsurgent convoys were foremost among important indicators of success (Hosmer and Crane, 1962), and changes in these outcomes may signal important shifts in insurgent strategy according to key decisionmaking factors. Subsequently, systems analysts at the DoD during the Vietnam War were fixated on attack data. A number of commentators have noted the disastrous use of enemy body count as an MOE of COIN operations at the time (Army, 2006; Corson, 1968; Murray, 2001; Long, 2006). Despite this improper use of one attack data metric, it must be acknowledged that changes in attack types have accurately pinpointed changes in insurgent strategy, such as with the insurgent group Farabundo Marti para la Liberacion Nacional in El Salvador in the early 1980s (U.S. General Accounting Office, 1991) as well as with the North Vietnamese in the late 1960s (Thayer, 1985). Thus, well-crafted attack metrics may shed considerable insight into group decisionmaking as an outcome.

Lessons Learned with Outcome Measures

Despite past analytical successes with attack data, past failures may be informative about the use of metrics by analysts and policymakers. Thayer's treatise *War Without Fronts* demonstrated that the operational environment must be considered in tandem with threat levels; he noted the destructive lack of awareness in the U.S. Army that clear seasonal attack trends were evident in the rainy seasons (Thayer, 1985). Attack trending based on calendar events has also been present in Operation Iraqi Freedom; attack levels have consistently spiked during Ramadan and during national elections (U.S. Congress, 2008). It is also intuitive that attack data will differ depending on CT/COIN operations and shifts in policy, such that the concept of "net assessment"—or the overall balance of insurgent and counterinsurgent forces—should be taken into consideration when assessing trends in attacks. Furthermore, analysis of allied strategic operations in World War II showed that secondary effects should be taken into account when analyzing the effectiveness of military operations (Roche and Watts, 1991). Thus, it may not be sufficient to use enemy attacks as an analytic measure alongside known calendar effects and CT/COIN actions; second- or third-order effects may be present and quantifiable and should also be taken into account. In this light, it must be acknowledged that, although attacks reflect group processes, they are higher-order outcomes that depend also on other processes and are useful in certain limited contexts and lack the precision of a true MOE.

Decisions

Some group decisions and outcomes are less easily observable, beyond simple attack planning, but may be interesting in helping us understand the choices being made by the decisionmaking unit. The decisonmakers in the group may have to decide whether to solicit funds from outside the immediate group structure or to generate their own revenues. Such decisions may be observable at certain points in time, given that relevant intelligence is being collected and becomes available. If it is known that a group decisionmaking unit is accepting funds from a higher node of the organization, it may indicate a lower risk tolerance, because generating revenue independently may increase the odds of capture. Other group decisions may be of interest to policy-makers, but it is difficult to attribute changes in these outcomes to particular policies because of the complex processes at play.

Organizational Structure

With some important outcomes of decisionmaking considered, I now examine metrics for important factors that contribute to group decisionmaking processes. The first key factor to consider is that of organizational structure; this feature of terrorist groups feeds into the other decisionmaking and behavioral factors. A group's organizational structure is difficult to gauge, and, as al-Qaeda has demonstrated, different structural elements may exist simultaneously. Appraising group structure is difficult in CT/COIN because of the clandestinity of such organizations; the potential for denial and deception makes high-quality multisource intelligence all the more necessary to accurately characterize structural forms. Developing a metric that captures group organization is difficult as a result of the artificial nature of the dichotomy between formal and networked organizations, but composite proxy metrics that combine aspects of group organization may be assembled from ordinal scales. Analysis of the literature indicates that organizations should be categorized on such criteria as division of labor, levels of decisionmaking authority within the organization, geographic specificity in operational command, and singularity of decision processes (that is, authoritarian versus participatory). These elements of group organization could be scaled from 1 (lowest) to 5 (highest), such as in the HES. Strict guidelines should be developed so that specific criteria are attributed to each step in the scale, and these ratings should be vetted extensively. Also, such a composite proxy metric would need to be constantly reassessed and evaluated for validity as new intelligence and analysis become available.

Resource Stocks

Stocks of weapons, money, and labor are clearly important factors for group decisionmaking that may be directly measurable and natural, and intelligence relating to these factors is should be of interest to CT/COIN forces. Although full-scale insurgencies make collection of such "order of battle" data more easily attainable, collection of these data for small clandestine terrorist cells can be extremely difficult. Internal group dynamics, risk tolerance, and organizational structure will determine how resources are allocated, and measures of resource allocation across time can, perhaps, be mapped against CT/COIN efforts to understand how operations may affect group decisions. Information regarding group stocks will become available only intermittently, and document intelligence may be of great interest, because it will often be the richest source of information about resource stocks. These measures may simultaneously be MOPs and MOEs, depending on how tightly they can be linked to strategic goals. Measures of validity will probably be quite high when evaluating the effectiveness of counterfinancing and counterrecruitment, but reliability is a major concern, and measure responsiveness will be moderate, because the measured concepts involve fairly complex processes with multiple inputs.

Support Factors

Developing useful metrics to assess support for terrorism or insurgency is difficult because of the scarcity of information resulting from the usually clandestine nature of such support. As mentioned previously (see the chapter by Paul), polled expressions of support may not directly map to the end-state of support being given, and the actual dispensation of support to terrorist or insurgent groups is often difficult to observe. Thus, without precise and dependable intelligence on the support being provided, CT/COIN analysts must rely on either natural or constructed proxy measures to gauge the outcome of dispensed support. Further, the presence or absence of a factor that contributes to support for terrorism is often even harder to observe and identify, thus forcing the analyst to rely on direct qualitative measures and proxy measures.

Measures of support must focus on local politics but should also capture support from global movements. For support, the desired end-state is the prevention or deterrence of support for terrorism or insurgency. The elements that I consider in subsequent paragraphs are outcomes, desire to contribute, social pressure, and propaganda. (See Table B.4.)

Outcomes

The outcome of dispensed support can appear in many forms: from manpower at the most concrete, to simple passivity at the most

Table B.4
Support Measures

Concept Being Measured	Measure Specification	Measure Type	Comments
Support outcomes	Ordinal support measure (e.g., manpower, funding, equipment, sanctuary)	MOO	Moderate to difficult data collection; reliability and sensitivity may be questionable
Desire to contribute	Availability of services (e.g., police, hospitals, schools, electricity, water)	MOP	Moderate data-collection difficulty; validity is questionable; not directly connected to end-state
Desire to contribute	HES desire measures (e.g., government response to population demands)	MOP	Difficult data collection; validity and reliability are questionable; not directly connected to end-state
Desire to contribute	Freedom of movement and freedom to meet	MOE	Moderate data-collection difficulty; reliability is questionable; direct connection to end-state
Social pressure	Population's willingness to talk about insurgency	MOE	Moderate data-collection difficulty; reliable and responsive; direct connection with end-state
Propaganda	Existence and strength of insurgent PSYOP	MOP	Easy data collection; reliability is questionable; may not directly connect to end-state
Propaganda	Existence and strength of counterinsurgent PSYOP	MOP	Easy data collection; reliability is questionable; may not directly connect to end-state

amorphous. The observation of insurgent activity within a given area could correspond with no support being given at all or with all possible types of support being present. Likewise, areas with no direct evidence of insurgency may also run the gambit from providing ample support to providing none at all. Where it is expected that support is being given to insurgents or terrorists, measures should be available to monitor the changes in flows of support over time. Intelligence collectors should be sensitized to the possible types of support classified by Metz and Millen (2004) (manpower, funding, equipment/supplies, sanctuary, intelligence, passivity), and qualitative indicators could be constructed to capture estimated levels of support that collectors believe exist in an area. As was done in the HES, geographic areas could be scored in given intervals along these six support dimensions and the scores could be double-checked against document intelligence. For example, captured documents found in Sinjar, Iraq, in 2007 that disclosed the source nations of al-Qaeda in Iraq foreign fighters could be used to reassess of the scoring of these indicators (Felter and Fishman, 2007), because they may provide an empirical check. Also, valid measures of financial support may be available through such methods as those pioneered in the HES by recording the presence of insurgent taxation in an area (Sweetland, 1968). This is clearly a MOO for both support and group decisionmaking, but it may also be a proxy MOE on the factors of intimidation and group legitimacy.

Desire to Contribute

Of the key factors that are relevant or contribute to support, the core chain of factors relating to the desire to contribute to resistance or to action by proxy may be approachable through constructed measures. The U.S. COIN field manual lists a number of indicators relating to government legitimacy that could be operationalized through collection of open-source information, including the availability of government services, the presence or absence of associations, and freedom of movement (Army, 2006). The HES contained three measures of administrative and political activities that are relevant to the desire to contribute to resistance within a more limited geographic area: level of government management, government response to popular aspirations, and strength of counterinsurgent information or PSYOP activities (Sweetland, 1968). Low or falling scores on these measures may proxy for increases in the populations' propensity to want to contribute to resistance.

Social Pressure

The second core factor chain that may be amendable to policy influence is that of social pressure. A direct way that insurgent or terrorist groups leverage social pressures on communities is through intimidation. Although levels of intimidation may seem to be a somewhat nebulous concept, it may in fact be directly measurable through a constructed measure. Sweetland's suggested "willingness to talk" to COIN forces about insurgent activity may be a valid and reliable indicator of the level of local intimidation if properly constructed (Sweetland, 1968). This measure would be sensitive and valid because it would indicate a population's response to CT/COIN activity or insurgent weakness, and it seems entirely feasible that a similar ordinal measure could be built for the areas of operation in the gobal war on terror. David Galula similarly noted that the best indicator of COIN success is the free flow of intelligence from the population (Hosmer 1962) and the DoD reports on measuring security and stability in Iraq present the number of insurgent caches found in a given period (U.S. Congress, 2007, 2008). Although measures of cache finds may be also be informative, they are more difficult to link to the end-state because they have more process inputs, such as insurgent material and counterinsurgent operations.

Propaganda

Group propaganda may also be of note for CT/COIN analysts when trying to track and understand shifts in support. The issue of assessing the effectiveness of PSYOP is particularly nettlesome because the mere existence of PSYOP does not necessarily indicate shifts in population behaviors, such as dispensing support. Thus, simple measures, such as the HES's indicators of ongoing insurgent or counterinsurgent PSYOP, constitute only MOPs because the performance of PSYOP missions may or may not affect the population's willingness to support the insurgency. Many have noted that the understanding of how to develop MOEs for individual PSYOPs is still in its infancy at the DoD despite being faced with the problem since the Vietnam era (Sammons, 2004). MOEs for PSYOP campaigns more generally may be even more difficult to develop, because psychological reactions to either insurgent or counterinsurgent propaganda may be difficult to parse out from the effects of other types of operations. Regardless, MOEs for PSYOP should be developed during the operational planning phase and, likewise, MOEs should be developed to monitor the effects of insurgent propaganda to better analyze the propensity of populations to give support to insurgents or terrorists.

End-of-Terrorism Factors

The factors that lead to the end of terrorism include many of those that are also relevant to root causes, as well as to other factors that have been covered in previous sections of this monograph (see the chapter by Gvineria). In this appendix, the desired end-state is the demise of terrorism or the demise of the terrorist or insurgent group. In this section alone, I seek to identify metrics that reflect the demise of terrorism or a terrorist group's ability to function, which may comprise a set of metrics that would inform policymakers about the effectiveness of CT/COIN efforts. First, I examine metrics that have a decline in group activity as an outcome, then I examine metrics of the key factors that may lead to a decline in the group's status. The elements that I consider in subsequent paragraphs are end of terrorism, strength of organization, capabilities, and support of population. (See Table B.5.)

End of Terrorism

The outcome of an end of terrorism can manifest in many ways. However, the decline of terrorism should present itself as a reduction in the number or severity of enemy-initiated attacks on civilian or security force targets, but a combination of attack data must be considered based on local conditions. Further, the number of enemy caches found should be increasing as we have seen in Iraq since early 2007 (U.S. Congress, 2007, 2008), as should the flow of intelligence tips to

Table B.5
End-of-Terrorism Measures

Concept Being Measured	Measure Specification	Measure Type	Comments
End of terrorism outcome	Attack data (e.g., number, types)	MOO	Easy or moderate data collection; should be analyzed alongside calendar events, CT/COIN action, and qualitative indicators of second- or third-order effects
End of terrorism outcome	Flow of intelligence or enemy caches	MOO	Moderate data collection; should be analyzed alongside CT/COIN action and qualitative indicators of second- or third-order effects
End of terrorism outcome	Social ferment (e.g., numbers or intensity of demonstrations, desertion rate from military, rise of subversive groups)	MOE	Moderate data collection difficulty; reliable and responsive; well-connected to end-state
Strength of organization	Freedom of operation	MOE	Easy data collection; reliability is a concern but responsive; well-connected to end-state
Strength of organization	Measures of regime control (e.g., taxes collected per capita, teacher assignments)	MOE	Moderate data-collection difficulty; reliable and responsive; direct connection to end-state
Capabilities	Order of battle (e.g., organization structure, weapons, training, leadership)	MOP/ MOE	Difficult or very difficult data collection; sensitivity is questionable; connection with end-state may be mediated by many other factors
Capabilities	Financial activity (e.g., revenue sources, expenditure types)	MOO	Very difficult data collection; reliability is questionable; other indicators of group processes are also necessary
Capabilities	Attack data (e.g., types, size, targets, numbers of attacks)	MOO	Easy or moderate data collection; should be analyzed alongside calendar events, CT/COIN action, and qualitative indicators of second- or third-order effects

Table B.5—continued

Concept Being Measured	Measure Specification	Measure Type	Comment
Support	Ordinal support measure (e.g., manpower, funding, equipment, sanctuary)	MOO	Moderate to difficult data collection; reliability and sensitivity may be questionable
Support	HES desire measures (e.g., government response to population demands)	MOP	Difficult data collection; validity and reliability questionable; not directly connected to end-state
Support	Population's willingness to talk about insurgency	MOE	Moderate data-collection difficulty; reliable and responsive; direct connection with end-state

security forces (Sunderland, 1964). Although indicators of reconstruction should be considered as important MOPs, the social ferment indicators delineated above should be closely monitored to gauge the effectiveness of COIN/CT operations.

Strength of Organization

The strength of the insurgent or terrorist organization may be measured by their freedom of operation or their control over the population. It may be argued that some terrorist groups do not *seek* to control territory, but it is certain that these groups need to maintain areas where operational security is nearly assured. Also, with a lack of territorial control comes clandestinity, which may increase the group's disposition to commit more spectacular attacks. Thus, measures for the group's freedom of operation should be evaluated alongside attack data. Upward shifts in measures of government territorial control may indicate that the government is increasing its authority over an area previously controlled by insurgents. Indicators such as the amount of government tax receipts or the ability to place administrative personnel (such as teachers) may allow for a better understanding of government control (see Fall, 1956, 1966). A similar indicator has been used in Operation Iraqi Freedom; namely, the percentage of battlespace the Iraqi Security Force is responsible for (Jones, 2006). However, this indicator should be monitored alongside the insurgent group's freedom of operation to properly assess the strength of organization.

Capabilities

The capabilities of the insurgent organization must also be well understood to determine the status of COIN/CT efforts. Indicators for capabilities include the enemy order of battle, which assess the enemy's quality and quantity of personnel, leadership, training, and materiel. Attrition in the enemy order of battle may signal a decline in the group's capabilities, but the quality of these data and the intelligence they are based on should be routinely critiqued. Further, the ability to raise funds should indicate the group's potential to operate at a given level over the medium to long term. If a group's finances are well diversified, it may indicate that COIN/CT forces will be less successful in attacking the group's lines of revenue generation, thus revealing robust capabilities to operate. Attack data may also show how capable the group is through the complexity of its attacks or the mix of attack types, and these data should be checked against the assessment of the group's order of battle.

Support of Population

A comprehensive set of measures on insurgent support was described above. I assess that the most general support measures to apply to the decline of terrorism are observed support, the desire to contribute, and social pressure. Observed support may be captured by ordinal indicators based on intelligence assessments and should include manpower, weapons, materiel, and funding. External or foreign support may be extremely difficult to gauge, particularly when borders are porous or banks are not well regulated. Captured documents, such as those mentioned above from Sinjar, Iraq (Felter and Fishman, 2007) and those recovered in February 2008 from the raid on Raul Reyes's FARC camp in Ecuador, demonstrate that document intelligence may be the best source for understanding the profile of insurgent support. The desire to contribute may be measured by the freedom of movement in the area, as well as by the indicators described above from the HES. Collecting these data would be difficult and doing so would present methodological challenges, but the effort may yield useful intelligence on how insurgents are interacting with the population. Finally, monitoring social pressure to contribute support may be of great interest, because it may allow military operators to properly target recruitment or financier agents of the insurgency. The indicators described by Galula (Hosmer and Crane, 1962) and Sweetland (1968) that capture the population's willingness to talk about the insurgent group may best reflect the existence of significant social pressure.

Bibliography

Army—*see* U.S. Department of the Army.

Artley, W., D. J. Ellison, and Bwell Kennedy, "Establishing an Integrated Performance Management System," *The Performance-Based Management Handbook*. Washington, D.C.: U.S. Department of Energy, 2001.

Bullock, R. K., *Theory of Effectiveness Measurement*, dissertation, Wright-Patterson Air Force Base, Air Force Institute of Technology, September, 2006.

Byman, D., "Measuring the War on Terrorism," *Current History*, 2003, pp. 411–416.

Chiarelli, P. W., and Patrick R. Michaelis, "Winning the Peace: The Requirement of Full-Spectrum Operations," *Military Review*, Vol. 85, No. 4, 2005, pp. 57–70.

Corson, W. R., *The Betrayal*, New York: Norton & Co. Inc., 1968.

Council, N. S., *National Strategy for Victory in Iraq*, Government Printing Office, Washington, D.C., 2005.

Darilek, R., *Measures of Effectiveness for the Information Age Army*, Santa Monica, Calif.: RAND Corporation, 2001. As of July 18, 2008:
http://www.rand.org/pubs/monograph_reports/MR1155/

Davis, Paul K., *Analytic Architecture for Capabilities-Based Planning, Mission-System Analysis, and Transformation*, Santa Monica, Calif.: RAND Corporation, 2002. As of July 18, 2008:
http://www.rand.org/pubs/monograph_reports/MR1513/

Defense Systems Management College, *Systems Acquisition Manager's Guide for the Use of Models and Simulation,* Ft. Belvoir, Va.: Defense Systems Management College, 1994.

DSMC—*See* Defense Systems Management College.

Eisenstadt, M., "Assessing the Iraqi Insurgency (Part II): Devising Appropriate Analytical Measures," Washington Institute for Near East Policy, Policy Watch, Vol. 3, 2005.

Eisenstadt, M., and J. White, "Assessing Iraq's Sunni Arab Insurgency," Washington Institute for Near East Policy, Policy Focus, Vol. 50, 2005.

Fall, B., "Indochina—The Last Year of the War. Communist Organization and Tactics," *Military Review*, Vol. 3, No. 11, October 1956.

―――, "Insurgency Indicators," *Military Review*, Vol. 3, No. 11, April 1966.

Felter, J., and Brian Fishman, *Al-Qa'ida's Foreign Fighters in Iraq: A First Look at the Sinjar Documents*, West Point, N.Y.: Combating Terrorism Center, 2007.

Geisler, E., *The Metrics of Science and Technology*, Westport, Conn.: Quorum Books, 2000.

Hosmer, S., and S. Crane, *Counterinsurgency: A Symposium, April 16–20, 1962*, Santa Monica, Calif.: RAND Corporation, 1962. As of December 25, 2008: http://www.rand.org/pubs/reports/R412-1/

Hunt, R. B., *Pacification: The American Struggle for Vietnam's Hearts and Minds*, Boulder, Colo.: Westview Press, 1995.

Jones, D. D., *Understanding Measures of Effectiveness in Counterinsurgency Operations*, Vol. 69, Ft. Leavenworth, Kan.: School of Advanced Military Studies, 2006.

Keeney, R. L., *Value-Focused Thinking: A Path to Creative Decision Making*, Cambridge, Mass.: Harvard University Press, 1992.

Kelley, Charles, Paul K. Davis, Bruce W. Bennett, Elwyn D. Harris, Richard O. Hundley, Eric V. Larson, Richard Mesic, and Michael Douglas Miller, *Metrics for the Quadrennial Defense Review's Operational Goals*, Santa Monica, Calif.: RAND Corporation, 2003. As of December 25, 2008: http://www.rand.org/pubs/documented_briefings/DB402/

Kirkwood, C. W., *Strategic Decision Making*, Duxbury Press, Belmont, Calif., 1997.

Krepinevich, B., *Are We Winning in Iraq?* Washington, D.C., Committee on Armed Services, 2005.

Long, A., *On "Other War": Lessons from Five Decades of RAND Counterinsurgency Research*, Santa Monica, Calif.: RAND Corporation, 2006. As of December 22, 2008: http://www.rand.org/pubs/monographs/MG482/

Metz, Steven, and Raymond B. Millen, *Insurgency and Counterinsurgency in the 21st Century: Reconceptualizing Threat and Response*, Carlisle, Pa.: Army War College, 2004.

Murray, W. S., "A Well to Measure," *Parameters,* Vol. 31, No. 3, 2001, p. 14.

O'Neill, Bard, *Insurgency & Terrorism: Inside Modern Revolutionary Warfare,* Dulles, Va.: Brassey's Inc., 2000.

Pape, R., *Bombing to Win: Air Power and Coercion in War,* Ithaca, N.Y.: Cornell University Press, 1996.

Sammons, D. H. J., *PSYOP and the Problem of MOE for the Combatant Commander,* Newport, R.I.: Naval War College, Joint Military Operations Department, 2004.

Sproles, N., "Identifying Success Through Measures," *PHALANX,* Vol. 30, No. 4, 1997, pp. 16–31.

———, "Formulating Measures of Effectiveness," *Systems Engineering,* Vol. 5, 2002, pp. 253–263.

Sunderland, R., *Antiguerrilla Intelligence in Malaya, 1948–1960,* Santa Monica, Calif.: RAND Corporation, 1964. As of July 18, 2008:
http://www.rand.org/pubs/research_memoranda/RM4172/

Sweetland, A., *Item Analysis of the HES (Hamlet Evaluation System),* Santa Monica, Calif.: RAND Corporation, 1968. As of December 22, 2008:
http://www.rand.org/pubs/documents/D17634/

Tanham, G. K., "Indicators of Incipient Insurgency," Santa Monica, Calif.: RAND Corporation, unpublished, 1988.

Thayer, T., *War Without Fronts: The American Experience in Vietnam,* Boulder, Colo.: Westview Press, 1985.

U.S. Congress, *Measuring Stability and Security in Iraq,* Washington, D.C., October 2005.

———, *Measuring Stability and Security in Iraq,* Washington, D.C., October 2006.

———, *Measuring Stability and Security in Iraq,* Washington, D.C., October 2007.

———, *Measuring Stability and Security in Iraq,* Washington, D.C., March 2008.

U.S. Department of the Army, *Counterinsurgency Field Manual,* Washington, D.C., 2006.

U.S. Department of Defense, *Dictionary of Military and Associated Terms,* Joint Publication 1-02, April 12, 2001, As of October 17, 2008:
http://www.dtic.mil/doctrine/jel/new_pubs/jp1_02.pdf

———, *Measuring Security and Stability in Iraq,* Washington, D.C., 2008, p. 60.

U.S. General Accounting Office, *El Salvador: Military Assistance Has Helped Counter But Not Overcome the Insurgency,* NSIAD-91-166, Washington D.C., 1991.

Endnotes

[1] MOP is also used for "measure of performance," as when characterizing a system's efficiency, such as the number of information operations conducted by an infantry unit within a specific time period. The Department of Defense (DoD) defines MOP (JP 1-02) more narrowly as "a criterion to assess friendly actions that is tied to measuring task accomplishment."

[2] Various definitions have been suggested. Sproles (1997) defines MOEs as quantitative or qualitative metrics that seek to determine how well a system tracks against its purpose or its normative behavior. DoD's definition in (JP 1-02) is quite broad: "tools used to measure results achieved in the overall mission and execution of assigned tasks." I use the definition given in Bullock (2006), which defines effectiveness measurement as the difference between a given system state (for example, lots of terrorism) and some reference state, such as a desired end-state (for example, no terrorism). System attribute measures can then be developed such that they yield a system state-space that can be characterized as a metric space, and differences in system states relative to the reference state can be monitored over time to gauge the effectiveness of certain tasks.

[3] The relationships are not always mutually exclusive, since a single measure can sometimes serve as both a MOP and a MOE (Keeney, 1992).

[4] This concept is identical to the idea of *amplitude* presented in Bullock (2006).

[5] See, for example, Davis (2002) and Kelley, Davis, Bennett, Harris, Hundley, and Larson (2003).

[6] I am grateful to RAND analyst William Rosenau for pointing this out.